Aging of Industrial Polymers 1

to Ian and Nico
to Christelle

SCIENCES

Materials Science, Field Director – Jean-Pierre Chevalier

Polymers, Subject Head – Jean-Pierre Chevalier

Aging of Industrial Polymers 1

Aging Mechanisms

Coordinated by

Emmanuel Richaud

iSTE

WILEY

First published 2025 in Great Britain and the United States by ISTE Ltd and John Wiley & Sons, Inc.

ISTE Ltd
27-37 St George's Road
London SW19 4EU
UK

www.iste.co.uk

John Wiley & Sons, Inc.
111 River Street
Hoboken, NJ 07030
USA

www.wiley.com

Library of Congress Control Number: 2025940262

British Library Cataloguing-in-Publication Data
A CIP record for this book is available from the British Library
ISBN 978-1-78945-189-4

ERC code:
PE5 Synthetic Chemistry and Materials
PE5_15 Polymer chemistry
PE11 Materials Engineering
PE11_4 Engineering of polymers and plastics

The manufacturer's authorized representative according to the EU General Product Safety Regulation is Wiley-VCH GmbH, Boschstr. 12, 69469 Weinheim, Germany, e-mail: Product_Safety@wiley.com.

Contents

Foreword

Samuel FOREST
CNRS, Académie des sciences, Mines Paris – PSL, France

A foreword to a book on polymer aging written by a researcher who specializes in modeling and simulation in the mechanical metallurgy field might be perceived as incongruous. However, Jean-Pierre Chevalier and Emmanuel Richaud were certainly aware of my training at the *école de l'homme-matériaux* led by André Pineau. I have therefore been close to polymer scientists for nearly 40 years, starting with Jean-François Agassant, who taught me that polymers were on their way to solving most of the materials problems. I became increasingly aware of this subject due to the passionate speech by Jean-Marc Haudin on these materials and, naturally, the speech by Jacques Verdu at the ENSAM, to whom this book is rightfully dedicated. Therefore, I gladly and confidently accepted the invitation addressed by Emmanuel Richaud to write the foreword to his book entitled *Aging of Industrial Polymers 1*.

The complexity of the mechanical behavior of polymers is a challenge for a modeler. These materials often combine, in equal measure, elasticity, viscosity, plasticity and damage caused by complex, cyclic and multiaxial loads, encountered in many industrial structural applications. Moreover, these materials are sensitive to the environment, particularly to humidity and temperature, this being inherent to their behavior.

Times have changed, and some limits of these materials have been reached in terms of environmental pollution and use of carbon resources. However, polymers and plastic materials remain ubiquitous and are often irreplaceable. New solutions

Aging of Industrial Polymers 1,
coordinated by Emmanuel RICHAUD.

for their elaboration, recycling and biodegradation are already available, and even more ambitious prospects are expected to be developed in the future.

Similar to metals, polymers are subjected to aging, but in a specific way. Light and water are their typical aging agents. This book is dedicated to synthetic organic polymers, in the form of thermoplastic and thermosetting elastomers.

While reading this beautiful book, and due to my propensity for metallurgy, I identified what polymers have in common with metallic materials in terms of aging. These include some mechanisms of brittle or ductile fracture or brittle–ductile transition under aging. The representativeness of accelerated aging tests is essential for the study of polymer aging, similar to metals and alloys in the nuclear industry, for example, subjected to radiation, to the aggression of fluids and to thermo-mechanical fatigue. I have also discovered many specific traits of polymers under their multiple forms, such as the fatigue durability of elastomers or the probabilistic modeling of the degradation of polymer chains.

The essential association of mechanics and chemistry, required for aging control of structural polymers, is similar to the corrosion of metals and alloys, a threat that is regularly present in aging structures. Once again, there is no escape without a strong coupling between mechanical stress, diffusion and chemical and physical reactions, mixing solids and fluids.

This extensive subject is justifiably covered by two volumes, the first volume being dedicated to the physical and chemical bases of polymer degradation mechanisms according to the theoretical, experimental and numerical perspectives, while the second volume approaches the specificities of various classes of industrial polymers.

In cooperation with nearly 40 co-authors, Emmanuel Richaud accomplished a synthesis, unique in the French publishing landscape and even beyond it, on a very important subject for both engineering and society. This is generally dealt with by numerous, and sometimes contradictory scientific articles.

The authors are active in universities, research institutes and large companies all over France, covering a broad range of domains: chemistry, physics, mechanics, environment and industry.

The first four chapters implement the kinetic model of aging by free radical chain oxidation. They describe the aging mechanisms of polymers in the presence of water and under ionizing radiation. Polymer stabilization involves the modification of their structure to increase their resistance to external factors or the addition of additives that slow down the aging mechanisms. This refers to the basically irreversible chemical aging of polymers. As for the physical aging by structural relaxation in the polymers, it leaves the chemistry unchanged and can therefore be reversible. This concerns the study of glass transition, phase changes and memory effect, as well as the consequences on the viscoelasticity of the concerned polymers. Three chapters are explicitly dedicated to the impact of chemical aging on the mechanical properties of polymers, such as elasticity, viscoelasticity, plasticity and fracture, but also fatigue of elastomers. The prediction of durability of polymers relies on detailed knowledge on microstructure–properties relations.

The second volume relates to the specificities of important classes of industrial polymers. It starts by the study of photooxidation and thermal oxidation of polyethylene, thanks, on the one hand, to tracers of oxidation phenomena and aging follow-up methods, and, on the other hand, to tests of aging accelerated by increased light intensity. Next come PolyVinyl Chloride (PVC), polyamides, polychloroprenes, polyurethanes, polyesters (Polyhydroxyalkanoates (PHA) Polylactic acid (PLA), possibly bio-sourced and biodegradable), which can be destabilized by thermo-oxidation or hydrolysis, for example, in contact with seawater. It is also important to note the polymers in the form of oils, gums or resins, such as polysiloxanes whose thermal degradation is accompanied by depolymerization, while the photochemical degradation of oils is rather associated with cross-linking.

The focus then is on polymer matrix composites such as the organic matrix structural composites (OMC) for automotive, aerospace and shipping applications, in the presence of aggressive environments. An experimental and numerical approach is used to study the impact of thermo-oxidative or hydrolytic aging on the mechanical behavior of carbon fiber composites and thermosetting matrix composites. The organic synthetic fibers used for lifelines (nets, webs and slings) provide an exceptional resistance to humid environments and to creep.

The range is completed by nano-composites and copolymers, multilayer polymer films characterized by Raman and infrared microscopy.

An original chapter is dedicated to polymer combustion, with the development of flame retardants and formulations whose performance in terms of fire resistance must be maintained in time. The volume ends with a chapter dedicated to the development of "high-temperature" polymers for the electronic and aerospace industry.

There are undoubtedly some omissions, such as phase changes, particularly for thermal insulation or thermal control, for which aging is an obstacle to diffusion. Natural, bio-sourced polymers and their degradation in the natural environment (pollution, a specific case in which degradation must be accelerated instead of being stabilized!) are not widely covered. However, covering such a broad topic might require a third volume.

Hopefully, this Foreword has convinced the reader that this book is a valuable resource in terms of variety of materials, fundamental concepts of physics and chemistry, inquiry methods, industrial applications and prospects of improvement of the elaboration processes and mechanical properties. Finally, I wish to thank the authors for this invaluable work that will serve in the future as a reference for students, scientists, engineers and all curious daily users.

July 2025

Introduction

Emmanuel RICHAUD
PIMM, CNRS, CNAM, ENSAM, Paris, France

Given their mechanical, optical and electromagnetic properties in combination with their light weight and great ease of use, polymers can be employed in a wide variety of applications. The main driving force for the use of polymers and polymer matrix composites is the weight gain allowed by the high levels of performance indices rigidity/density and tensile strength/density. In the food industry field, they are used for transparent (flexible or rigid) packaging with good barrier properties that allow a better preservation of food products. Being excellent electrical insulators, the solutions they offer cover a broad range of fields, from general electricity to microelectronics. In terms of costs (in the transportation field, mainly of fuel savings), 1 kg weight reduction is associated with a cost saving of approximately $500 for a Boeing 747 and $30,000 for a space shuttle.

Obviously, other indicators should be taken into account for a sustainable use of polymer materials and organic matrix composites. Industry requirements are indeed very high in terms of performance, and also of supply management, operation, maintenance and service life duration.

This leads to high standards related to the durability and aging control of these materials. Consequently, engineers and even out spacing between words degradation mechanisms, the consequences on their properties and the solutions brought by a better formulation. Furthermore, the design of their systems must take into account the incompatibilities between material and its conditions of use.

Aging of Industrial Polymers 1,

To address the above, this book is structured to cover the following:

– fundamental studies on the aging mechanisms;

– degradation of various categories of polymers in their conditions of use;

– case studies on real systems.

Written by a large group of researchers and engineers in the aging community, this book is a tribute to the memory of Professor Jacques Verdu (1941–2018) at the *École nationale supérieure des Arts et Métiers*, who played a major role in the study of these phenomena and inspired many vocations in this field, particularly among the co-authors of this book.

At the end of this Introduction, I would like to express my gratitude to Jean Pierre Chevalier for entrusting me with the coordination of this book and for his invaluable help in finalizing the manuscript. My warmest thanks go to Samuel Forrest, who honored me by writing the Foreword to this book.

In memoriam

Figure I.1. *Jacques Verdu*

After graduating the *École nationale supérieure de chimie* in Montpellier, Jacques Verdu started his career at the *Centre des matériaux plastiques* of ENSAM in 1966, where he became Assistant Professor in 1971. After having defended his thesis related to the study of PVC photo-oxidation in 1976, elaborated under the supervision of Professor Sigwalt, he became a Full Professor in 1981. He established the *Laboratoire de transformation et de vieillissement des polymères*, which he led until 2005. The research he conducted in this laboratory was mainly related to the study of the mechanisms and kinetics of oxidative aging, humid aging and to the study of the relations between structure and properties in thermosetting networks. These efforts resulted in the publication of around 300 articles, some 20 books and chapters, and also to many awards, such as the Montgolfier medal of the *Societé pour l'encouragement de l'industrie française* (1999), the Nessim Habif Prize of the *Société des ingénieurs Arts et Métiers* (2002), medal of the Polymer Institute of the Slovak Academy of Sciences (2003), medal of the City of Lyon (2004) and Réaumur medal of SF2M (2005). During his career, Jacques Verdu supervised about 50 PhD students, including the coordinator of this book, who would like to express the most sincere gratitude to his exceptional mentor.

1

Mechanistic and Kinetic Aspects of Oxidative Aging

Emmanuel RICHAUD
PIMM, CNRS, CNAM, ENSAM, Paris, France

1.1. Introduction

The first observations of hydrocarbons getting oxidized date back to the late 19th century (for example, Engler and Weissberg (1898)). According to Hock et al.'s (1950) works, in the first half of the 20th century, hydroperoxides could be isolated as products of the oxidation of liquid hydrocarbons. Moreover, Semenov (1936) demonstrated that free radical oxidation was a branched chain mechanism.

In 1946, Bolland and Gee (1949) proposed an adaptation of these works to polymers. In the following years, studies on free radical oxidation allowed a better comprehension of hydroperoxide decomposition reactions, for example, thanks to the works of Bolland and Gee (Bateman et al. 1953), of termination reactions thanks to the works of Russell (1957) and of propagation reactions leading to the formation of hydroperoxides thanks to the works of Howard et al. (1968).

Moreover, the works of Niki (1973) and Gillen et al. (1995) laid the foundations for the kinetic modeling of oxidative processes.

The objective of this first chapter is to review some of the key elements required for understanding oxidative aging problems.

Aging of Industrial Polymers 1,
coordinated by Emmanuel RICHAUD. © ISTE Ltd 2025.

1.2. Oxidation mechanisms

1.2.1. *General mechanism*

1.2.1.1. *Initiation*

There are two main paths for the initiation of polymer oxidative aging mechanisms.

– Extrinsic initiation is induced, for example, by UV or ionizing radiation.

– Intrinsic initiation induced by structures present in the polymer structure can be grouped into three families:

- homolysis of polymer bonds: these occur at temperatures significantly higher than those of polymer aging under use conditions;

- decomposition of impurities that were initially present in the material (denoted by X), generally at quite low concentrations;

- unimolecular or bimolecular decomposition of hydroperoxides, which are, as we shall see in this chapter, the species produced during the propagation stages.

The relative concentrations of these two last types of species are schematically represented in Figure 1.1.

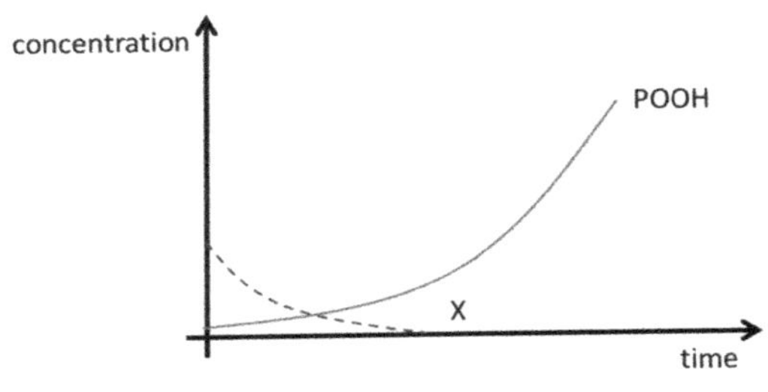

Figure 1.1. *Concentration of "extrinsic" species (X) and "intrinsic" species (hydroperoxides POOH) participating at the aging initiation*

From a thermochemical perspective, the energy required for the homolytic dissociation of a C–C or C–H bond is on the order of 350 or 400 kJ mol^{-1}, whereas that of a peroxide bond is on the order of 150 kJ mol^{-1}. It can thus be understood that under service conditions, thermooxidative aging is exclusively initiated by hydroperoxide decomposition. Two possible paths are schematically represented in Figure 1.2 for polyethylene.

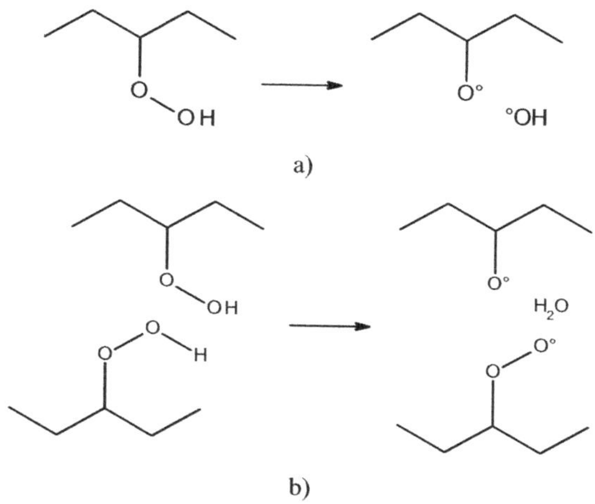

Figure 1.2. *(a) Mechanisms of decomposition of hydroperoxides via (a) a unimolecular path and (b) a bimolecular path*

The two initiation modes can, in principle, coexist, but they are competing with one another, according to the expression of their rate:

$$r_u = \frac{d[\text{POOH}]}{dt} = -k_u[\text{POOH}]$$

$$r_b = \frac{1}{2}\frac{d[\text{POOH}]}{dt} = -k_b[\text{POOH}]^2$$

Therefore, the two rates are equal when:

$$[\text{POOH}]_C = \frac{k_u}{2k_b}$$

In other words, unimolecular initiation predominates (i) at low concentrations, and (ii) when POOH is very unstable (k_u high), which appears to be, for example, the case of polyamides and other polymers in which the attracting inductive effects due to neighboring heteroatoms destabilize hydroperoxides. Importantly, the energy required for the initiation of the unimolecular reaction is always higher than that of the bimolecular reaction, with the latter being favored at low temperatures.

As will be explained in what follows, the decomposition of alkoxy PO° is not a stage of initiation, stricto sensu, but it is part of a series of reactions that are kinetically comparable to a reaction of initiation. There are several paths for the decomposition of alkoxy radical:

– cage decomposition, possible only if the alkoxy radical is not tertiary (Figure 1.3(a));

– β-scission (Figure 1.3(b));

– the removal of one hydrogen atom leading to the formation of an alcohol (Figure 1.3(c)).

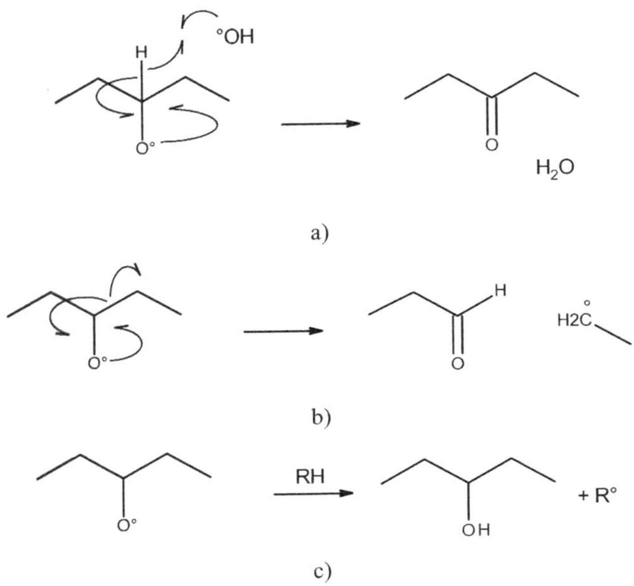

Figure 1.3. *Decomposition of an alkoxy (note: RH is a chain involving a methylene group CH2 and R° corresponding to the methyl radical –CH°–)*

In the end, initiation is a reaction involving the decomposition of at least one hydroperoxide, which leads to:

– The formation of stable oxygenated products (alcohols, ketones and aldehydes). This is, in fact, a simplified view: aldehydes are oxidized, for example, into carboxylic acids and other products (esters) eventually appear. Consequently, the comparison of the infrared spectra of a new or oxidized polymer reveals several peaks in the region centered on 1,700 cm^{-1}. Figure 1.4 illustrates, for example, the case of an oxidized PE.

– The appearance of chain scissions or volatile compounds that can be detected by, for example, gas chromatography–mass spectrometry (GC–MS) (Philippart et al. 1995).

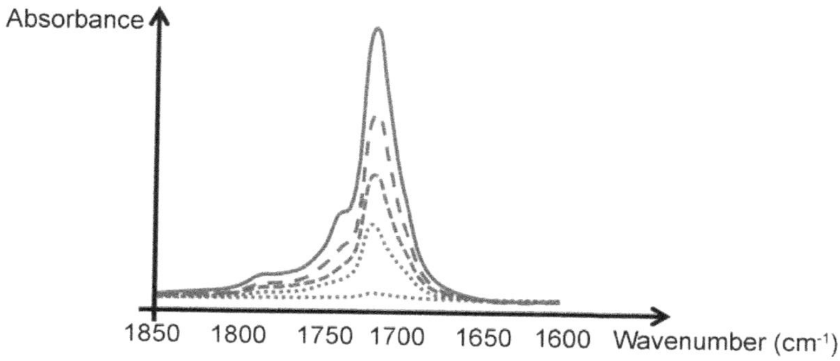

Figure 1.4. *Schematic representation of the IRTF spectrum in the carbonyl region for new and thermo-oxidized PE (various aging times ranging between 0 and 50 h at 100°C) (adapted from Gardette et al. (2013))*

1.2.1.2. *Propagation*

During the propagation stages, radical species react with non-radical compounds, leading to a new radical species. The two propagation stages shared by all of the polymerization mechanisms are as follows:

– The reaction between an alkyl radical and oxygen (Figure 1.5(a)). O_2 being, in reality, a diradical, this reaction is very rapid, and its rate constant is among the highest of the kinetics, namely, on the order of 10^8 L mol^{-1} s^{-1}.

– The reaction in which a POO° radical removes an unstable C–H, which leads to the formation of a hydroperoxide and an alkyl (Figure 1.5(b)). Although semiempirically, it is possible to link the rate constant of this reaction to the C–H bond dissociation energy (BDE) (Korcek 1972). Although only approximate, this relation can be used to identify the main oxidizable sites in the polymers and the preferred position of the hydroperoxides (Figure 1.6) and to establish a first ranking based on the "intrinsic" polymer stability: for example, aromatic polymers are significantly more stable than polymers containing aliphatic hydrogen (PE, PP), which in turn are more stable than dienic elastomers (that contain allylic hydrogens), and polymers carrying C-H in the α position of heteroatoms (polyamides, polyethers and epoxies).

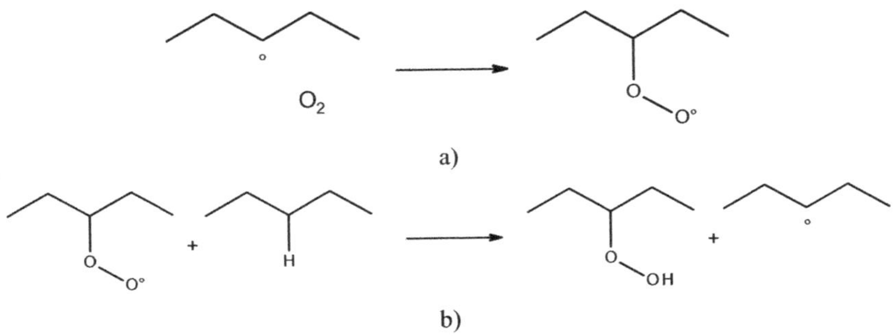

Figure 1.5. *Stages of propagation (illustrated for polyethylene)*

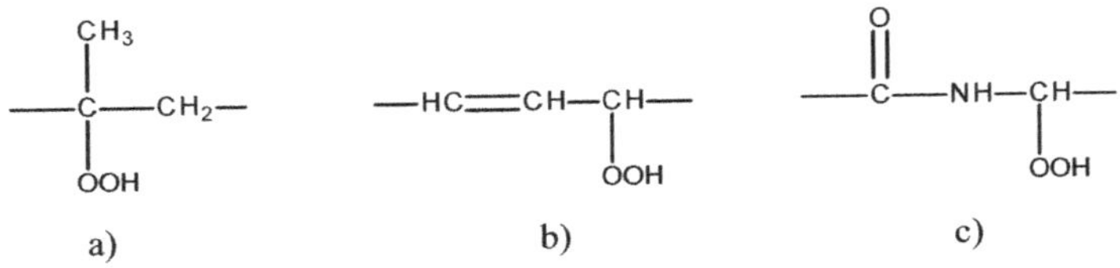

Figure 1.6. *(a) Structure of hydroperoxides predominating in polypropylene; (b) dienic elastomers; (c) polyamides*

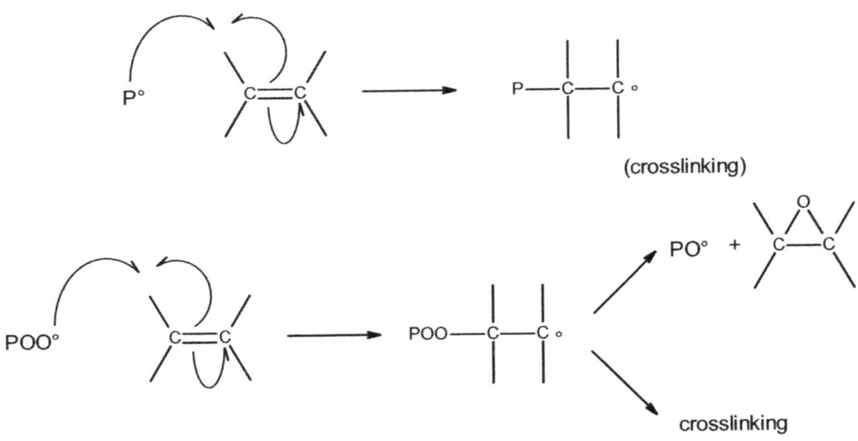

Figure 1.7. *Reactions of propagation involving the opening of double bonds*

Two other reactions are possible in non-saturated polymers: it is the opening of double bonds by radicals (Figure 1.7). These reactions are responsible for the appearance of an insoluble gel in the case of oxidation of non-cross-linked polydienes or an increase in the rubbery module in the case of cross-linked elastomer materials (Delor-Jestin et al. 2000; Le Gac 2016).

1.2.1.3. *Termination*

Termination corresponds to the stage in which radicals react together to form inactive species. There are three types of processes involving P° and POO° radicals, whose proportion depends on the external concentration of oxygen (Richaud et al. 2006), as illustrated in Figure 1.8:

$P^\circ + P^\circ \rightarrow$ inactive products

$P^\circ + POO^\circ \rightarrow$ inactive products

$POO^\circ + POO^\circ \rightarrow$ inactive products

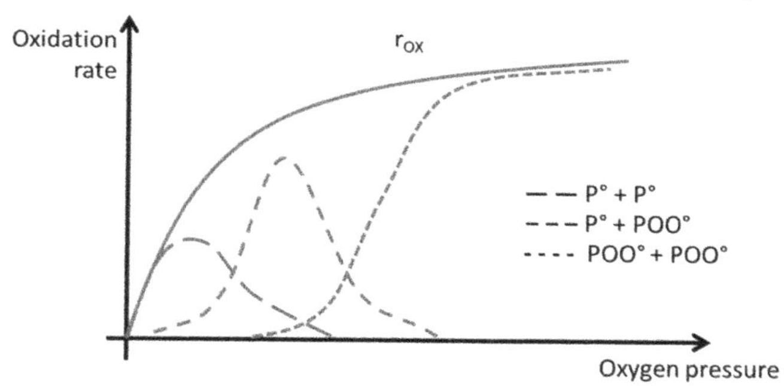

Figure 1.8. *Global oxidation rate and termination rates represented as a function of the external oxygen pressure*

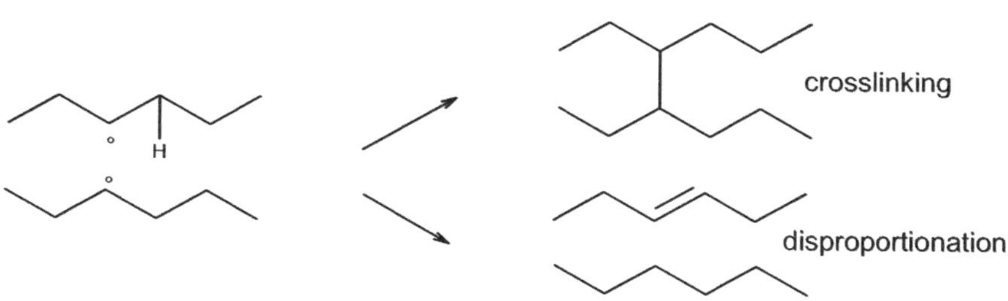

Figure 1.9. *Termination between two alkyl radicals*

From a more mechanistic perspective, terminations can be cross-linking paths, such as the termination reaction between two alkyls (Figure 1.9).

The relative proportion of these two reactions was quantified based on PE radiolysis experiments in an inert atmosphere, for which we have the following mechanism:

- Initiation: polymer (PH) + hν → P° + ½ H_2 (r_i).
- Propagation: P° + PH → PH + P° (k_p).
- Termination by coupling: P° + P° → P–P (+ x) (k_c).
- Termination by disproportionation: P° + P° → double bond + PH (k_d):

$$\frac{dx}{dt} = k_C[P°]^2 = \frac{r_i}{2} \cdot \frac{k_C}{k_C + k_d}$$

The termination between two peroxy radicals POO° + POO° has also been extensively researched. The result is that radicals start by coupling to form a tetroxy POOOOP, but this is unstable and leads to the formation of a pair known as a caged pair of radicals:

$$\text{POOOOP} \rightarrow [\text{PO}° + O_2 + °\text{OP}]_{cage} \rightarrow [\text{PO}° + °\text{OP}]_{cage} + O_2$$

There are several ways to terminate this pair of radicals, such as follows:

- the formation of a dialkyl peroxide POOP (which is itself unstable);
- the disproportionation by Russell reaction;
- the decomposition of each alkoxy is as follows:

$$[\text{PO}° + °\text{OP}]_{cage} \rightarrow 2\text{PO}° \rightarrow 2\text{P}° + 2\gamma_{CO}\ \text{P=O}$$

1.2.2. *Kinetic aspects*

For a linear polymer whose aging leads exclusively to chain breakages, these must reach a concentration given by the following equation to reach embrittlement (Chapter 5):

$$s_C = 2 \cdot \left(\frac{1}{M_{WC}} - \frac{1}{M_{W0}}\right)$$

Let us consider, for example, the case of polyethylene, whose average molar mass leading to brittle behavior amounts to 70 kg.mol^{-1}. In the case of a polymer of average molar mass of about 200 kg.mol^{-1}, we have s_C = 0.018 mol.kg^{-1}. In other words, in many situations, embrittlement is reached at a low advance rate, and the appearance of "chemical tracers" essentially confirms the nature of the degradation mechanism but does not anticipate degradation of the functional properties.

Figure 1.10 presents a classical case, showing curves of the kinetics of antioxidant consumption (AH), the formation and decomposition of hydroperoxides (POOH) and carbonyls (P = O) and the decrease in mechanical properties. Antioxidants disappear (by migration processes but also by capturing the radicals created by radical decomposition of unstable species present in the polymer). The increase in the concentration of the latter and hydroperoxides, to which they lead, is progressive and increasingly easier, as the antioxidant concentration decreases with the exposure time.

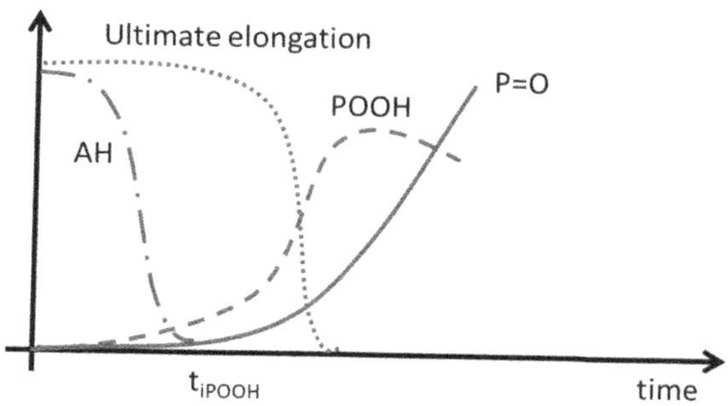

Figure 1.10. *Schematic representation of the aging kinetics of a polyolefin*

It is important to note the sigmoidal shape of the curve of appearance of carbonyls and the self-accelerated shape of the hydroperoxidation curve (with an induction period (t_{iPOOH})), and also a peak reached after an exposure period, at which point the rate of carbonyl appearance is also maximal. This period of induction (for carbonyls and hydroperoxides) is, in reality, typical for cases in which the kinetic chain length:

$$\Lambda = \frac{k_3[\mathrm{PH}][\mathrm{POO}^\circ]}{k_1[\mathrm{POOH}]}$$

is great, which means that the propagation rate is initially considerably higher than the initiation rate. It is, for example, the case of polypropylene, polyethylene and many hydrocarbon elastomers, in which hydroperoxides can be considered "relatively" stable.

On the other hand, there are cases (for example, polyamides) in which peroxides are unstable and lead to self-slowing curves, in which the period of induction does not exist in the case of unstabilized materials.

1.2.3. *Control by oxygen diffusion*

In the thick part, oxidation generally occurs only in the surface layers (Figure 1.11). Several criteria can be used to measure the depth of the oxidized layer. For example, we can study aging for samples of various thicknesses (for example, by mass loss) and then plot the mass loss as a function of the inverse of thickness.

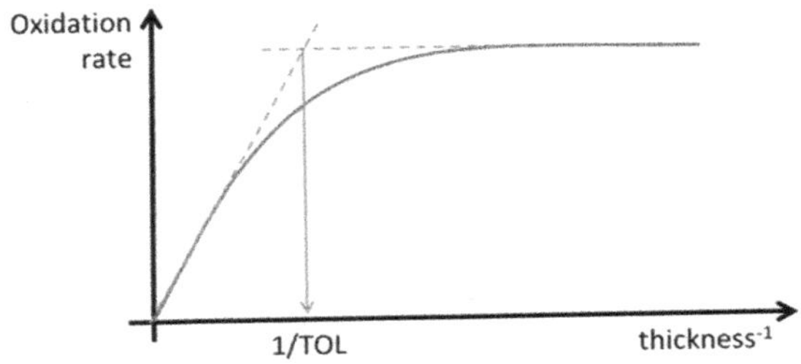

Figure 1.11. *Initial mass loss rate as a function of the inverse of thickness (according to Colin et al. (2001))*

This figure shows (i) a quasilinear increase in the mass loss rate (which is linked to the oxidation rate) in the range of high thickness; and (ii) a plateau where the oxidation rate is constant when the thickness is below a critical value. The value of the critical thickness separating the two regimes is generally quite close to the thickness of the surface zone whose aspect changes in relation to oxidation (such as the bleaching induced by fissuring), which can be observed, for example, by optical microscopy (Figure 1.12).

These results are explained by the expression of the oxygen consumption rate:

$$\frac{\partial[O_2]}{\partial t} = D_{O2} \cdot \frac{\partial^2[O_2]}{\partial x^2} - r_{OX}([O_2]$$

where $r_{OX}([O_2])$ is the quantity of oxygen reacting in time.

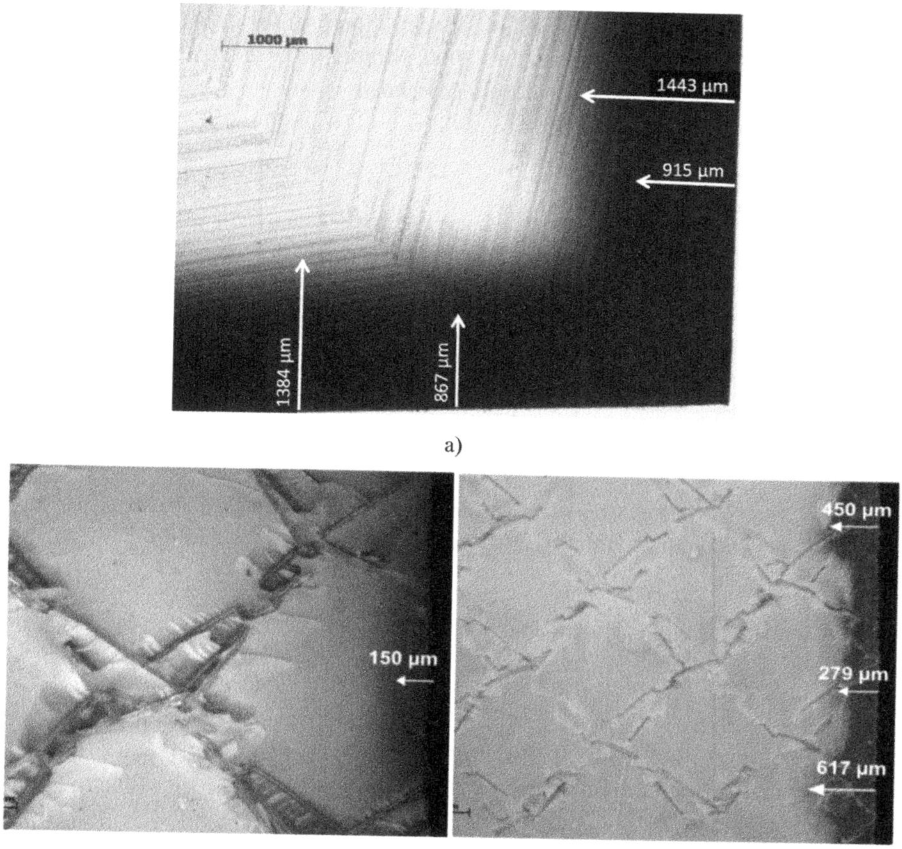

b) c)

Figure 1.12. *Optical microscope image of the thickness of the oxidized layer in an epoxy resin (Richaud et al. 2019) and image of the oxidized layer advancementby cracking in an adhesive (Delozanne et al. 2023). For a color version of this figure, see www.iste.co.uk/richaud/industrialpolymers1.zip*

Generally, oxidation is not homogeneous throughout the thickness, since oxygen is consumed mainly in the superficial layers by oxidative reactions, and only a small quantity migrates toward the core. The depth of the oxidized layer stabilizes, in principle, around a value that depends only on the external conditions of temperature and partial oxygen pressure. The thickness of oxidized layer (TOL) can be approximated as follows (Audouin et al. 1994):

$$TOL^2 = D_{O2}[O_2]/r_{OX}$$

Literature abounds with examples of concentration profiles in oxidation products or in mechanical properties (Celina et al. 2000; Minervino et al. 2013; Richaud et al. 2019) (Figure 1.13), the two being closely correlated.

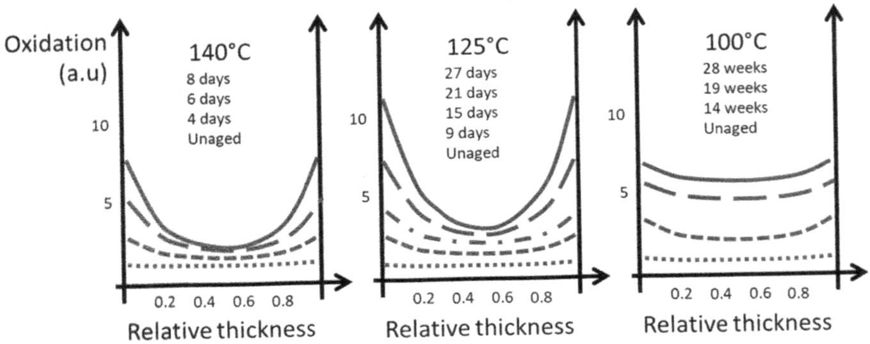

Figure 1.13. *Carbonyl concentration and module profiles in an elastomer at three aging temperatures (Celina et al. 2000)*

A general trend is clearly shown by the results in Figure 1.13: the depth of the oxidized layer decreases with temperature.

This can be easily explained as follows: the diffusion activation energy is generally lower (< 50 kJ mol^{-1}) than that of oxidation processes (> 70 kJ mol^{-1}). Observing that the thickness of the oxidized layer decreases with temperature is therefore not surprising.

When relying on predictions of oxidized layers at working temperatures, based on observations conducted under accelerated aging, the following should be taken into account:

– The oxidized layer thickens when the temperature decreases (therefore, there is a risk of underestimating its value at low temperature from an accelerated aging test).

– The thickness of the oxidized layer depends on the oxygen pressure in $[O_2]^{1/2}$.

In practice, the oxidized layer may be brittle enough to get cracked, providing additional access to the oxygen, which leads to oxidized layers going deep into the material and advancing step by step (Figures 1.12(b) and (c)).

1.3. Kinetic modeling

1.3.1. *Introduction*

The non-empirical method for predicting service life can be schematically summarized as follows:

– analytical study of the degradation mechanism;

– development of a kinetic scheme containing the essential kinetic stages of the mechanism established at stage 1;

– identification of its kinetic parameters and extrapolation values under the expected conditions of use to predict the evolution of the chemical structure under such conditions;

– the use of the structure–property relationships allows the estimation of the rate at which physical (for example, mechanical) properties vary based on the rate of structural changes (this point being dealt with in Chapter 4);

– identification of an end-of-life criterion in relation to the specifications.

This section aims to present several key elements of stages 2 and 3 to elaborate a kinetic model for thermo-oxidative aging. The model output data are, for example, in the case of oxidative aging: disappearance of functions that were initially present in the polymer (double bonds), increase in the concentration of stable species formed during aging (carbonyls), of reaction intermediates (radicals and hydroperoxides), and any other marker of oxidative aging (mass changes). Another important output data point is the appearance of chain scissions and/or cross-linking, which can be used to predict the evolution of the polymer architecture, based on which the residual mechanical properties of the material can be estimated.

The model input data must account for the conditions of use of the material, such as the aging temperature (allowing the estimation of the rate constants of the mechanism), oxygen concentration, formulation (presence of antioxidants) and thickness (in relation to diffusion phenomena, namely, the possible appearance of an oxidation gradient and/or the migration of antioxidants out of the material).

1.3.2. *Standard scheme*

This section will show how a simple scheme can be derived from mechanisms such as those presented in the first part of this chapter. Then, it will be used to predict degradation kinetics and also several elements for the estimation of rate constants. First, two examples of simplified models (based on the hypothesis of an initiation that is either purely unimolecular or purely bimolecular) will be presented. These can be dealt with analytically and serve as a basis for more complex (but closer to practical cases) cases for which numerical tools will be needed.

1.3.2.1. *Analytical models*

Under standard aging conditions (with the exception of aging in the presence of ionizing radiation), hydroperoxides play a key role in the stage of initiation (creation of radicals) (Backstrom 1934), since the O–O bond is relatively weak and its homolysis produces two radicals. Two main mechanisms are possible.

– Unimolecular decomposition: the initial reaction is the "spontaneous" decomposition of hydroperoxide:

$$POOH \rightarrow PO^\circ + {}^\circ OH \qquad k_u$$

Alkoxy groups, on the one hand, and hydroxyl groups, on the other hand, are considered to decompose extremely rapidly, and the three following reactions occur nearly instantaneously after their decomposition reaction:

$$HO^\circ + PH \rightarrow P^\circ + H_2O \qquad k_{1a}$$

$$PO^\circ + PH \rightarrow P^\circ + POH \qquad k_{1b}$$

$$PO^\circ \rightarrow P{=}O + P^\circ \qquad k_{1c}$$

This leads to:

$$\frac{d[P^\circ]}{dt} = 2k_u[\text{POOH}]$$

$$\frac{d[\text{P=O}]}{dt} = \frac{k_{1c}}{k_{1c} + k_{1b}[\text{PH}]} k_u[\text{POOH}]$$

In other words, these four reactions can be grouped so that unimolecular initiation can be schematized in one kinetically equivalent stage:

$$POOH \rightarrow 2P^\circ + \gamma_1 P{=}O + \gamma_2 s \qquad k_{1u}$$

– Bimolecular decomposition: in the case of bimolecular initiation, we have, more simply:

$$POOH + POOH \rightarrow PO^\circ + H_2O + POO^\circ \qquad k_b$$

$PO° + PH \rightarrow P° + POH \qquad k_{1b}$

$PO° \rightarrow P{=}O + P° \quad k_{1c}$

Similarly, the hypothesis of a stationary state leads to:

$$\frac{d[\mathrm{P{=}O}]}{dt} = \frac{k_{1c}}{k_{1c} + k_{1b}[\mathrm{PH}]} k_b[\mathrm{POOH}]^2$$

The bimolecular initiation can also be considered in the form of the following kinetically equivalent reaction:

$POOH + POOH \rightarrow P° + POO° + \gamma_1 P{=}O + \gamma_2 s \qquad k_b$

Finally, a simplified model would be, in the case of "pure" unimolecular initiation, composed of reactions (1u), (2), (3) and (6) and, in the purely bimolecular case, of reactions (1b), (2), (3) and (6):

(1u) $POOH \rightarrow 2P° + \gamma_1 P{=}O + \gamma_2 s$

(1b) $POOH + POOH \rightarrow P° + POO° + \gamma_1 P{=}O + \gamma_2 s$

(2) $P° + O_2 \rightarrow POO°$

(3) $POO° + PH \rightarrow POOH + P°$

(6) $POO° + POO° \rightarrow$ inactive products

In reality, this is a simplified view of the oxidation process. Other reactions should be taken into account for an accurate representation of aging. These will be presented in what follows.

The simple unimolecular diagram can be solved analytically, and the essential steps are presented below (Audouin et al. 1995).

The first step is to write the equations related to the appearance and consumption of the reactive species:

$$\frac{d[\mathrm{POOH}]}{dt} = -k_u[\mathrm{POOH}] + k_3[\mathrm{POO°}][\mathrm{PH}]$$

$$\frac{d[\mathrm{P°}]}{dt} = 2k_u[\mathrm{POOH}] - k_2[\mathrm{P°}][\mathrm{O_2}] + k_3[\mathrm{POO°}][\mathrm{PH}]$$

$$\frac{d[\mathrm{POO}^\circ]}{dt} = k_2[P^\circ][\mathrm{O}_2] - k_3[\mathrm{POO}^\circ][\mathrm{PH}] - 2k_6[\mathrm{POO}^\circ]^2$$

Using a simplifying hypothesis, namely, that of the stationary state (which is actually debatable), we deduce the following:

$$\frac{d[\mathrm{P}^\circ]}{dt} + \frac{d[\mathrm{POO}^\circ]}{dt} = 0$$

Therefore, at equilibrium, $k_u[\mathrm{POOH}]_{\infty u} = k_6[\mathrm{POO}^\circ]_{\infty u}^2$.

However, applying the same principle to the balance equations for hydroperoxides formation and decomposition yields:

$$k_u[\mathrm{POOH}]_{\infty u} = k_3[\mathrm{POO}^\circ][\mathrm{PH}]$$

Combining the last two equations then leads to the hydroperoxide concentration in the stationary state:

$$[\mathrm{POOH}]_{\infty u} = \frac{{k_3}^2[PH]^2}{2k_u k_6}$$

This reflects the equilibrium between their formation and their decomposition. The POOH conservation equation then becomes:

$$\frac{d[\mathrm{POOH}]}{dt} = -k_u[\mathrm{POOH}] + k_3\sqrt{\frac{k_u}{k_6}}[\mathrm{PH}][\mathrm{POOH}]^{1/2}$$

and is integrated by writing: $y = [\mathrm{POOH}]^{1/2}$:

$$\frac{dy}{dt} = -\frac{k_u}{2}\left(y - [\mathrm{POOH}]_\infty{}^{1/2}\right)$$

or, considering at:

$$t = 0, [\mathrm{POOH}] = [\mathrm{POOH}]_0$$

and writing: $\delta = 1 - ([\mathrm{POOH}]_0/[\mathrm{POOH}]_\infty)^{1/2}$

$$y = [\mathrm{POOH}]_\infty{}^{1/2} \cdot \left(1 - \delta.\exp\left(-\frac{k_u t}{2}\right)\right)$$

$$[\text{POOH}] = [\text{POOH}]_\infty \cdot \left(1 - 2\delta.\exp\left(-\frac{k_u t}{2}\right) + \delta^2.\exp(-k_u t)\right)$$

The expression of the induction period can then be deduced. This expression is, in fact, very close to the duration after which the mechanical properties disappear:

$$t_{i\text{POOH}} = \frac{0.386}{k_u}$$

In the case of a bimolecular diagram, the same reasoning leads to the equation of hydroperoxide consumption:

$$\frac{d[\text{POOH}]}{dt} = -2k_b[\text{POOH}]^2 + k_3\sqrt{\frac{k_b}{k_6}}[\text{PH}][\text{POOH}]$$

Also, considering:

$$[\text{POOH}]_{\infty b} = \frac{k_3[\text{PH}]}{2\sqrt{k_b k_6}}$$

$$[\text{POOH}] = \frac{[\text{POOH}]_\infty}{1 + \frac{[\text{POOH}]_\infty - [\text{POOH}]_0}{[\text{POOH}]} \cdot \exp(-2k_b[\text{POOH}]_\infty t)}$$

The induction period can be expressed by the following equation:

$$t_{i\text{POOH}} = \frac{1}{k_3[\text{PH}]}\left[\ln\left(\frac{[\text{POOH}]_\infty - [\text{POOH}]_0}{[\text{POOH}]_0}\right) - 2\right]$$

It can be easily verified that these models can be used to suitably reproduce the shape of the POOH accumulation curves already presented (Figure 1.10) by imposing rate constants reflecting the reactivity of the considered substrate. In simple cases, it is possible to estimate the induction time $t_{i\text{POOH}}$, which is a first approximation of the polymer lifetime under service conditions.

1.3.2.2. *Numerical models*

In reality, the above-presented models have primarily a heuristic value, while they only partially represent the complexity of phenomena occurring in most practical cases. Let us consider, for example, the oxidative aging of polypropylene.

During air aging, even when considering only the surface of a thick part, other termination reactions involving P° radicals should be integrated:

P° + P° → inactive product

P° + POO° → inactive product

Hence, for pure polypropylene (without an antioxidant, for which other equations should still be added), the "simplest" diagram contains at least equations 1u, 1b, 2, 3, 4, 5 and 6. These can clearly be solved only using numerical solvers, such as the ODE function in MATLAB.

Similarly, in the case of non-saturated elastomers (polychloroprene and EPDM), the role of double bonds shall be taken into account since it contributes not only to the high sensitivity of these materials to oxidation but also to the appearance of crosslinking reactions (Coquillat et al. 2007):

P° + > C=C < → cross-linking

POO° + > C=C < → cross-linking

For most polymers, aging is considered heterogeneous at the macroscopic scale, as it is accompanied by oxygen diffusion-based control. In fact, it reacts faster during oxidation processes and does not diffuse in the material. Thus, the oxidation rate decreases with depth (as presented in the previous section). The oxygen consumption equation then becomes:

$$\frac{\partial[O_2]}{\partial t} = D_{O2} \cdot \frac{\partial^2[O_2]}{\partial x^2} - k_2[\mathrm{P}^\circ][O_2] + k_6[\mathrm{POO}^\circ]^2$$

Finally, antioxidants, which are almost always added to the material, are another source of complication for kinetic models. For phenols (for example, Chapter 3), two additional balance equations must be (at least) integrated (Richaud et al. 2011b):

POO° + AH → POOH + A° k_{S1}

POO° + A° → inactive product k_{S2}

This leads to modifications of the differential equations presented above. There are many examples in which the model was completed to describe the stabilization process, possibly including the disappearance of the antioxidant through chemical paths (reaction with deleterious species) and physical paths (evaporation, diffusion), but their presentation is beyond the introductive scope of this chapter (Richaud et al. 2009).

1.3.3. *Strategy for estimating rate constants and orders of magnitude*

One of the main difficulties associated with the model is the identification, and even the prediction, of rate constants. We present here a brief overview of the corresponding methods, accompanied by several research perspectives.

1.3.3.1. *Initiation constants*

There are at least two methods for estimating the initiation constants related to the decomposition of solid polymers:

– the study of the thermolysis of model hydroperoxides in (liquid) solution;

– thermolysis, when "stable" enough POOH is formed, but which requires the capacity to precisely monitor the in situ peroxide concentration, which was described in a recent article (Huang et al. 2020).

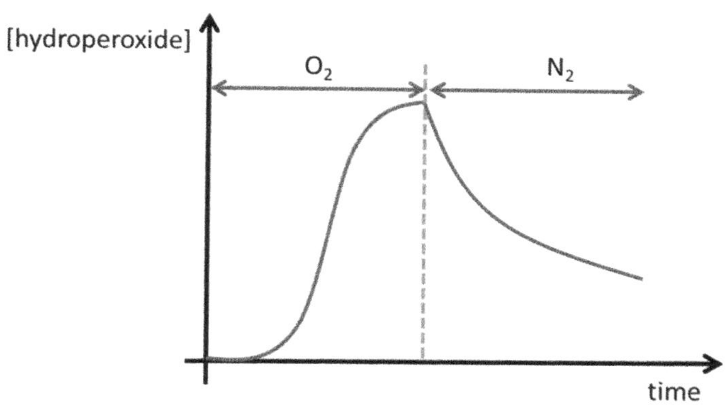

Figure 1.14. *Schematic representation of a method for estimating initiation constants from the decrease in hydroperoxides under nitrogen*

Ideally, if hydroperoxides decompose through a unimolecular process, the results corresponding to the part under nitrogen are as follows:

$$\ln[\text{POOH}] = \ln[\text{POOH}]_0 - k_u$$

Also, if decomposition follows a purely bimolecular process:

$$\frac{1}{[\mathrm{POOH}]} = \frac{1}{[\mathrm{POOH}]_0} + 2k_b t$$

However, the method becomes complicated because the two processes often overlap.

1.3.3.2. *Propagation constants*

This reaction was studied in particular by the Ingold team (Howard 1968). These authors proposed a relationship between the rate constant and the C–H BDE as a function of the (secondary or tertiary) nature of the POO° radical:

$$\log k_3\ {}^{t}\mathrm{POO°}\ (30°\mathrm{C}) = 15.2 - 0.2 \times \mathrm{BDE}$$

$$\log k_3\ {}^{s}\mathrm{POO°}\ (30°\mathrm{C}) = 16.2 - 0.2 \times \mathrm{BDE}$$

$$E_3\ {}^{t}\mathrm{POO°} = 0.55 \times (\mathrm{BDE} - 65)$$

$$E_3\ {}^{t}\mathrm{POO°} = 0.55 \times (\mathrm{BDE} - 62.5)$$

The first one could be completed by finer approaches integrating the calculation of bond energies that are not based on analogies with simple model compounds but, on the contrary, integrate environmental and stereochemical configuration effects, taking into account the nature of the states of transition of the reaction.

1.3.3.3. *Termination constants and link with molecular mobility*

Figure 1.8 provides, in principle, a strategy for estimating termination rate constants.

Quite schematically, it can be used to identify a "critical pressure" on which the oxygen pressure effect becomes minor, beyond which it can be proven that aging kinetics depend only on the initiation kinetics (see above) of k_3 and k_6. This latter parameter is thus identified, and we can then use the tests at increasingly lower pressures to identify the values of k_4 and k_5.

The dependence of these reactions on macromolecular mobility (with calculations depending on the gap between aging temperatures and those expressing polymer relaxations) is, in our opinion, a path that has only been partially explored (Colin et al. 2020).

1.4. Conclusion

This chapter summarizes the mechanisms involved in one of the large families of aging processes: thermo-oxidative processes. As the reader will note in what follows, they are relatively similar to those involved in aging under UV radiation. In addition to an approach to the study of mechanisms, a kinetic approach is presented and allows the establishing of a link between the observation of mechanisms and their transposition into a rate law to advance toward service life prediction. The latter will be completely developed in the following chapters, taking into consideration important aspects for industrial materials: the addition of antioxidants and other additives slowing down aging (Chapter 4) and the transposition of these phenomena with respect to the reduction in mechanical properties (Chapter 5).

1.5. References

Audouin, L., Langlois, V., Verdu, J., de Bruijn, J.C.M. (1994). Role of oxygen diffusion in polymer ageing: Kinetic and mechanical aspects. *Journal of Materials Science*, 29(3), 569–583.

Audouin, L., Gueguen, V., Tcharkhtchi, A., Verdu, J. (1995). "Close loop" mechanistic schemes for hydrocarbon polymer oxidation. *Journal of Polymer Science Part A Polymer Chemistry*, 33(6), 921–927.

Backstrom, H. (1934). Der Kettenmechanismus bei der Autoxydation von Aldehyden. *Zeitschrift für Physikalische Chemie.*, 25B, 99–121.

Bateman, L., Hughes, H., Moris, A. (1953). Hydroperoxide decomposition in relation to the initiation of radical chain reactions. *Discussion of Faraday Society*, 14, 190–199.

Bolland, J.L. and Gee, G. (1949). Kinetic studies in the chemistry of rubber and related materials. II. The kinetics of oxidation of unconjugated olefins. *Transactions of the Faraday Society*, 42, 236–243.

Celina, M., Wise, J., Ottesen, D.K., Gillen, K.T., Clough, R.L. (2000). Correlation of chemical and mechanical property changes during oxidative degradation of neoprene. *Polymer Degradation and Stability*, 68(2), 171–184.

Colin, X., Marais, C., Verdu, J. (2001). A new method for predicting the thermal oxidation of thermoset matrices: Application to an amine crosslinked epoxy. *Polymer Testing*, 20(7), 795–803.

Colin, X., Essatbi, F., Delozanne, J., Moreau, G. (2020). Towards a general kinetic model for the thermal oxidation of epoxy-diamine networks. Effect of the molecular mobility around the glass transition temperature. *Polymer Degradation and Stability*, 181, 109314.

Coquillat, M., Verdu, J., Colin, X., Audouin, L., Nevière, R. (2007). Thermal oxidation of polybutadiene. Part 2: Mechanistic and kinetic schemes for additive-free non-cross-linked polybutadiene. *Polymer Degradation and Stability*, 92(7), 1334–1342.

Delor-Jestin, F., Lacoste, J., Barrois-Oudin, N., Cardinet, C., Lemaire, J. (2000). Photo-, thermal and natural ageing of ethylene–propylene–diene monomer (EPDM) rubber used in automotive applications. Influence of carbon black, crosslinking and stabilizing agents. *Polymer Degradation and Stability*, 67(3), 469–477.

Delozanne, J., Montana, J.S., Guinault, A., Desgardin, N., Cuvillier, N., Richaud, E. (2023). Thermal ageing of bonded assemblies. Effect of adhesive curing degree. *Journal of Adhesion*, 99(6), 911–929.

Engler, C. and Weissberg, J. (1898). Ueber Activirung des Sauerstoffs. *Berichte der deutschen chemischen Gesellschaft*, 31(3), 3046–3059.

Gardette, M., Perthue, A., Gardette, J.L., Janecska, T., Földes, E., Pukánszky, B., Therias, S. (2013). Photo- and thermal-oxidation of polyethylene: Comparison of mechanisms and influence of unsaturation content. *Polymer Degradation and Stability*, 98(11), 2383–2390.

Gillen, K.T., Wise, J., Clough, R.L. (1995). General solution for the basic autoxidation scheme. *Polymer Degradation and Stability*, 47(1), 149–161.

Hock, H., Depke, F., Knauel, G. (1950). Autoxydation von Kohlenwasserstoffen, XI. Mitteil: Über tertiäre Hydroperoxyde. *Chemische Berichte*, 83(3), 238–244.

Howard, J.A., Schwalm, W.J., Ingold, K.U. (1968). Absolute rate constants for hydrocarbon autoxidation VII. Reactivities of peroxy radicals toward hydrocarbons and hydroperoxides. *Advances in Chemistry*, 75, 6–23. doi: 10.1021/ba-1968-0075.ch002.

Huang, J., Minne, W., Drozdzak, R., Recher, G., Le Gac, P.Y., Richaud, E. (2020). Thermal oxidation of poly(Dicyclopentadiene) – Decomposition of hydroperoxides. *Polymer Degradation and Stability*, 174, 109102.

Korcek, S., Chenier, J.H.B., Howard, J.A., Ingold, K.U. (1972). Absolute rate constants for hydrocarbon autoxidation. XXI. Activation energies for propagation and the correlation of propagation rate constants with carbon–hydrogen bond strengths. *Canadian Journal of Chemistry*, 50(14), 2285–2297.

Le Gac, P.Y., Celina, M., Roux, G., Verdu, J., Davies, P., Fayolle, B. (2016). Predictive ageing of elastomers: Oxidation driven modulus changes for polychloroprene. *Polymer Degradation and Stability*, 130, 348–355.

Minervino, M., Gigliotti, M., Lafarie-Frenot, M.C., Grandidier, J.C. (2013). The effect of thermo-oxidation on the mechanical behaviour of polymer epoxy materials. *Polymer Testing*, 32(6), 1020–1028.

Niki, E., Decker, C., Mayo, F.R. (1973). Aging and degradation of polyolefins. I. Peroxide-initiated oxidations of atactic polypropylene. *Journal of Polymer Science: Polymer Chemistry Edition*, 11(11), 2813–2845.

Philippart, J.L., Posada, F., Gardette, J.L. (1995). Mass spectroscopy analysis of volatile photoproducts in photooxidation of polypropylene. *Polymer Degradation and Stability*, 49(2), 285–290.

Richaud, E., Farcas, F., Bartoloméo, P., Fayolle, B., Audouin, L., Verdu, J. (2006). Effect of oxygen pressure on the oxidation kinetics of unstabilised polypropylene. *Polymer Degradation and Stability*, 91(2), 398–405.

Richaud, E., Monchy Leroy, C., Colin, X., Audouin, L., Verdu, J. (2009). Kinetic modelling of stabilization coupled with stabilizer loss by evaporation. Case of dithioester stabilized polyethylene. *Polymer Degradation and Stability*, 94(11), 2004–2014.

Richaud, E., Colin, X., Monchy-Leroy, C., Audouin, L., Verdu, J. (2011a). Diffusion-controlled radiochemical oxidation of bisphenol A polysulfone. *Polymer International*, 60(3), 371–381.

Richaud, E., Fayolle, B., Verdu, J. (2011b). Polypropylene stabilization by hindered phenols – Kinetic aspects. *Polymer Degradation and Stability*, 96(1), 1–11.

Richaud, E., Guinault, A., Baiz, S., Nizeyimana, F. (2019). Epoxidized linseed oils based networks. Case of thermal degradation. *Polymer Degradation and Stability*, 166, 121–134.

Russell, G.A. (1957). Deuterium-isotope effects in the autoxidation of aralkyl hydrocarbons. Mechanism of the interaction of peroxy radicals. *Journal of American Chemical Society*, 79, 3871–3877.

Semenoff, N. (1936). Chemical kinetics and chain reactions. *Nature*, 137, 127–128.

2

Polymer Aging in the Presence of Water

Emmanuel RICHAUD
PIMM, CNRS, CNAM, ENSAM, Paris, France

2.1. Introduction

The aging of polymers in the presence of water is a practical case concerning the degradation induced by improper drying of the material before its processing, its degradation under conditions of use and, obviously, the "postend-of-life" degradation, with the formation of microplastics, which has become an unavoidable problem.

This chapter begins with a synthesis of the (reversible) physical aging mechanisms, involving water penetration into polymers, followed by a presentation of the (irreversible) chemical aging mechanisms and associated stabilization paths. For each of them, we summarize aspects related to thermodynamics (the quantity of water actually dissolved in the polymer), kinetics (water diffusion and the rate at which it may cause chain scissions) and consequences for the mechanical and thermomechanical properties.

2.2. Physical aging of polymers in the presence of water

First, it is important to note that "physical" aging corresponds to the case where the chain architecture (average molar mass of chains for linear polymers and molar mass between cross-linking nodes for networks) remains constant. The variations in properties under the aging effect are considered reversible in this context. It is typically the case of a polymer in the presence of water, which migrates and promotes polymer swelling (plasticization). In this type of aging, two parameters should be taken into account: solubility, which reflects the state of equilibrium ("what quantity of water can get into the polymer?"), and diffusivity, which expresses the duration of the transient state ("at what speed will water penetrate the material?"). These two aspects will be addressed in this first part, which will also present several examples of lesser mechanical properties caused by water absorption.

2.2.1. *Water solubility in the polymers*

When an initially dry polymer is put in contact with water, the latter migrates to increase its chemical potential. Although counterexamples will be presented further, the most general curve showing the variation in the absorbed mass as a function of time is as follows.

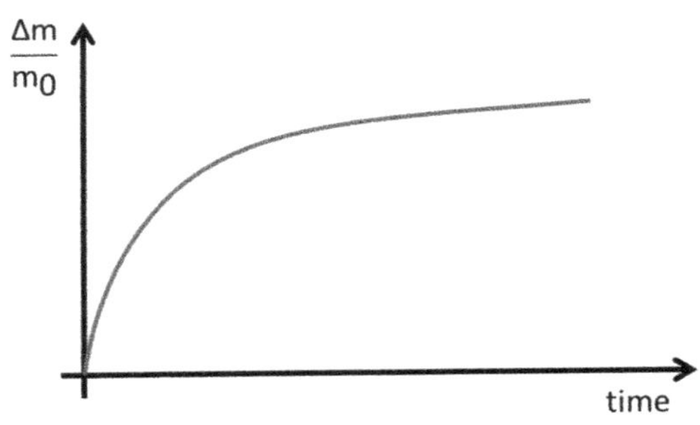

Figure 2.1. *Mass gain during aging in the presence of water*

The following can be noted:

– a transient regime (more or less complex) that depends on the diffusion mechanism and kinetics;

– a plateau at long times indicates the affinity of the polymer for water.

The interaction between a polymer and water is described using sorption isotherms, giving the quantity of water absorbed as a function of the partial pressure of water. Three possible cases are depicted in Figure 2.2.

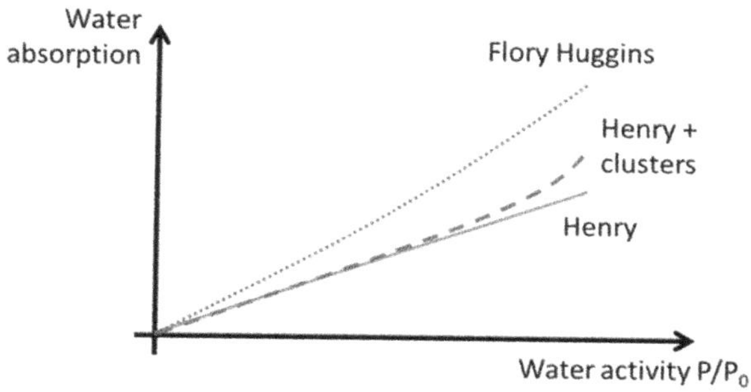

Figure 2.2. *Typical shape of sorption isotherms*

1) Henry's law, which is generally valid for ideal gas (small molecules at low pressure):

$$c = S \times P \quad [2.1]$$

in which c is the water concentration at equilibrium, S is the coefficient of solubility expressed in m^3 (gas)/m^3 (polymer)/Pa and P is the given pressure of water vapor that follows van't Hoff's law:

$$S = S_0 . \exp\left(-\frac{H_S}{RT}\right) \quad [2.2]$$

$$P = P_0 . \exp\left(-\frac{\Delta H_{vap}}{RT}\right) \quad [2.3]$$

H_S can be positive (endothermal polymer–water interaction) or negative (exothermal polymer–water interaction).

ΔH_{vap} is the heat of vaporization of the solvent (here, water). It is closely linked to cohesive energy: ΔH_{vap} is relatively low for low-polarity liquids, such as hydrocarbons, and high for liquids with hydrogen bond (e.g. $\Delta H_{vap} = 43$ kJ.mol^{-1} for water).

To summarize, the solvent concentration at equilibrium varies according to the following expression:

$$c = c_0 . \exp\left(-\frac{H_S + \Delta H_{vap}}{RT}\right) \quad [2.4]$$

Thus,

– if H_S is positive, then $\Delta H_{solub} > 0$, and the concentration of the solvent at equilibrium increases with temperature (ΔH_{vap} is still positive);

– this remains true if H_S is negative, but $\Delta H_{vap} > -H_S$;

– on the other hand, if the H_S is negative, with a high absolute value so that: $|H_S| > \Delta H_{vap}$, then c decreases when the temperature increases.

2) If Henry's law is not verified and the isotherm has a convex shape, it can be described using Henry's law at low levels of activity, completed by an additional term at high levels of activity describing the appearance of clusters (water aggregates) (Lacuve 2019):

$$V = a.H + B.a^n \qquad [2.5]$$

where B and n can be identified from the experimental results.

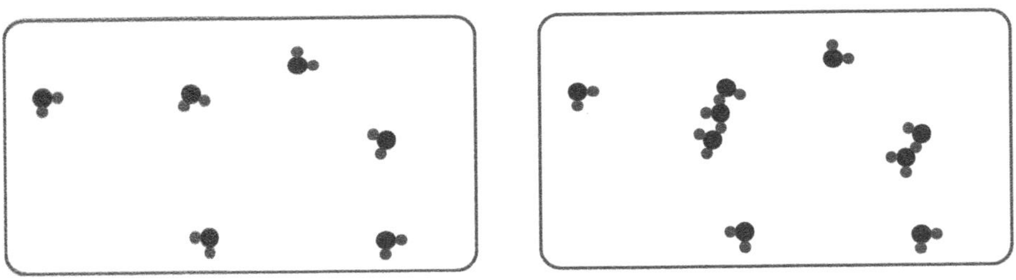

Figure 2.3. *Schematic representation of the appearance of water clusters. For a color version of this figure, see www.iste.co.uk/richaud/industrialpolymers1.zip*

Water clusters (schematically represented in Figure 2.3) appear when water–water interactions are stronger than polymer–polymer interactions. They are detectable using a theoretical approach of sorption isotherms in Figure 2.2 using Zimm–Lundberg method (Zimm 1966; Lundberg 1969):

$$f_{ZL} = -(1 - v)\left[\frac{\partial a/v}{\partial a}\right]_{T,P} - 1 \qquad [2.6]$$

The average size of the clusters is then given by:

$$MCS = 1 + v.f_{ZL} \qquad [2.7]$$

MCS is equal to 1 for an isolated water molecule and 2 for a cluster composed of two water molecules (Davis and Elabd 2013). Equation [2.6] expresses that the presence of clusters is not associated with higher or lower values of the final solubility but rather with the convexity of the sorption isotherm reflected by the term ∂(a/v)/∂a. The size of the clusters may reach 4, including for low-polarity polymers, such as PS, and even more for polychloroprene (Le Gac et al. 2014).

3) Finally, a third shape of the isotherm is given by the Flory–Huggins law:

$$\ln(a) = \ln(v) + (1 - v) + \chi(1 - v)^2 \quad [2.8]$$

where χ is the water–polymer interaction parameter given by equation [2.9] and where $\delta_{polymer}$ and δ_{water} are the water solubility and polymer solubility parameters, respectively, the values of which are summarized in Table 2.1:

$$\chi = \frac{V_m}{RT} \times (\delta_{polymer} - \delta_{water})^2 \quad [2.9]$$

	δ (J.cm^{-3})
PE	15.8–17.1
PP	16.8–18.8
PS	17.4–19
PVC	19.2–22.1
PTFE	12.7
PMMA	18.6–26.2
PET	19.9–21.9
PA6	27.8

Table 2.1. *Solubility parameters for several polymers*

It can be deduced that:

1) apolar polymers (PE, PP, etc.) have a low affinity for water;

2) moderately polar polymers (polyesters) have a moderate affinity;

3) highly polar polymers (epoxies, polyamides) have a high affinity.

At this stage, it should be noted that there is no link between the quantity of water that can penetrate the polymer and its sensitivity to hydrolysis. For example, polyesters (unsaturated polyesters cured with styrene or thermoplastic polyesters such as PET or PBT) absorb less water than epoxies, but water absorption in epoxy matrices is considered reversible, whereas in polyesters, it leads to a hydrolysis that is irreversible.

2.2.2. *Diffusivity of water into polymers*

The simplest method for the experimental study of diffusion in a polymer consists of immersing the polymer and monitoring the mass gain through regular weighing. The results are schematically represented in Figure 2.4. Importantly, if the polymer undergoes hydrolysis, a mass loss phenomenon is superimposed and induces a decrease in the curves of mass gain. A commonly used method to differentiate the various mechanisms of water diffusion is to track the gain in reduced mass $(m - m_0)/(m_\infty - m_0)$ as a function of the square root of the immersion time.

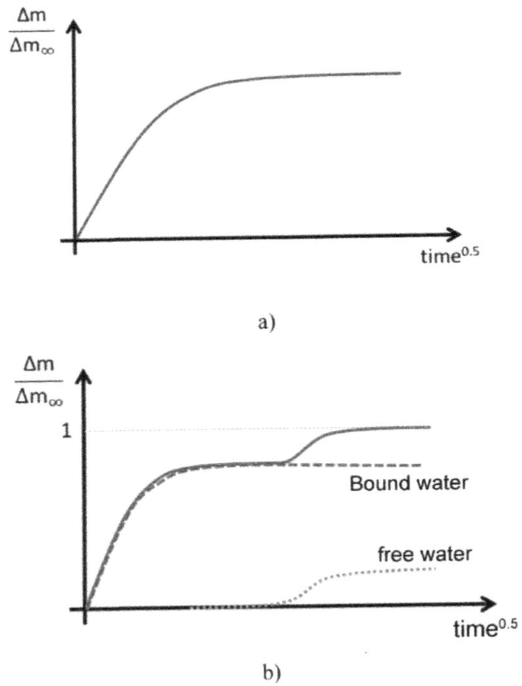

Figure 2.4. *(a) Shape of the sorption curves in the case of a diffusion mechanism described by Fick's law or (b) Langmuir's law, with a schematic representation of the diffusion of free and bound water (Popineau et al. 2005)*

In the case of Fickian diffusion, the analytical solutions of diffusion equations were developed by Crank (1979). For a thin plate whose thickness is 2e, where water is initially at concentration c_0 and whose surfaces are maintained at concentration c_1, the mass gain is given by:

$$\frac{m(t) - m_0}{m_\infty - m_0} = 1 - \frac{8}{\pi^2} \cdot \sum_{n=0}^{\infty} \frac{1}{(2n+1)^2} \cdot exp\left(-\frac{D.(2n+1)^2.\pi^2.t}{e^2}\right) \quad [2.10]$$

where m_∞ corresponds to the mass absorbed at saturation, that is, at the solubility threshold of the penetrant in the polymer.

The plot of this equation is an accurate reproduction of the shape of the experimental results, such as those in Figure 2.4. The greater the number of terms of the sum, the more accurate the shape of the curve is at short times. Equation [2.10] yields an approximate solution as follows:

1) At low levels of water absorption ($m/m_\infty < 0.6$):

$$\frac{\Delta m(t)}{\Delta m_\infty} = \frac{4}{e} \cdot \sqrt{\frac{D.t}{\pi}} \quad [2.11]$$

2) At high levels of water absorption:

$$\frac{\Delta m(t)}{\Delta m_\infty} = 1 - \frac{8}{\pi^2} \cdot \exp\left(-\frac{D\pi^2}{e^2} t\right) \quad [2.12]$$

For samples whose geometry is far from ideal, the diffusion coefficient of edge effects must be corrected when the thickness is not sufficiently small compared with the length (L) and width (l). Shen and Springer (1976) thus propose a geometric correction to estimate the real coefficient D_{real} from an apparent diffusion coefficient D_{app} deduced from sorption curves:

$$D_{\text{real}} = \frac{D_{\text{app}}}{(1 + \frac{e}{L} + \frac{e}{l})^2} \quad [2.13]$$

In summary:

1) If the reduced mass gain curve features a plateau at long times and a linear dependence on $t^{1/2}$ at short times, the diffusion is referred to as Fickian diffusion (the term "case II" is sometimes employed) (Masaro and Zhu 1999), and the water diffusion coefficient is estimated from the initial slope.

2) If the reduced mass gain curve features a “double” plateau at long times and a linear dependence on $t^{1/2}$ at short times, then it is referred to as Langmuir diffusion, and the diffusion coefficient of bound water at polar sites can be estimated on the basis of the initial slope.

3) If the mass gain varies as a function of t, this is referred to as case I Fickian diffusion, which is characterized by the fact that the time required for penetrant motions is much shorter than that related to the mobility of chain segments [2.9].

The diffusion coefficient depends on the penetrant concentration: indeed, the penetrant, which is still a small molecule, results in its own small volume, which facilitates diffusion. The variations in D with the penetrant concentration can be described by an empirical law:

$$D = D_{c=0}.\exp(\gamma.c) \quad [2.14]$$

This type of apparent law was developed in more complex theories, such as that of Vrentas and Duda (1977), who consider that the solvent, which diffuses in its own polymer, results in its own free volume and that diffusion is self-facilitated.

Finally, the water diffusivity usually follows the Arrhenius law, in which the apparent activation energy is of the order of 50 $kJ.mol^{-1}$ (Roggero et al. 2021; Siedlaczek et al. 2022):

$$D = D_0.\exp(-E_D/RT) \quad [2.15]$$

2.2.3. The consequences of physical aging by water absorption on the mechanical properties of polymers

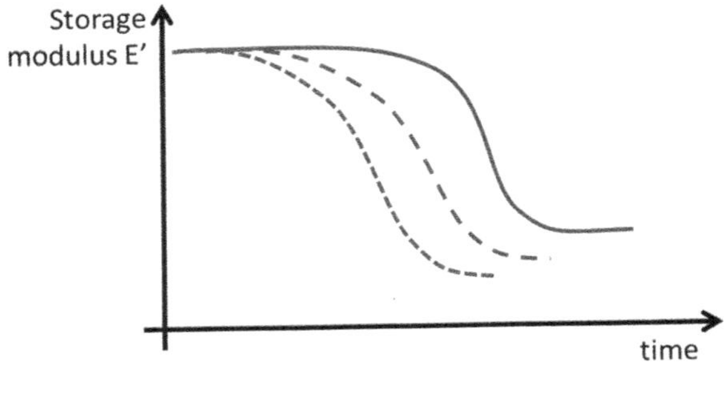

Figure 2.5. *Schematic representation of the effect of water on the thermomechanical properties of a polymer*

Water is a small molecule that promotes polymer swelling, thus playing a role as a plasticizer. Figure 2.5 shows the main consequences of water diffusion in a polymer for its elastic properties.

It mainly induces:

1) A decrease in the glass transition temperature (Golaz et al. 2013; Alessi et al. 2014) that is caused by the fact that T_g of the water (134 K) is much lower than that of a thermosetting resin. By considering that free volumes add up, the law referred to as the Gibbs–DiMarzio law is obtained, which is very often employed in a first approximation:

$$\frac{1}{T_{g1,2}} = \frac{w_1}{T_{g1}} + \frac{w_2}{T_{g2}} = \frac{1}{T_{g1}} + w_2\left(\frac{1}{T_{g2}} - \frac{1}{T_{g1}}\right) \qquad [2.16]$$

Moreover, the higher the glass transition temperature of the unaged network (T_{g1}) is, the greater the decrease in $T_{g1,2}$ with increasing water content, as we have:

$$\frac{dT_{g1,2}}{dw_2} = \frac{\frac{1}{T_{g1}} - \frac{1}{T_{g2}}}{\left[\frac{1}{T_{g1}} + w_2\left(\frac{1}{T_{g1}} - \frac{1}{T_{g2}}\right)\right]^2} \approx T_{g1}{}^2 . \left(\frac{1}{T_{g1}} - \frac{1}{T_{g2}}\right) \qquad [2.17]$$

2) A plasticization of the material (Figure 2.6), with an increase in elongation at break and a decrease in yield stress, given, for example, by Kambour's law:

$$\sigma_Y = C. (T_g - T) + \sigma_{Y0} \qquad [2.18]$$

with C ~ 0.5–1 MPa K^{-1} (Cooke et al. 1998; Li and Strachan 2011). In other words, if water induces a decrease in the T_g, then the yield stress will also decrease (Le Gac et al. 2017).

3) In the specific field of glued assemblies, a decrease in performance indicates adhesion (such as the adherence energy illustrated in Figure 2.7 or the assembly break stress) (Bordes et al. 2009). Microscopic observations generally reveal that the break of a "healthy" glued assembly is cohesive (within the adhesive), and after water diffusion, the assembly becomes adhesive, that is, at the substrate–adhesive interface.

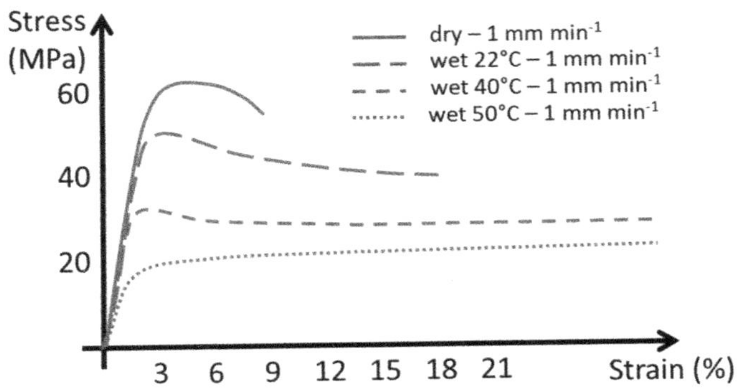

Figure 2.6. *Stress–strain curves of an epoxy resin before and after various humid agings (Gibhardt et al. 2023)*

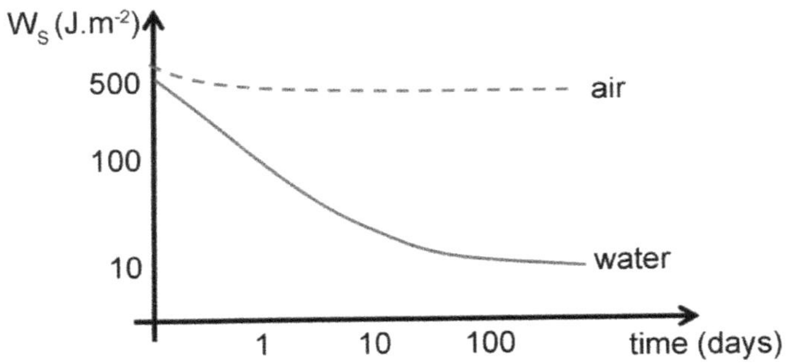

Figure 2.7. *Loss of adhesion properties of a bonded assembly during humid aging*

2.3. Chemical aging in the presence of water (hydrolysis)

Water diffusion is, in principle, a reversible phenomenon. However, some chemical groups are known to be sensitive to water and react irreversibly by hydrolysis. These factors lead to chain scissions that have a serious impact on the mechanical properties.

This section presents several theoretical approaches for describing the chain scission rate induced by hydrolysis. This approach will also facilitate the understanding and prediction of the damage to mechanical properties.

2.3.1. *General case*

In the case of polymers, hydrolysis is the reaction between an ester, amide or imide group and water molecules present in the polymer. At this stage, two cases should first be identified:

1) Hydrolysis affects a "side" group, as is the case, for example, in PMMA (Semen and Lando 1969; Ali et al. 2015) (Figure 2.8), where hydrolytic aging is assumed to modify first of all the polarity of the material.

2) Hydrolysis affects a bond of the main chain, which is the main reason for polymer failure. The first stage to predict its appearance involves predicting the rate of a hydrolysis reaction. This value is calculated from the concentration of hydrolyzable groups $[E]$ (esters or amides) in water molecules $[W]_0$ and the concentration of chain scissions:

$$[E] = [E]_0 - s \qquad [2.19]$$

$$\frac{ds}{dt} = k_H.([E]_0 - s).[W]_0 \qquad [2.20]$$

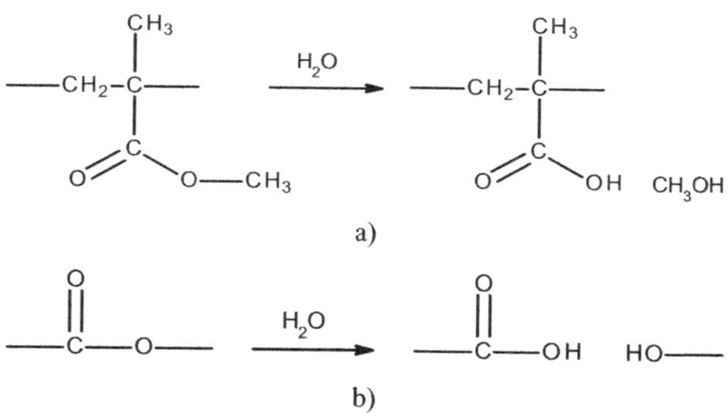

Figure 2.8. *Hydrolysis of a lateral group in the case of (a) PMMA and (b) the main chain*

Here, k_H is the rate constant of the hydrolysis reaction. The integration of this equation yields:

$$s = [E]_0.[1 - \exp(-k_H.[W]_0.t)] \qquad [2.21]$$

For a mechanism of exclusive chain scission, the following can be written:

$$\frac{1}{M_N} - \frac{1}{M_{N0}} = s \qquad [2.22]$$

If the end-of-life criterion $M_N = M_F$ is applied to the molar mass (Chapter 5), the lifetime t_F can be deduced (Bellenger et al. 1995). Notably, the lifetime depends on three terms:

1) k_H expresses the reactivity and is exclusively dependent on temperature.

2) $[W]_0$ depends on humidity, which resulted from a previous study on water sorption (section 2.2.1).

3) $[E]_0$ is exclusively a function of the polymer structure.

Some studies, for example, on the hydrolysis of unsaturated polyesters, have shown that maleate and fumarate esters are more unstable than isophthalates and orthophthalates. Isophthalate esters are themselves more stable than orthophthalate esters (Ashbee et al. 1967; Bélan et al. 1996).

2.3.2. Some practical cases of hydrolysis

In reality, many practical cases lead to a greater complexity of the kinetic laws of hydrolysis, either for chemical reasons or for physical reasons. It is worth citing the following:

1) Steric effects can be noted in unsaturated polyesters. If the hydrolysis rate in the networks ($r_{network}$) and that of the corresponding prepolymer ($r_{prepolymer}$) are compared, the following can be noted:

$$R_{network} < (1 - sty)_{prepolymer} \qquad [2.23]$$

Here, "sty" is the styrene mass fraction. This is valid even when the rate of styrene considered here as a non-reactive diluent is taken into account. In other words, it would induce steric hindrance, slowing hydrolysis reactions (equation [2.21]).

2) The reactions characteristic of chain ends that would be more reactive than the ester groups at mid-chains, as described for unsaturated polyesters in equation [2.21] (Figure 2.9).

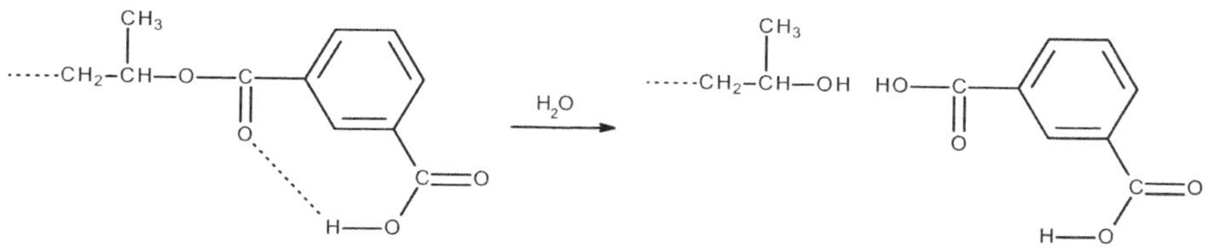

Figure 2.9. *Decomposition of chain ends in polyesters*

3) The existence of a phenomenon of reversible hydrolysis, which has been particularly described in the case of polyamide 11, for which the following balance equation can be proposed:

$$E + W \underset{k_R}{\overset{k_H}{\leftrightarrows}} \text{Amine} + \text{Acid} + s$$

If Am and Ac denote the carboxylic acids or amine chain ends created by the hydrolysis reaction and s the chain scissions, we obtain:

$$\frac{ds}{dt} = k_H[E][W] - k_R[Am][Ac] \qquad [2.24]$$

This is solved by choosing the following conditions:

At $t = 0$, $[E] = [E]_0$ and $[Am]_0 = [Ac]_0 = 1/M_{n0}$ [2.25]

At t $[Am] = [Ac] = \frac{1}{M_{n0}} + s$ [2.26]

The hydrolysis rate can be written as follows:

$$\frac{ds}{dt} = k_H([E]_0 - s)[W]_0 - k_R\left(\frac{1}{M_{n0}} + s\right)^2 \qquad [2.27]$$

The fact that hydrolysis is reversible leads to a value of the molar mass at equilibrium. Assuming that chains are long $[E]_0 >> M_{N0}^{-1}$, the value of the average molar mass at equilibrium can be calculated:

$$\frac{1}{M_{ne}} = \sqrt{\frac{k_H}{k_R}[E]_0[W]_0} \qquad [2.28]$$

This leads to:

$$M_n(t) = M_{ne}\cdot\left[\frac{\left(M_{n0}{}^{-1} + M_{ne}{}^{-1}\right) + \left(M_{ne}{}^{-1} - M_{n0}{}^{-1}\right).\exp(-K.t)}{\left(M_{n0}{}^{-1} + M_{ne}{}^{-1}\right) - \left(M_{ne}{}^{-1} - M_{n0}{}^{-1}\right).\exp(-K.t)}\right] \quad [2.29]$$

with:

$$K = 2M_{ne}.k_H.\,[E_0][W]_0 \quad [2.30]$$

It can also be shown that:

$$\left(\frac{dM_n(t)}{dt}\right)_{t\to 0} = \frac{M_{n0}{}^{-2} + M_{ne}{}^{-2}}{M_{n0}{}^{-2}}\cdot k_R \quad [2.31]$$

$[E]_0$ can be readily calculated from the structure of the repetitive constitutional unit. A strategy for estimating rate constants can be deduced from experimental results:

1) The measure of the molar mass at M_{ne}, that is, at k_H/k_R.

2) On the basis of the values of M_{N0} and M_{ne}, the decrease in molar mass at $t = 0$ gives access to k_R.

3) Based on the values of M_{ne}, k_R, $[W]_0$ and $[E]_0$, k_H can be calculated.

4) The existence of autocatalytic phenomena. In polyesters, each hydrolysis reaction produces a carboxylic acid that can itself catalyze the following reactions. This leads to the following rate equations, which are employed, for example, in studies on polyester hydrolysis (Antheunis et al. 2010; Richaud et al. 2014):

$$\frac{ds}{dt} = k_0.([E]_0 - s).w + k_{cat}.([E]_0 - s).s.[W] \quad [2.32]$$

5) Water solubility variations are induced by chemical structure modifications related to hydrolysis, which is the case when a (moderately hydrophilic) ester is transformed into an alcohol group and a carboxylic acid. Equation [2.21] can then be written as follows:

$$[W] = [W]_0 + a.s \quad [2.33]$$

where s is the number of chain scissions.

6) The effects of water diffusion. This phenomenon is, in fact, due to diffusion–reaction coupling, meaning that only the water molecules that do not react

at the surface can diffuse to the deeper layers of the polymer. This results in the following balance equation:

$$\frac{\partial[W]}{\partial t} = D_w \frac{\partial^2[W]}{\partial z^2} - r([W]) \quad [2.34]$$

where:

a) D_w is the diffusion coefficient of water in the polymer;

b) z is the depth of the layer considered in the thickness;

c) $r([W])$ is the water consumption rate.

In the case of nonreversible hydrolysis, at a low conversion rate ($[E] \sim [E]_0 =$ constant), $r(W) = K.[W]$ (pseudo-first order) can be written. In this case, the thickness of the degraded layer is quasi-constant and of the order of l_D (Audouin et al. 1994):

$$l_D = (D/K)^{1/2} \quad [2.35]$$

l_D follows the Arrhenius law, with an activation energy E_{lD}:

$$E_{lD} = \frac{1}{2}(E_D - E_K) \quad [2.36]$$

Since, in general, $E_D < E_K$, the thickness of the degraded layer decreases when the temperature increases.

7) Acid-catalyzed hydrolysis. In the case of a weak acid (e.g. a chain end created or not created by a previous hydrolysis event), the hydrolysis rate constant could therefore be formulated similarly to the kinetics of catalyzed polycondensation:

$$k_H = k_{H2O} + k_{H+}.[H_3O^+] + k_{HO-}.[HO^-] \quad [2.37]$$

For example, in the case of a pure acid catalysis, we would have:

$$\frac{ds}{dt} = k_H[E][W][H_3O^+] \quad [2.38]$$

$[H^+]$ is expressed using the laws of equilibrium (Merdas et al. 2003):

$$H_2O \leftrightarrows H^+ + HO^-$$

$$AH \leftrightarrows H^+ + A^-$$

which lead to:

$$[A^-] + [HO^-] = [H^+] \quad [2.39]$$

$$[H_3O^+].[HO^-] = K_e \quad [2.40]$$

Knowing that:

$$K_A = \frac{[H_3O^+][A^-]}{[AH]} \quad [2.41]$$

we deduce:

$$[H_3O^+] = \frac{K_A[AH]}{[H_3O^+] - {}^{K_e}/_{[H_3O^+]}} \quad [2.42]$$

or:

$$[H_3O^+] = \sqrt{K_A[AH] + K_e} \quad [2.43]$$

Finally, with:

$$\gamma = K_a/K_e \quad [2.44]$$

$$\frac{ds}{dt} = k_H[E][W]K_e\sqrt{1 + \gamma[AH]} \quad [2.45]$$

A catalytic effect can be observed if $\gamma.[AH] >> 1$.

2.3.3. Consequences for mechanical properties

The consequences of hydrolytic aging are those observed in the general case of polymers experiencing chain scissions, a subject that will be discussed in further detail in Chapter 4.

For the time being, Figure 2.10 shows that the ultimate elongation decreases continuously. The other quantities (ultimate stress, yield stress and elastic modulus) remain unchanged overall and feature no strong variations unless the material becomes highly degraded.

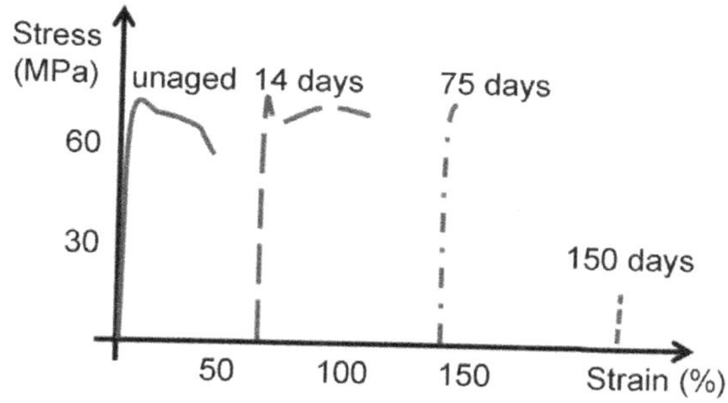

Figure 2.10. *Stress–strain curves of PA 6 hydrolyzed at 120°C (adapted from Deshoulles et al. 2021) (note that curves have been shifted for further clarity)*

2.3.4. *Stabilization against hydrolytic aging*

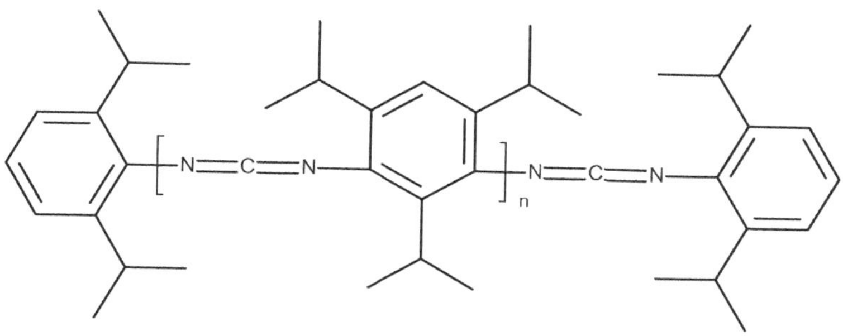

a)

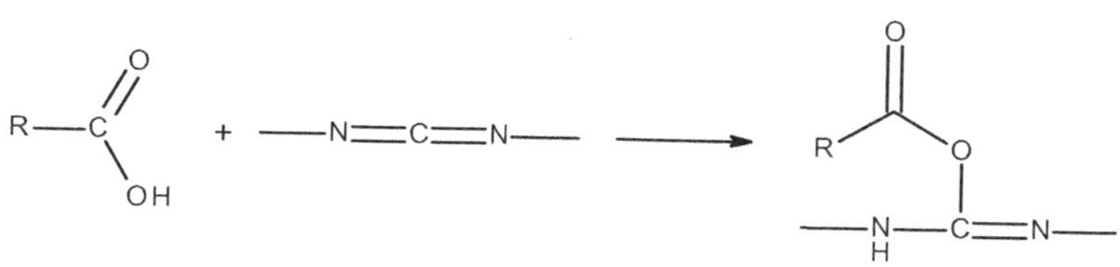

b)

Figure 2.11. *(a) General structure of carbodiimides; (b) possible mechanism of stabilization*

There are essentially two main paths for countering hydrolytic aging:

1) by blocking access to water via, for example, the use of a hydrophobic coating (this may generate further technical and economic constraints);

2) by limiting the water reactivity of the polymer.

If the focus is on the second path, there are two possibilities:

1) The choice of internal stabilization, that is, comonomers decreasing hydrophily. For example, the results of Bellenger on the effects of the diol employed for the synthesis of prepolymers from polyesters are important (Bélan et al. 1997).

2) The choice of stabilizers that limit hydrolysis reactions, such as carbodiimides, as illustrated in Figure 2.11, and their stabilization mechanism. These molecules react in particular with carboxylic acid groups, which may, for example, block carboxylic acids and the associated phenomena of autocatalysis (Yang et al. 2008), whose efficiency is illustrated, in the case of hydrolysis of TPU, by Bardin et al. (2020).

2.4. Conclusion

This chapter presents the essential elements of polymer aging in the presence of water:

1) On the one hand, (reversible) physical aging leads to a plasticization phenomenon. The models describing the states of equilibrium (water solubility in the polymer) and transient states (diffusion) are summarized.

2) On the other hand, a (irreversible) chemical mechanism involving (total or balanced, autocatalyzed) hydrolysis reactions can be slowed down by the action of stabilizers. These are unavoidable phenomena that lead to the total destabilization of the part. They must be particularly controlled during the formation of polymers, where incomplete drying may lead to brittle parts even before the beginning of their aging under “service” conditions.

Notably, the literature is essentially dedicated to the study of these phenomena while ignoring their correlation with other aging mechanisms. However, this approach is not representative of real cases, in which humid aging occurs under stress, or of materials that may already present a sensitive level of thermo-oxidation.

Water absorption in a thermo-oxidized material (containing polar functions) may be greater than that in a non-thermo-oxidized material. The study of such couplings is thus necessary in most practical cases.

2.5. References

Alessi, S., Pitarresi, G., Spadaro, G. (2014). Effect of hydrothermal ageing on the thermal and delamination fracture behaviour of CFRP composites. *Composites Part B: Engineering*, 67, 145–153.

Ali, U., Karim, K.J.B.A., Buang, N.A. (2015). A review of the properties and applications of poly (methylmethacrylate) (PMMA). *Polymer Reviews*, 55(4), 678–705.

Antheunis, H., van der Meer, J.C., de Geus, M., Heise, A., Koning, C.E. (2010). Autocatalytic equation describing the change in molecular weight during hydrolytic degradation of aliphatic polyesters. *Biomacromolecules*, 11(4), 1118–1124.

Ashbee, K.H.G., Frank, F.C., Wyatt, R.C. (1967). Water damage in polyester resins. *Proceedings of the Royal Society A*, 300(1463), 415–419.

Audouin, L., Langlois, V., Verdu, J., Bruijn, J.C.M. (1994). Role of oxygen diffusion in polymer ageing: Kinetic and mechanical aspects. *Journal of Materials Science*, 29(3), 569–583.

Bardin, A., Le Gac, P.Y., Cérantol, S., Simon, G., Bindi, H., Fayolle, B. (2020). Hydrolytic kinetic model predicting embrittlement in thermoplastic elastomers. *Polymer Degradation and Stability*, 171, 109002.

Bélan, F., Bellenger, V., Mortaigne, B., Verdu, J., Yang, Y.S. (1996). Hydrolytic stability of unsaturated polyester prepolymers. *Composites Science and Technology*, 56(7), 733–737.

Bélan, F., Bellenger, V., Mortaigne, B., Verdu, J. (1997). Relationship between the structure and hydrolysis rate of unsaturated polyester prepolymers. *Polymer Degradation and Stability*, 56(3), 301–309.

Bellenger, V., Ganem, M., Mortaigne, B., Verdu, J. (1995). Lifetime prediction in the hydrolytic ageing of polyesters. *Polymer Degradation and Stability*, 49(1), 91–97.

Bordes, M., Davies, P., Cognard, J.-Y., Sohier, L., Sauvant-Moynot, V., Galy, J. (2009). Prediction of long term strength of adhesively bonded steel/epoxy joints in sea water. *International Journal of Adhesion and Adhesives*, 29(6), 595–608.

Cooke, W.D., Mayr, A.E., Edwards, G.H. (1998). Yielding behaviour in model thermosets – II. Temperature dependence. *Polymer*, 39(16), 3725–3733.

Crank, J. (1979). *The Mathematics of Diffusion*. Oxford Science Publications, Oxford.

Davis, E.M. and Elabd, Y.A. (2013). Water clustering in glassy polymers. *The Journal of Physical Chemistry B*, 117, 10629–10640.

Deshoulles, Q., Le Gall, M., Dreanno, C., Arhant, M., Stoclet, G., Priour, D., Le Gac, P.Y. (2021). Origin of embrittlement in polyamide 6 induced by chemical degradations: Mechanisms and governing factors. *Polymer Degradation and Stability*, 191, 109657.

Gibhardt, D., Krauklis, A.E., Doblies, A., Gagani, A., Sabalina, A., Starkova, O., Fiedler, B. (2023). Time, temperature and water aging failure envelope of thermoset polymers. *Polymer Testing*, 118, 107901.

Golaz, B., Michaud, V., Lavanchy, S., Månson, J.A.E. (2013). Design and durability of titanium adhesive joints for marine applications. *International Journal of Adhesion and Adhesives*, 45, 150–157.

Lacuve, M., Colin, X., Perrin, L., Flandin, L., Notingher, P., Tourcher, C., Ben Hassine, M., Tanzeghti, H. (2019). Investigation and modelling of the water transport properties in unfilled EPDM elastomers. *Polymer Degradation and Stability*, 168, 108949.

Le Gac, P.Y., Roux, G., Davies, P., Fayolle, B., Verdu, J. (2014). Water clustering in polychloroprene. *Polymer*, 55, 2861–2866.

Le Gac, P.Y., Arhnant, M., Le Gall, M., Davies, P. (2017). Yield stress changes induced by water in polyamide 6: Characterization and modeling. *Polymer Degradation and Stability*, 137, 272–280.

Li, C. and Strachan, A. (2011). Molecular dynamics predictions of thermal and mechanical properties of thermoset polymer EPON862/DETDA. *Polymer*, 52(13), 2920–2928.

Lundberg, J. (1969). Clustering theory and vapor sorption by high polymers. *Journal of Macromolecular Science Part B: Physics*, 3(4), 693–710.

Masaro, L. and Zhu, X.X. (1999). Physical models of diffusion for polymer solutions, gels and solids. *Progress in Polymer Science*, 24(5), 731–775.

Merdas, I., Thominette, F., Verdu, J. (2003). Hydrolytic ageing of polyamide 11 – Effect of carbon dioxide on polyamide 11 hydrolysis. *Polymer Degradation and Stability*, 79(3), 419–425.

Popineau, S., Rondeau-Mouro, C., Sulpice-Gaillet, C., Shanahan, M.E. (2005). Free/bound water absorption in an epoxy adhesive. *Polymer*, 46(24), 10733–10740.

Richaud, E., Gilormini, P., Coquillat, M., Verdu, J. (2014). Crosslink density changes during the hydrolysis of tridimensional polyesters. Macromolecular. *Theory and Simulations*, 23(5), 320–330.

Roggero, A., Caussé, N., Dantras, E., Villareal, L., Santos, A., Pébère, N. (2021). In situ study of the temperature activated kinetics of water sorption in an epoxy varnish. *Polymer*, 213, 123206.

Semen, J. and Lando, J.B. (1969). The acid hydrolysis of isotactic and syndiotactic poly(methyl methacrylate). *Macromolecules*, 2(6), 570–575.

Shen, C.H. (1976). Moisture absorption and desorption of composite materials. *Journal of Composites Materials*, 10(1), 2–20.

Siedlaczek, P., Sinn, G., Peter, P., Wan-Wendner, R., Lichtenegger, H.C. (2022). Characterization of moisture uptake and diffusion mechanisms in particle-filled composites. *Polymer*, 249, 124799.

Vrentas, J.S. and Duda, J.L. (1977). Diffusion in polymer – Solvent systems. 1. Reexamination of the free volume theory. *Journal of Polymer Science Polymer Physics Edition*, 15(3), 403–439.

Yang, L., Chen, X., Jing, X. (2008). Stabilization of poly(lactic acid) by polycarbodiimide. *Polymer Degradation and Stability*, 93(10), 1923–1929.

Zimm, B.H. and Lundberg, J.L. (1966). Sorption of vapors by high polymers. *The Journal of Physical Chemistry*, 60(4), 425–428.

3

Polymers Under Ionizing Radiations: Introduction and Basic Principles

Yvette NGONO
CIMAP, UNICAEN, Caen, France

3.1. Introduction

Before the topic of polymer behavior under ionizing radiation is explored, it is worth becoming briefly familiar with these materials. This section focuses on synthetic organic polymers, that is, those synthesized from chemical compounds resulting from petrochemistry. There are natural polymers (DNA, collagen, proteins, cellulose or hemicellulose, starch, etc.), and active research on the development of biosourced polymers (from starch, for example) is underway. Natural and biosourced polymers are beyond the scope of this chapter.

There is a wide variety of synthetic polymers. Among them, one of the best known and most commonly used polymers is polyethylene (PEX). Packaging and other equipment containing it are marked by the PE symbol. It may be high-density PE (HDPE; bottles, tubes, etc.) or low-density PE (LDPE; bags, package coating, etc.), for example. Several other polymers quite widely used in everyday objects include polypropylene (PP), polystyrene (PS), polyvinyl chloride (PVC) and polyethylene terephthalate (PET).

All synthetic polymers are denoted by an abbreviation that usually starts with the letter P, which stands for *poly* (the Greek word for "many"). Polymers are

Aging of Industrial Polymers 1,
coordinated by Emmanuel RICHAUD.

sometimes represented by the monomers composing them. For example, ABS (acrylonitrile–butadiene–styrene) is a symbol that can be seen on shock-absorbing equipment.

	Abbreviation	RU	Class
Polyethylene	PE	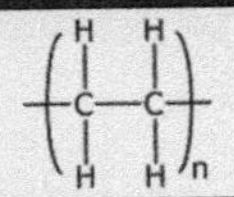	Thermoplastic
Polypropylene	PP	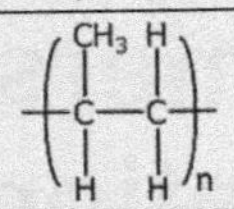	Thermoplastic
Polystyrene	PS	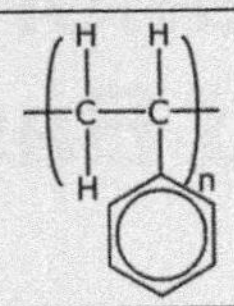	Thermoplastic
Polyvinyl chloride	PVC	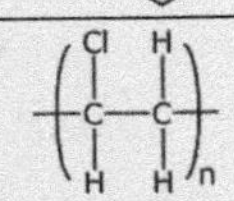	Thermoplastic
Polyethylene terephthalate	PET	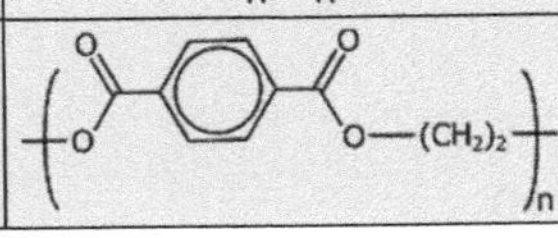	Thermoplastic

Table 3.1. *Several commonly used polymers. RU = repeating unit*

Polymers are macromolecules. They consist of chains formed by the repetition of one or more types of repeating units (RUs; chemical unit that repeats in the chain), known as monomers, linked by covalent bonds. An intuitive image is that of a string of beads (Figure 3.2(a)). The various beads (monomer or RU) are connected by a thread (covalent bond). The difference is that, unlike the string of beads, which is designed to be fastened (assuming it is used as a necklace), polymer chains remain open (in the vast majority of cases) and can be very long (thousands of RUs). As for the monomer, it is a more or less complex chemical molecule[1].

Depending on the number of different monomers forming the polymer chain, there can be homopolymers (a single RU) or copolymers (at least two different RUs). In the case of a copolymer composed of two monomers, A and B monomers

1 For further information on the synthesis of polymers, please see the file proposed by the author on Futura-Sciences (ref): https://www.futura-sciences.com/sciences/dossiers/physique-vulgarisation-polymeres-synthetiques-709.

can be organized alternately, in a random or statistical manner, or by blocks. For the latter, the term "block copolymer" is often used.

~A-B-A-B-A-B-A-B-A-B-A-B-A-B-A-B-A-B~

a) Alternating copolymer

~A-B-B-A-B-A-A-A-B-A-B-B-A-A-B-A-B-B~

b) Statistical copolymer

~[A-A-A-A][B-B-B][A-A-A-A][B-B-B][A-A-A-A]~

c) Block copolymer

Figure 3.1. *Possible configurations for a two-monomer copolymer (Aubrit 2017). For a color version of this figure, see www.iste.co.uk/richaud/industrialpolymers1.zip*

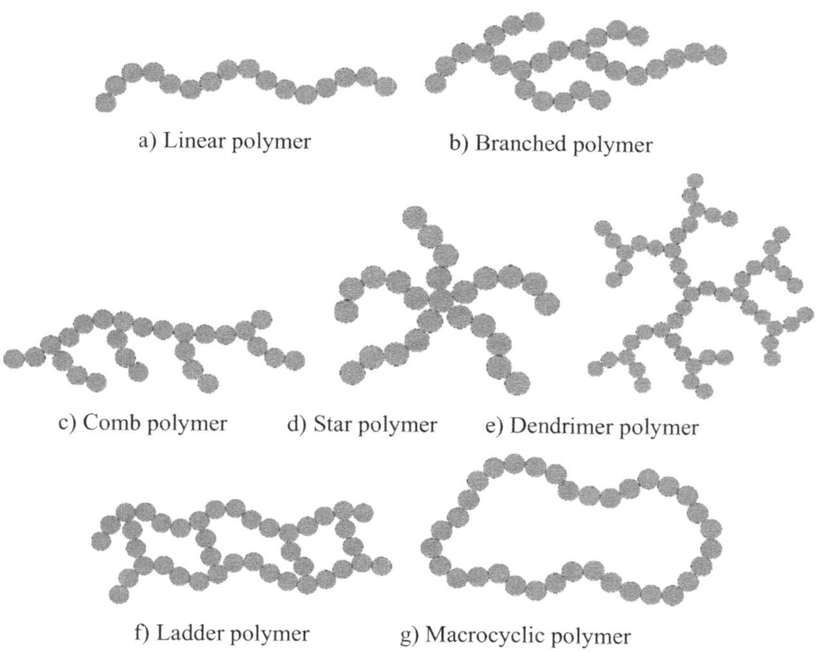

Figure 3.2. *Possible configurations of macromolecular chains (Aubrit 2017)*

The polymer chain, also known as the "main chain" or "backbone", can be purely linear or have side chains (or branches). The layout of these branches can be comb-like or star-like (Figure 3.2). Three-dimensional development of the chains is also possible through the formation of covalent bonds between various chains. The resulting materials are then referred to as cross-linked materials. The chain is no longer similar to the above-mentioned string of beads but rather to a fishing net with smaller or larger meshes (Figure 3.3).

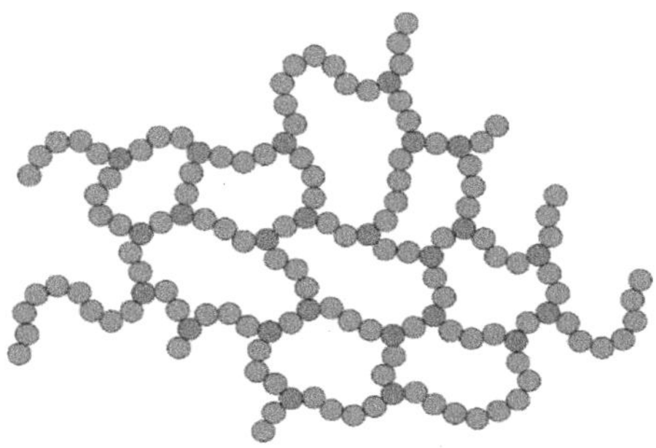

Figure 3.3. *Configuration of a cross-linked polymer material (Aubrit 2017)*

Polymers can be classified in various ways. One of the most widespread classifications identifies three main classes of polymers.

– Thermoplastics: polymers in this class become malleable when the temperature increases, and this effect is reversible.

– Thermosets harden under the effect of temperature (the polymer is generally synthesized during this stage), and the effect is irreversible. Once obtained, these polymers are infusible (do not melt) and insoluble. Epoxy resins after curing are part of this polymer class.

– Elastomers can be reversibly deformed. Some of these elastomers are thermoplastics.

Polymers have several levels of organization and therefore various characteristic temperatures related to relaxations or transitions, which influence the mobility of reactive species in the material:

– Molecular (groups composing the monomer).

– Macromolecular (organization of monomers along the chain).

– Supra-macromolecular (organization of chains with respect to one another). This level of organization defines the degree of crystallinity of the polymers, which ranges from complete disorder (amorphous structure) to three-dimensional and long-distance order (crystalline structure). In many cases, polymers are semi-crystalline materials: the material is composed of crystallites (highly organized chains) immersed in an amorphous matrix. The degree of crystallinity influences, for example, gas permeability, chain mobility or mechanical properties.

Irrespective of the imposed external constraints, the behavior of polymers under irradiation is influenced by all of this organization, on which the various modes of mobility depend. Mobility may concern side groups or side chains, chain segments, entire chains and chains with respect to one another.

The (mechanical, electrical, etc.) properties of materials are characterized at the macroscopic level, and for this reason, they are referred to as "macroscopic properties". In the field specific to this book, the change in macroscopic properties is caused by the accumulation of defects at the molecular and macromolecular scales.

The term "defect" covers any chemical modification occurring in the RU or in the macromolecular chains. To estimate the stability of a polymer material under ionizing radiations, it is essential to consider the building blocks: the monomer and the chemical groups composing it and the links between these elementary units. Depending on these chemical groups, a polymer involving the same type of atoms can be more or less stable under irradiation depending on whether these atoms are organized in saturated or unsaturated aliphatic structures or in aromatic structures. These notions will be considered further in this chapter.

After World War II, polymer production and use intensified in all sectors, including the nuclear power industry (electric power generation driven by the fission of uranium 235 nuclei). In this sector, polymers are irradiated[2] under normal operating conditions (cables or paints in nuclear power plants) or following contamination with radioactive nuclei during their usage. For example, this is the case for materials used to protect users and surfaces (PVC gloves and tablecloths) during nuclear fuel production or during spent fuel reprocessing.

Polymers can also be intentionally irradiated for obtaining new materials (membranes for filtration, obtained by nanoscale modifications under ion beams, for example), during medical equipment sterilization, to debug artifacts or to improve some macromolecular properties of standard materials.

In daily life, an increasingly common material obtained by irradiation is cross-linked PEX. This material, which has gradually replaced copper for water supply pipes in residential buildings, is obtained by the irradiation of PEX tubes (above mentioned PE) with electron beams.

---

2 Exposed to radiation.

Importantly, the polymer irradiation conditions are related to the desired application. The deposited doses range from several impacts for membrane structuring at the nanometric scale to several megagrays for some applications in the nuclear power industry, passing through several kilograys for sterilization. This notion of dose, which is very important in the study of polymer behavior under ionizing radiation, will be examined further.

The following section presents the various types of ionizing radiation, the interactions between ionizing radiation and polymers, the resulting defects and the factors influencing the behavior of a given material exposed to ionizing radiation.

Polymers can be subjected to various types of ionizing radiation. These include γ-rays, X-rays, electron beams or swift heavy ions (SHIs in the following). Given that the action of neutrons on the behavior of synthetic polymers has not been extensively studied, it will be covered only to a little extent in this chapter. Several principles for approaching the interaction between neutrons and synthetic polymers are presented below.

There are various factors that explain why studies on neutron–polymer interactions are not very common. Importantly, in addition to those present in close vicinity to particle accelerators, polymers in the nuclear environment are especially subjected to thermal neutrons. The energy of the latter (several dozen millielectron-volts) is too low to induce direct ionizations through elastic collisions with the atoms of the target.

Thermal neutrons interact with the target via nuclear reactions. The two most common nuclear reactions are (n,γ[3]) and (n,p[4]). The (n,γ) reaction is effective for various nuclei that are common in synthetic polymers, with cross-sections[5] varying by several orders of magnitude between the highest and the lowest (45 barns for ^{35}Cl, 210^{-4} barns for ^{16}O and 0.34 barn for ^{1}H for thermal neutrons of 25 meV[6]). The (n,p) reaction at these energies is only possible if no energy is required to overcome the Coulomb barrier. This is the case for two nuclei common in polymers: ^{14}N and ^{35}Cl.

3 When the target nucleus absorbs the neutron, γ radiation is emitted.

4 When the target nucleus absorbs the neutron, a proton is emitted and a new nucleus is formed. Moreover, this is accompanied by a recoil atom.

5 The cross-section represents the probability of interaction of a given particle for a given reaction. Its unit in the International System is m^2, but it is often expressed in barn (10^{-28} m^2).

6 Database of the National Nuclear Data Center: https://www.nndc.bnl.gov/sigma/basicresults.jsp.

Therefore, the action of thermal neutrons on the polymers can be reduced to the action of γ-rays resulting from the (n,γ) reaction, to that of protons from the (n,p) reaction, or to that of recoil nuclei, when applicable. The predominant process depends on the cross-sections of the occurring nuclear reactions, the energy of the emitted protons or γ radiation, and the energy of the recoil nuclei. Consequently, from a physics standpoint, interactions involving thermal neutrons can be simulated by γ-rays or ion beams.

Determining the dose deposited by thermal neutrons is a much more complex process. Indeed, it takes into account not only the cross-section of the involved nuclear reactions but also the atomic fraction of the nuclei forming the polymer.

The behavior of polymers under ionizing radiation is normally studied using γ-rays or electron beams under an inert atmosphere at room temperature. We will maintain this aspect and present the behavior under an ion beam or an oxidative atmosphere as a specificity.

However, first, an overview of ionizing radiation types should be provided, including the definition of what ionizing radiation is, how it is obtained and how its effects on matter are quantified. The vocabulary specific to this field will also be introduced. The first of the terms used is irradiation, which is the action of exposing a material to radiation.

3.2. Overview of ionizing radiation

3.2.1. *Definition*

Radiation can be defined as the propagation of energy in the form of electromagnetic waves or particles, in vacuum or in a material. On the basis of this definition, radiation can be divided into two main classes: electromagnetic radiations and particle radiations. The visible light is part of electromagnetic radiation, whereas electron beams are part of particle radiations.

During their interaction with matter, radiations deposit energy following processes that depend on the radiation type (electromagnetic or particle) and on its energy. The resulting species depend on the deposited energy and the main process involved. We focus herein on ionizing radiation.

Radiation is referred to as "ionizing" when its interaction with matter leads to electron removal from the target atoms. This results in the formation of ions.

Radiation can be indirectly ionizing (γ-rays, X-rays, vacuum UV and extreme UV) or directly ionizing (particle radiation, such as electrons or ions). Visible light and UV radiation of the solar spectrum (whose wavelengths are above 200 nm) do not have sufficient energy to ionize matter. Therefore, they are not part of ionizing radiation and will not be considered in the remaining document. However, low-energy UV radiation, though non-ionizing, contributes to the aging of polymers via excitation.

	Radiation	Notation	Ionizing profile	Production mode
Electromagnetic	Gamma	γ	Indirect via secondary electrons	Radionuclides (^{60}Co, ^{137}Cs)
	X-rays	XR	Indirect via secondary electrons	Interactions electrons accelerated under high vacuum/metallic target[a]
	Vacuum-UV radiation (200–10 nm)	VUV-EUV	Indirect via secondary electrons	Discharge lamps[7], synchrotrons
Particle	Electrons or positrons	β	Direct	Radionuclides, accelerators
	Alpha (helium ions)	α or ^4He	Direct	Radionuclides, accelerators
	Ion beams	Various ions	Direct	Accelerators

Table 3.2. *Ionizing radiation commonly used in the study of polymers under radiation and the main modes of production. [a]Usually tungsten*

Ionizing radiation (IR) results in the deposition in the polymer of an energy higher than that of any covalent bond forming the macromolecular chain. This energy can consequently be absorbed by any group present in the material. This trend is also one of the main differences between γ radiation, for example, and low-energy UV radiation (wavelengths above 200 nm), as the latter can be absorbed only by chemical groups whose excitation energy corresponds exactly to the energy provided by the radiation (this is referred to as the resonance condition).

7 The gas in the lamp (deuterium, xenon) depends on the sought-for energy.

3.2.2. *Sources of ionizing radiations*

Ionizing radiations are produced either by the natural decay of radioactive nuclei, also referred to as "radionuclides", or via accelerators (electrons (β-), ion beams, X-rays) (Table 3.2).

3.2.2.1. *Electromagnetic radiation*

Radionuclides form radioactive sources. These sources can be sealed or not. A radioactive source is characterized by the type of radiation emitted and its energy, its activity and its half-life or period.

The energy of radiation is generally *expressed* in multiples *of electron-volt* (eV). The electron-volt represents the energy acquired by an electron accelerated in vacuum by a voltage of 1 V. The relationship between electron-volt and Joule energy, the common unit of energy, is 1 eV = 1.6×10^{-19} J.

Activity corresponds to the *number of disintegrations per second*. It is expressed as becquerel (Bq). This very small unit is generally replaced by the curie (Ci); 1 Ci = 3.7×10^{10} Bq.

As for the *half-life*, it corresponds to the time at the end of which half of the atoms present at a given time t will have decayed. Therefore, it indicates the time required for the activity at a given time to decrease by half.

The most frequently used radionuclides are cobalt (^{60}Co) and cesium (^{137}Cs) for γ-rays. Cobalt (^{60}Co) sources are most commonly used in large irradiation installations. They are characterized by a gamma ray energy of 1.17 and 1.33 MeV. Cesium (^{137}Cs) sources, which are generally sealed, emit γ radiation of 0.66 MeV. The sources of α particles contain actinides, such as plutonium (^{241}Pu), for example.

3.2.2.2. *Particle radiation*

Continuous or pulsed electron beams, whose energy varies between 0.1 MeV (electrostatic accelerators) and 30 MeV (linear accelerators (LINACs)), are provided by accelerators.

Ion beam accelerators are classified according to the geometry of the accelerating part. This can be linear (LINAC, Van de Graaff, tandem) or circular (cyclotrons and synchrotrons). Linear accelerators are mainly used in the production of light ion beams (H^+-C^{6+}). As for cyclotrons, they are characterized by increased flexibility in terms of the number of ions accelerated and the energies reached. The Grand Accélérateur national d'ions lourds (GANIL), located in Caen, France, accelerates ions ranging from carbon to uranium. The highest energy reached for carbon ion beams is 95 MeV A^{-1} owing to its two cyclotrons. Spiral 2, the LINAC recently built on the same site is used, in its phase 1, to accelerate light ions (He, H^+, $^2H^+$) and heavier ions (A/Q < 3) up to energies of 14.5 MeV A^{-1} (Ganil 2022).

3.2.3. Quantifying energy deposition and radiation effects on matter

3.2.3.1. *Energy deposition*

The energy deposited by ionizing radiation in materials is generally quantified in terms of dose. The *dose* (D) represents the energy deposited per unit mass of material. It is expressed in gray. The gray represents 1 Joule of energy deposited on 1 kg of matter: $1\ \text{Gy} = 1\ \text{J.kg}^{-1}$.

Dose rate (D° or I) is used to quantify the rate at which the dose is deposited: it corresponds to the dose deposited per unit time. It is expressed in $Gy.s^{-1}$. As an example, the dose rates that can be reached with the above-mentioned ^{60}Co sources range between 0.5 and 10 kGy h^{-1}.

	Notation	Definition	Unit
Dose	D	Energy deposited per unit mass	Gray (Gy) = 1 $J.kg^{-1}$
Dose rate	I or D°	Dose deposited per unit time	$Gy.s^{-1}$
Fluence	φ	Number of particles entering the material per unit surface	$Particles.cm^{-2}$
Flux	Φ	Number of particles entering the material per unit surface and per unit time	$Particles.cm^{-2}.s^{-1}$
Stopping power*	LET	Energy deposited by one particle per unit length	$eV.nm^{-1}$
Radiochemical yield	G	Number of moles of defects per unit energy deposited	$Mol.J^{-1}$

Table 3.3. *Summary of the parameters used to quantify the energy deposited by radiation and their action. *Stopping power and LET are identical in the electronic interaction regime, as it is the case for beams used for the study of polymers under ionizing radiation, which is the focus of this chapter. There are different units*

Various systems, referred to as dosimeters, are used to quantify the dose deposited in the materials during irradiation. These systems depend on the radiation type considered. There is a wide range of systems for γ radiation dosimetry (Blanc 1997). Most of them are passive (their operation does not require a source of energy) and must be revealed after their irradiation. These include chemical systems, such as the Fricke dosimeter (the best known) or alanine, radiochromic films, calorimeters or ionization chambers. Radiochromic films are polymeric films whose color changes under radiation. This color variation as a function of dose is known because of previous calibrations.

For irradiations with accelerated particles, the quantifying parameters are the fluence, φ, and the flux, Φ, instead of the dose and the dose rate. The fluence corresponds to the number of particles entering the target per unit surface. It is expressed in particles.cm^{-2}. As for the *flux*, it represents the number of incident particles per unit surface and per unit time. It is expressed in particles.cm^{-2}.s^{-1}. It is related to the fluence according to the following equation:

$$\Phi = \frac{d\varphi}{dt} \qquad [3.1]$$

The fluence and flux are related to the dose and dose rate through the energy deposit per unit mass or LET per unit mass (in MeV.cm^2.mg^{-1}), respectively, following equations [3.2] and [3.3]. Linear energy transfer (LET) represents the energy deposited by a particle per unit of distance traversed:

$$D = 1.602 \times 10^{-7} \times LET \times \varphi \qquad [3.2]$$

$$\frac{dD}{dt} = 1.602 \times 10^{-7} \times LET \times \Phi \qquad [3.3]$$

In the presence of continuous particle beams, electrons or ions, irradiation is followed by the measurement of an electric current using a Faraday cup. As the latter is interceptive (stops the particles), it cannot be left in place during irradiation. Therefore, it must be associated with another system whose calibration prior to irradiation will make it possible to track the number of particles traversing the sample per unit time and in fine the fluence.

For electrons, this second system can be a diaphragm electrically insulated from the rest of the line. Under ion beams, the non-interceptive system is either an X-ray sensor for low doses or a secondary electron multiplier. In some irradiation facilities using particle beams, the Faraday cup is inserted over very short periods at a frequency selected to understate the perturbations caused by its insertion during irradiation.

Alanine dosimeters can also be used under continuous electron beams, but only for low doses and for verification purposes. Chemical dosimeters are used for pulsed electron beams.

3.2.3.2. *Radiation effects: the radiochemical yield*

The modifications induced in the material (defects) are quantified through the *radiochemical yield* (*G*). This factor represents the number of moles of defects created per unit energy deposited in the target. It also represents the efficiency with which the considered defects are formed.

The term "defect" may denote either the destruction (or consumption) of a group present in the material (B) prior to irradiation or the creation of a new group (A). In the following equations, n_A and n_B correspond to the numbers of moles of A created or B consumed, respectively, and E is the energy deposited in the material:

$$G_A = \frac{dn_A}{dE}; \; G_{-B} = -\frac{dn_B}{dE} \qquad [3.4]$$

The radiochemical yield is expressed in mol.J^{-1}. The former unit (100 eV)$^{-1}$ is, however, commonly encountered in the literature. It corresponds to the number of defects created per 100 eV deposited in the target. One unit can be transformed into the other using equation [3.5]:

$$G_{(\text{mol.J}^{-1})} = \frac{G((100\ \text{eV})^{-1})}{9.63 \times 10^6} \qquad [3.5]$$

3.3. Interactions between ionizing radiation and polymers: initial processes

The initial processes of interaction between radiation and matter depend on the nature of the radiation (electromagnetic or particles), and for a given radiation type, they will be a function of its energy.

3.3.1. *Photon–polymer interactions*

Gamma-ray photons, with their dual particle and electromagnetic nature, can interact with the atoms in the target through three processes, which are classified in ascending order of energy:

– Photoelectric effect: the incident γ photon, with sufficient energy $h.\upsilon$, transfers its entire energy through a single collision to an inner shell electron. The transferred energy is used to extract the electron whose bond energy is E_L and provide it a kinetic energy E_c so that $E_c = h.\nu - E_L$.

– Compton effect: the incident photon, with an energy $h.\upsilon$, transfers part of its energy to a free or loosely bound electron of the target. This electron is ejected with a kinetic energy equal to the energy transferred by the photon. During this process, the incident photon is deflected, and its energy, $h.\upsilon'$, is below the energy of the incident photon.

– The electron–positron pairs creation: this process only occurs for photons whose energy is strictly above 1,022 keV, which corresponds to the energy equivalent to the remaining mass of two electrons. When a photon of energy $E > 1{,}022$ keV enters the electric field surrounding a nucleus, it may transform into an electron–positron pair. The rapid annihilation of the positron by an electron of the target (neighboring electron) generates two photons of energy 511 keV, propagating along the same line but in opposite directions.

3.3.1.1. *Relative importance of various processes*

The above-described processes are quantified using attenuation coefficients, μ, obtained from single-energy photons hitting the target. The attenuation coefficient of a given material is related to the number of photons absorbed (dN) in the thickness of the material (dx) and to the number of incident photons, N, by equation [3.6], whose integration leads to equation [3.7]. The variation in the mass attenuation coefficient for each energy domain of the incident photon is represented in Figure 3.4:

$$dN = -\mu \times N \times dx \qquad [3.6]$$

$$N = N_0 \times e^{-\mu.x} \qquad [3.7]$$

Gamma rays used for studying the behavior of polymers result from the radioactive disintegration of ^{60}Co or ^{137}Cs nuclei. The maximal energy is 1.2 MeV, and the Compton effect is predominant. The irradiation of polymers that do not contain heavy atoms with γ radiation can therefore be considered isotropic irradiation with the electrons of the target.

In the presence of atoms with high atomic numbers, the boundary between the photoelectric effect and Compton scattering is shifted toward high energies, and depending on the atoms present, the photoelectric effect can no longer be ignored.

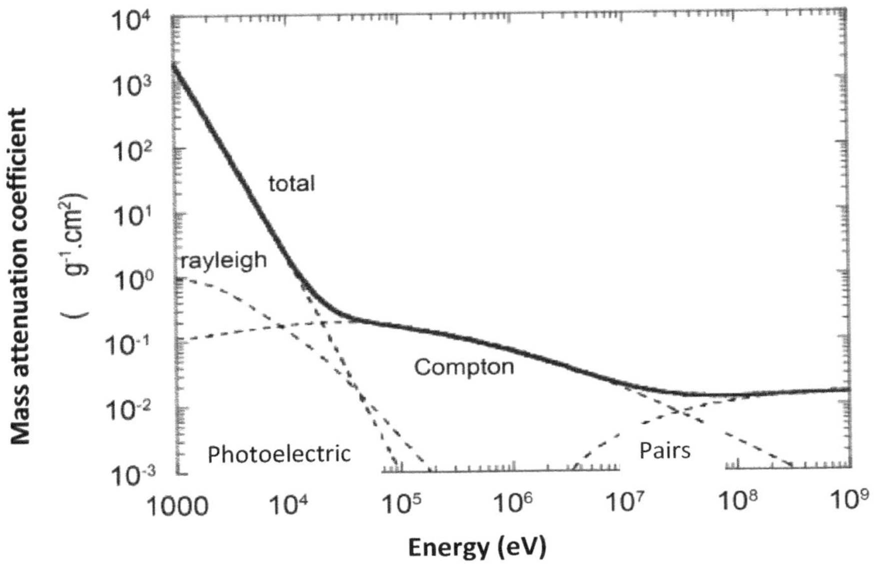

Figure 3.4. *Variation in the mass attenuation coefficient (by the material of density ρ) μ1/ρ depending on the photon energy. Case of polyethylene (Corbin 2001)*

3.3.2. *Interactions between particles and polymers*

3.3.2.1. *Stopping power*

When a projectile penetrates a material, it gradually loses energy. The energy deposit in the material is quantified by the stopping power $\left(-\frac{dE}{dx}\right)$ of the projectile. The stopping power represents the loss of energy ΔE of the projectile over a distance Δx traveling in the target. It can be written as follows:

$$LET = -\frac{\Delta E}{\Delta x} \quad [3.8]$$

Accelerated ions can interact with target atoms via two mechanisms: elastic collisions with their nuclei or inelastic collisions with their electrons. The total stopping power is therefore the sum of the nuclear (n) and electronic (e) stopping powers:

$$\left(-\frac{dE}{dx}\right)_{\text{total}} = \left(-\frac{dE}{dx}\right)_n + \left(-\frac{dE}{dx}\right)_e \quad [3.9]$$

The nature and velocity of a projectile determine the predominance of one or another of the two mechanisms of interaction (electronic or nuclear), as shown below.

The stopping power is generally expressed in keV.nm^{-1}.

Nevertheless, a common way to bypass density and directly compare several materials is to use the mass stopping power, expressed in MeV.cm^2.mg^{-1}, obtained by dividing the linear stopping power by the density of the target:

$$\left(-\frac{dE}{dx}\right)_{mass} = \frac{1}{\rho} * \left(-\frac{\Delta E}{\Delta x}\right) \quad [3.10]$$

In the energy field where electronic interactions predominate, the stopping power and LET are similar.

3.3.2.2. *The projected path of the particle into the target*

When an accelerated particle penetrates a material, its path can deviate by a series of impacts with the atoms of the medium. The real distance R_r over which a particle of energy E traverses a material depends on the atomic number and energy of the projectile on the one hand and on the atomic composition and target density on the other hand. It can be assessed by the following relation:

$$R_r = \int_0^E \left(\frac{dE}{dx}\right)^{-1} \times dE = \int_0^E (-LET)^{-1} \times dE \quad [3.11]$$

Instead of the actual path, it is more common to use the projected path along the direction of the incident beam R_p.

The path of electrons in the polymers is not rectilinear, as they deviate after each interaction. As long as the energy of the electron remains above 10 keV, the projected path is approximately $0.7 \times R_r$ (Rullier-Albenque 1993).

For high-energy ions, in the regime of electronic interactions, the projected and actual paths are considered equal: $R_r \simeq R_p$. When nuclear collisions are effective, the ion paths are no longer rectilinear.

Both the ion range in a given material and the resulting energy deposition are usually calculated using the SRIM software based on the works of Ziegler et al. (1985).

3.3.2.3. *Electron–polymer interactions*

When an electron interacts with matter, it may lose energy through various interaction processes, depending on the collision partner. It may:

– collide with electrons from the target atoms, causing ionization and/or excitation of said atoms;

– collide with atom nuclei. These interactions have a small contribution to the energy deposited by electrons in the material. However, the incident electron deviates from its path because of Coulomb interactions;

– be slowed down by the target and produce a Bremsstrahlung or generate a radiation referred to as Cerenkov. These two types of interactions lead to energy deposition that is negligible in the realm of energies used for polymer studies under ionizing radiation.

Most of the electron beams used to study polymers have an energy ranging between 0.5 and 2 MeV. In this range, interactions preferentially occur with the target electrons. The incident electrons will therefore transfer their energy through a series of electron collisions until reaching an energy equal to that of the electrons in the target: this phenomenon is referred to as thermalization.

The fate of the target electrons depends on the energy received during interactions with the incident electron. It may be sufficient for the electrons to be ejected from the parent atoms: *this is ionization*. In case the energy received by the target electrons is not sufficient to be ejected, they will be promoted to a higher energy level: *this is excitation*. Electrons ejected during ionization are referred to as "secondary electrons". They may, in turn, induce ionization and excitation phenomena if they have sufficient energy until their thermalization (Figure 3.5).

From the ionization density standpoint, there are different ionization sites, depending on the energy deposited by the incident electron. The ionization density increases from spurs, which contain only a few ionizations, to short tracks in which the ionization density is higher. The path of secondary electrons with energies above 5 keV is similar to that of the primary electron.

During polymer irradiation with electron beams produced by accelerators, the ionization sites are composed of 65% spurs, 25% short tracks and 10% blobs (Mozumder and Magee 1966). On the basis of this distribution, irradiations using accelerated electrons can be considered homogeneous to a first approximation.

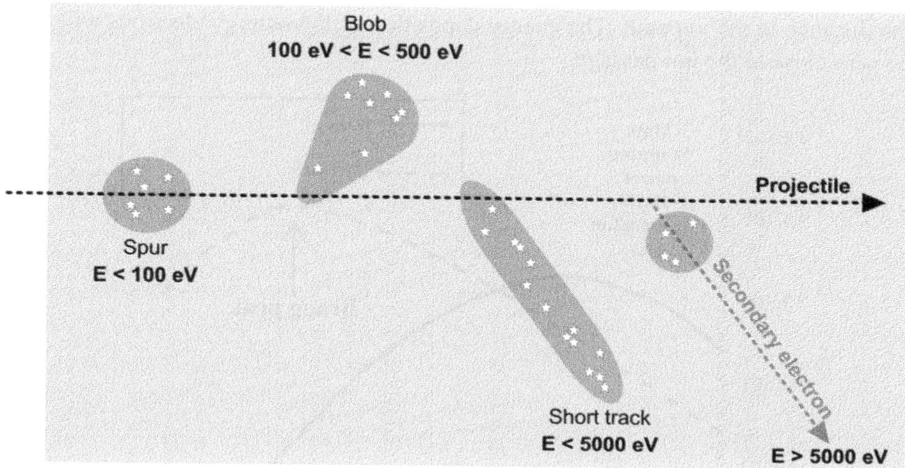

Figure 3.5. *Freely adapted representation of the distribution of ionization sites in the electron track according to the classification of Mozumder and Magee (Hickel 2000). E represents the energy deposited during the event*

3.3.2.4. *Interactions between swift heavy ions and polymers*

3.3.2.4.1. Spatial structure of energy deposition

Accelerated ions can interact either with the target electrons during interactions referred to as "electronic" or with the target nuclei during interactions referred to as "nuclear". The predominance of one of these processes depends on the energy of the incident ions (Figure 3.6).

The ion beams used for studying polymers (E > 1 MeV A^{-1})[8] preferentially lead to ion–electron collisions (or electronic interactions). Consequently, only the latter are presented in the following.

During electronic interactions, the incident ion transfers part of its kinetic energy to the target electrons that will be either ejected from their parent atoms, resulting in *ionizations*, or promoted to a higher energy level, resulting in *excitations*. The secondary electrons issued from the primary interaction between the ion and the target electrons have enough energy to induce other excitations and ionizations on their way until their thermalization. As already seen for accelerated electrons, the energy deposition by secondary electrons is inversely proportional to the square of

8 MeV/A energy of 1 MeV per nucleon or per unit of atomic mass.

the distance to the ion path. The energy deposition of low-energy electrons will thus be very close to the ion passage.

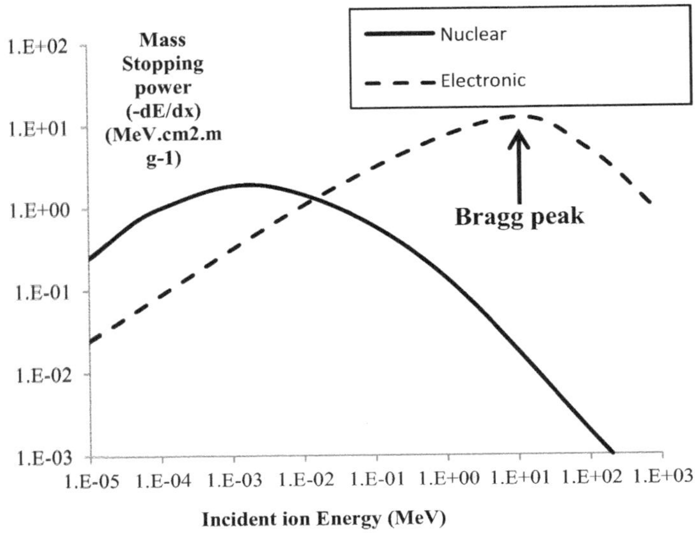

Figure 3.6. *Evolution of the nuclear and electronic mass stopping powers as a function of the energy of the incident ion. Case of ^{18}O ions on a polystyrene target (Ventura 2013)*

Owing to the very significant mass difference between electrons[9] and incident ions, the latter practically do not deviate from their path, which is rectilinear. Therefore, the ions deposit their energy through electronic processes along their path in a cylinder referred to as the *track*. The track can be schematically segmented into two regions: the *track core* and *the halo* or *penumbra*.

The track core, with a diameter of several nanometers, is characterized by a very high density of excitations and ionizations, induced, on the one hand, by the primary interaction between the ion and the target electrons and, on the other hand, by low-energy secondary electrons (by far the most numerous). The halo is a less dense area constituted by excitations and ionizations induced by high-energy secondary electrons. Its diameter is given by the range of these secondary electrons, which increases when the energy (or velocity) of the electrons increases (Gervais and Bouffard 1994).

9 It should be noted that an electron is approximately 2,000 times lighter than a proton and the lightest ion is H^+, which has a proton.

The spatial distribution of the energy deposition by heavy ions in the regime of electronic collisions is highly heterogeneous at the nanometric level. Furthermore, the radial distribution of the energy deposition results in a radial distribution of the dose (and of the dose rate[10]) around the ion passage.

Given their high atomic number, heavy ions induce high LETs. At a given energy, LET increases when the atomic number increases. For a given ion, the LET increases when the energy decreases, up to a maximum known as the Bragg peak, and then decreases. For such an ion, the same value of LET can be obtained with two energies positioned on each side of the Bragg peak (Figure 3.6).

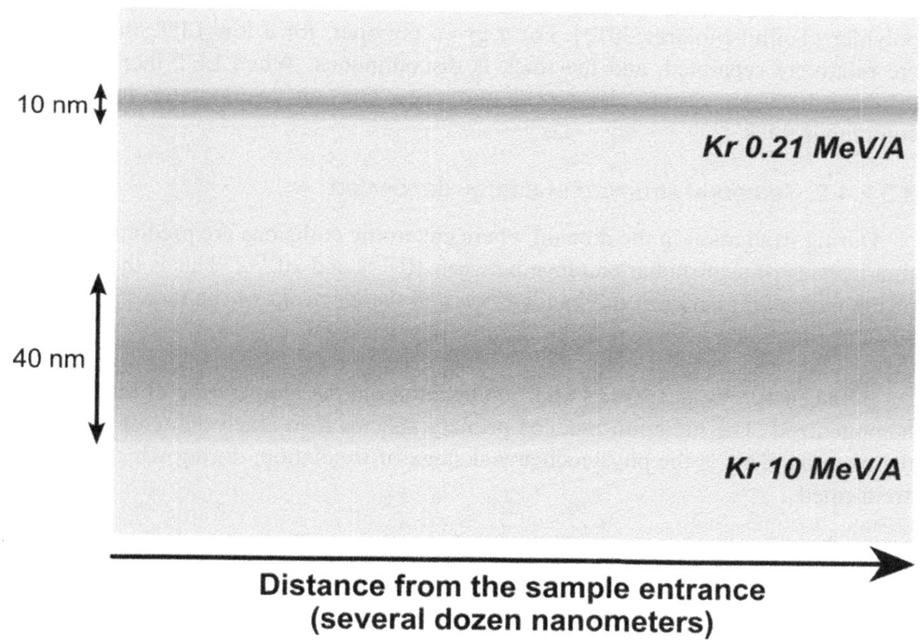

Figure 3.7. *Schematic representation of the track of two Kr ions in polyethylene: an upper image (Kr^{5+}, 0.21 MeV A^{-1}) and a lower image (Kr^{35+}, 10 MeV A^{-1}), leading to an identical stopping power of 4 keV.nm^{-1} (adapted from the works of Gervais (1993) and Gervais and Bouffard (1994)). For a color version of this figure, see www.iste.co.uk/richaud/industrialpolymers1.zip*

10 See equations [3.3] and [3.4] for the relation between dose, number of particles and energy deposition.

Therefore, LET is not sufficient for characterizing the energy deposition of accelerated ions in a given material. Two additional parameters should be considered: the ion energy and its charge state. For a given atom, the same stopping power can be obtained by adjusting the velocity–charge pair, as shown by the results of the Monte Carlo simulations presented in Figure 3.7. Another result of these works is that, for ions of the same atom leading to an identical stopping power, the track radius increases when the energy of incident ions increases.

The track (as presented in Figure 3.7) appears to be continuous. This is not always the case. The distribution of primary events, organized in the form of clusters, depends, on the one hand, on the LET value and, on the other hand, on the polymer (Toimil-Molares 2012). For a given polymer, for a low LET, the clusters are relatively separated, and the track is discontinuous. When LET increases, the distance between events decreases, and they end up overlapping to form a continuous track.

3.3.2.4.2. Temporal structure of energy deposition

During irradiation in the domain where electronic collisions are predominant, the first instants of postirradiation range between 10^{-17} s and 10^{-15} s. This is the duration of the interaction between the incident ion and the electrons of the target. This first stage of irradiation is known as the “atomic physics stage”.

Between 10^{-15} s and 10^{-12} s after ion penetration, the temperature of the target is homogenized. The radiation-induced primary species then reach the temperature of the material. This is the physicochemical stage of irradiation, during which radicals are formed.

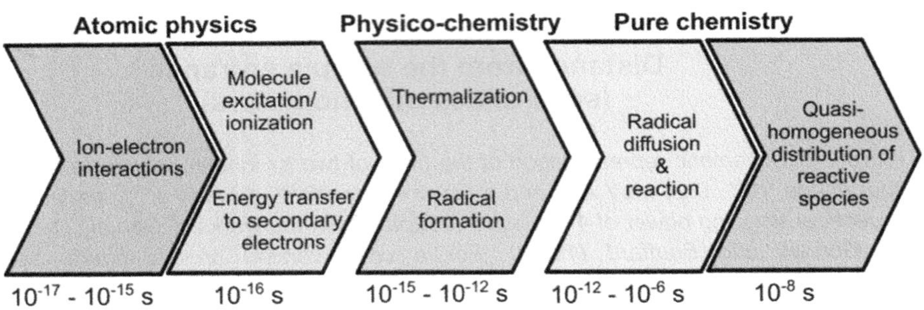

Figure 3.8. *Orders of magnitude of the characteristic times of events resulting from the ion–matter interaction. Figures normally used by the water radiolysis community. For a color version of this figure, see www.iste.co.uk/richaud/industrialpolymers1.zip*

Between 10^{-12} s and 10^{-6} s, the radicals are distributed in the material and react with the target atoms. This is the pure chemical stage of irradiation. The stationary state is considered to be reached after 10^{-6} s.

In solid materials, such as polymers, the radicals formed during the physicochemical stage can react very rapidly, within a few seconds, or even several years later if they are created in areas of very low mobility or under thermal conditions in which their mobility is significantly reduced.

3.3.3. *Differences and similarities between various types of ionizing radiation*

The main information resulting from the above section, presenting the interaction between ionizing radiation and polymers, in commonly used energy domains, is that ionizing radiation essentially interacts with the target electrons. This results either in excitations or ionizations. The secondary electrons resulting from ionization phenomena lead, in turn, to excitations and ionizations if their energy is sufficient through interaction with the target electrons.

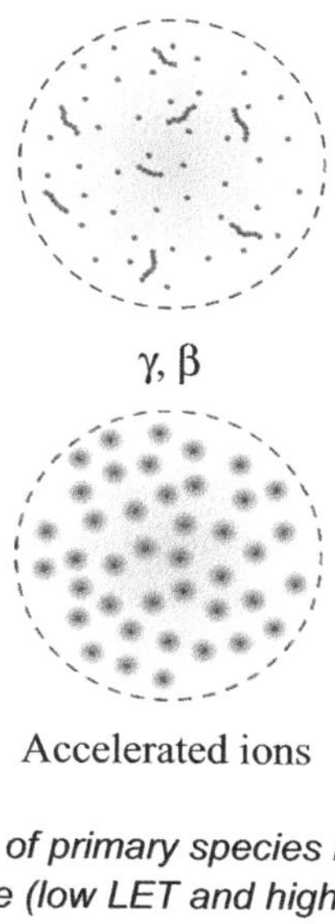

Figure 3.9. *Distribution of primary species in a polymer depending on the radiation type (low LET and high LET) (Aubrit 2017)*

Irradiations with electromagnetic radiations, such as γ- and X-rays, can be considered isotropic irradiations with the target electrons. Moreover, in some cases, γ-rays are used to simulate the effect of accelerated electrons and vice versa. Gamma radiation and accelerated electrons are, however, different in their range in the material. The ranges of electromagnetic radiation are longer than those of charged particles. The irradiated thicknesses will therefore be more significant with γ-rays than with electrons in the energy ranges applied to the study of polymers.

SHIs, in the energy domains used for studying polymers, also interact preferentially with the electrons of the target, inducing both excitation and ionization. However, because of the very large mass difference between incident ions and electrons, the energy is deposited very heterogeneously at the nanometric scale (Figure 3.9) in a cylinder centered on the passage of the ion referred to as the "latent track". The consequence of this distribution of ionizations and excitations is a radial distribution of the dose and the existence of very high doses and local dose rates.

As indicated above, accelerated ions induce very high LETs and therefore very high ionization density and very high excitations in the track. In the track core, owing to the very high number of ionizations and excitations per unit volume, compared with the volume occupied by regular RUs, the ionization and excitation of any RU in the track core is nearly compulsory. The high density of excitation and ionization in the track core leads to a specific evolution of primary and reactive species, with the development of cooperative phenomena and, as will be seen further on, the formation of specific defects in the polymers (Balanzat et al. 1994, 1995).

The above discussion concerns the physical aspect of the interaction between ions and matter. When the focus is on the chemical evolution of the material following irradiation, the migration (or diffusion) of species beyond the physical halo should be considered, which can be referred to as the "chemical halo". In the case of ion beam irradiation, the high heterogeneity in the distribution of primary events in the material at the nanometric scale leads to a highly heterogeneous distribution of defects formed in the material at the nanometric scale. In the case of electrons, over shorter or longer periods, the distribution of chemical defects is considered quasihomogeneous, as is the case for γ irradiation (Figure 3.9).

3.4. Evolution of polymers under ionizing radiation

The energy deposited during the interaction between radiation and polymers leads, after the stages presented below, to the formation of defects. The latter can be created in polymer chains (macromolecular defects) or be created as low-mass molecules that are released from the polymer as gas.

The initial stage involves the formation of primary species whose transfer or reaction at their creation sites will lead to radicals that generally constitute the predominant reactive species in polymers submitted to ionizing radiation.

3.4.1. *Primary species*

The ionization energies of chemical groups forming the RU in most polymers range between 10 and 15 eV. Consequently, the energy deposited by ionizing radiation used for studying these materials is high enough to extract electrons from any atom of the target. This leads to two species: the emitted electron and the resulting cation. In physics, this pair is commonly referred to as the electron–hole (e–h) pair. The first modification observed at very short times (Figure 3.8) in a polymer irradiated with ionizing radiation is therefore the appearance, at a local scale, of positive electric charges.

The ions created are usually in an excited (electronic or vibrational) state because of the large difference between the energy deposited by the radiation and the energy of the first ionization. In polymers devoid of groups with high affinity for electrons and because of the significant interaction between the electron and the cation, these two species recombine within very short periods, on the order of several picoseconds in PE in the molten state (Brede et al. 1989), to form highly excited species.

Excited species are also created directly during the interaction between radiation and the atoms of the target. During the excitation phenomenon (Figure 3.11), the groups that are initially at their fundamental level of energy are promoted to a higher electronic energy level.

3.4.2. *Evolution of primary species*

Under ionizing radiation, the target atoms are excited, and cations, resulting from the emission of secondary electrons, are formed. Excitations, cations and electrons are referred to as “primary species”. These species can either react rapidly near their creation site (recombination followed by the formation of radicals for the charged species or the formation of radicals from excitations) or migrate in the material before being trapped at adequate site. This is the energy transfer phenomenon.

3.4.2.1. *Energy transfers*

Energy transfer phenomena include species transfers and electronic excitation transfers. The notion of “species transfers” covers the transfer of isolated charges (electrons or holes) and the transfer of protons (hydrogen atoms, charged or not charged). It is very difficult, above the glass transition temperature[11], to differentiate transfers involving charged species with a mass from those involving electrons and holes.

Energy transfers are only effective in the presence of groups for which the stabilization of the charged species or of the transferred excitation is possible. The trapping groups differ from the chemical groups forming the polymer and depend on the type of transfer considered. Specifically, they either have an electron affinity that differs from that of the other groups in the RU of the polymer (charge trapping) (Zhang and Thomas 1996; Thomas 2007) or a lower excitation energy (excitation trapping) (Wündrich 1974). For example, it is difficult to trap electrons in a polymer, such as pure PE, which has no trapping groups. On the other hand, electrons can be more readily trapped in polymers that, similar to poly(methyl methacrylate) (PMMA), contain polar groups.

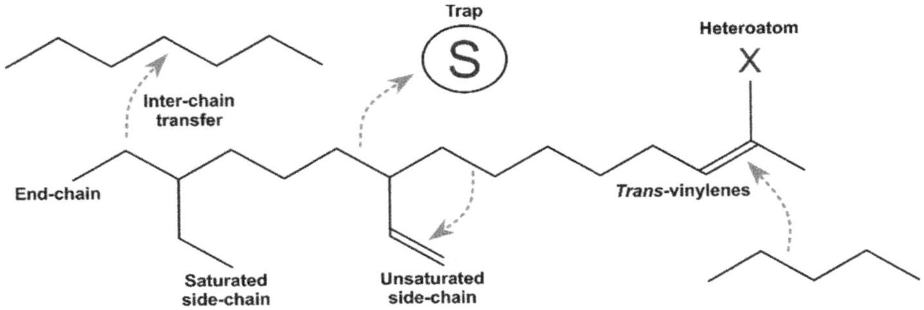

Figure 3.10. *Schematic representation of energy transfers in polymers under ionizing radiation (Ferry and Ngono 2020)*

These energy-accepting groups, also known as "energy traps", can be covalently bound to polymer chains or inserted in the polymer bulk (Figure 3.10). In PEX, the presence of unsaturation[12] (>C=C< or >C=O), issuing from the synthesis or deliberately inserted, modifies the behavior of the polymer, as it is stabilized through electronic excitation transfers from PE chains to these "alien" groups (Seguchi 2001;

11 Temperature above which the macromolecular chains present in the amorphous phase gain increased mobility with respect to one another and, for a given macromolecular chain, long-distance mobility is developed.

12 An unsaturation or unsaturated bond is formed by two atoms bound by at least two covalent bonds.

Ferry et al. 2012; Ventura et al. 2016). The introduction of benzene rings in common polymers, such as PEX or PP, has been used for several decades to improve their stability under ionizing radiation (Gardner and Harper 1965).

3.4.2.2. *Formation of radicals*

Excited molecules evolve toward the homolytic scission[13] of a covalent bond to give radicals (Figure 3.11).

In most cases, the broken bond is the pH bond, resulting in the formation of H° radicals and of the most stable P° radical. In the presence of halogens (chlorine, fluorine, for example, denoted by X in what follows), in addition to the C–H bonds, a high proportion of C–X bonds will also be broken. In some cases, depending on the chemical structure of the RU, low-mass radicals, such as CH_3°, can also be formed. The formed radicals are classified into two categories: free radicals (low mass radicals, such as H° radicals or CH_3° radicals, for example) and macroradicals (P°, radicals carried by the polymeric chain).

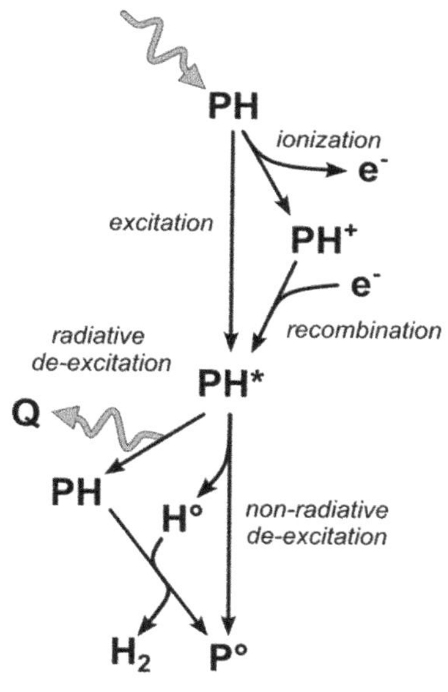

Figure 3.11. *Polymer under ionizing radiation: from energy deposition to the formation of radicals (Aubrit 2017)*

13 Each partner of the bond leaves with one of the electrons forming the bond.

Under low-LET irradiations (γ, β, α irradiations), at low doses, the chemical structure of the radicals reflects that of the chemical groups composing the polymer. At low doses and under inert conditions, the macroradicals formed in PEX are alkyl radicals (Ungar 1981). In irradiated PS, benzyl and cyclohexadienyl groups are the most frequently observed (Hedvig 1972).

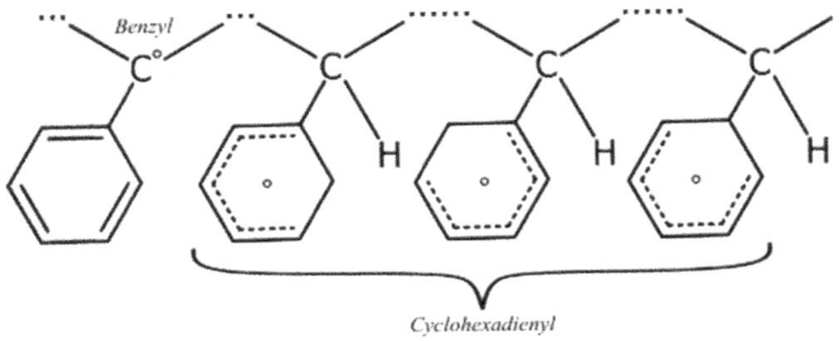

Figure 3.12. *The main radicals formed in the irradiated polystyrene under an inert atmosphere with low LET beams*

3.4.3. *Evolution of radicals: defect formation*

As noted above, very few defects are formed from primary species. The chemistry of polymers subjected to ionizing radiation is therefore predominantly led by radical reactions.

Radiation-induced radicals are highly reactive and can either recombine or react with the polymer in the neighboring environment of their sites of creation to yield new chemical groups referred to as *defects*. Radicals can also migrate along chains, essentially through H° or H° exchanges, even at temperatures as low as the temperature of liquid nitrogen (Waterman and Dole 1971), leading to the formation of defects far from their creation site.

The radioinduced defects in polymers can be gases (also known as radiolysis gases) or defects formed in the chain (also known as macromolecular defects) (Bovey 1958; Dole 1972). Gases and macromolecular defects are the counterparts of one another. For a thorough study of a given polymer behavior under ionizing radiation, both of them should be studied.

3.4.3.1. *Outgassing*

In polymers irradiated under an inert atmosphere (vacuum or inert gases, such as nitrogen or helium) at low doses with γ and β radiation, the composition of radiolysis gases is a perfect signature of the chemical structure of the RU. Among standard polymers (polyolefin without heteroatoms), dihydrogen (H_2) is the most efficiently created radiolysis gas. It is formed by the initial scission of a C–H bond in the material (Figure 3.11), and its formation is generally concomitant with that of alkene unsaturations. This also stands for irradiations under SHIs with low LET.

In polymers containing carbonyl groups (ketones, carboxylic acids, esters, etc.) in their RU, carbon oxides (CO, CO_2) are created quantitatively (Ferry et al. 2016).

Halogenated polymers, such as PVC and PEX tetrafluoride (PTFE, the commercial brand known as Teflon), are part of the least radiation-resistant polymers because of the fragility of carbon–halogen bonds. The gases emitted following the scission of such bonds contain hydrogen halides: HCl for PVC and HF for PTFE.

As noted in the introduction, the polymer backbone may contain longer or shorter side groups (Figure 3.2(c)). Under β and γ radiation, low-mass hydrocarbons are created by the scission of the covalent bond between the backbone and the side group. These hydrocarbons are then a signature of the side chain composition (Bowmer and O'Donnell 1977).

3.4.3.2. *Macromolecular defects*

The chemical structure of macromolecular defects created in polymers irradiated under an inert atmosphere (vacuum or inert gases) with β or γ beams is directly linked to that of RU. These macromolecular defects include:

– New bonds. These are generally unsaturations (>C=C< or >C=O, depending on the chemical structure of RU), created following the chemical reorganization of radiation-induced radicals. Chemical groups with simple bonds can also be formed by the reaction of macroradicals with a proton radical or by recombination of two macroradicals.

– Chain scissions, that is, scissions of bonds constituting the polymer backbone.

– Cross-linking, obtained by the setting of a covalent bond between two polymer chains (interchains) or between two atoms of the same chain (intrachain).

Chain scissions and cross-linkings have a significant influence on the length of polymer chains and some mechanical properties, such as the ultimate elongation or elastic limit. The same is true for solubility, which increases when the rate of chain scissions increases and decreases in the presence of cross-links. Three-dimensional structures, formed after cross-linkings, can be used to trap solvents in the resulting matrices. This is the principle of hydrogels obtained by cross-linking longer or shorter chains initially soluble in the respective solvent.

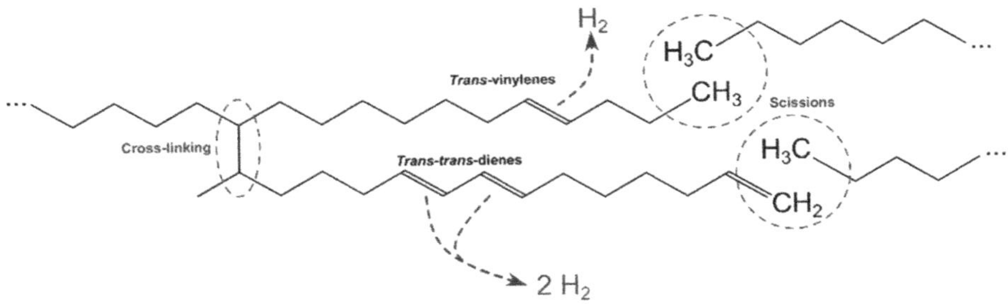

Figure 3.13. *Defects created in the PE irradiated in an inert atmosphere under electron beams (Ventura 2013). For a color version of this figure, see www.iste.co.uk/richaud/industrialpolymers1.zip*

Radiation-induced unsaturation types depend on the chemical structure of the RU. In polyolefines irradiated under an inert atmosphere, the main unsaturated defects are *trans*-vinylenes (TVs) in PEX, vinylidenes in PP and vinyls in polybutylene (Ngono-Ravache et al. 2007). Conjugated double bonds[14] can also be formed. In PE, the level of conjugation increases when the dose increases, leading to polyenes at high doses (Ungar 1981).

In the presence of heteroatoms (oxygen, nitrogen or sulfur, for example), unsaturations can contain these heteroatoms. In the case of polymers containing oxygen atoms, the unsaturated defects created include carbonyl groups (essentially ketones or aldehydes) (Dole 1972).

Importantly, defects are created simultaneously. Depending on the irradiation conditions, the chemical structure of RU and the organization at various scales, some defects may become predominant over others.

In this principle, chain scissions and cross-linkings coexist in polymers subjected to ionizing radiation under an inert atmosphere. However, one or the other may become predominant. Table 3.4 presents an experience-rooted empirical rule that can be used to classify polymers according to their tendency to undergo cross-linkings or chain scissions.

---

14 Two unsaturated bonds are referred to as conjugated if they are separated by a single simple bond.

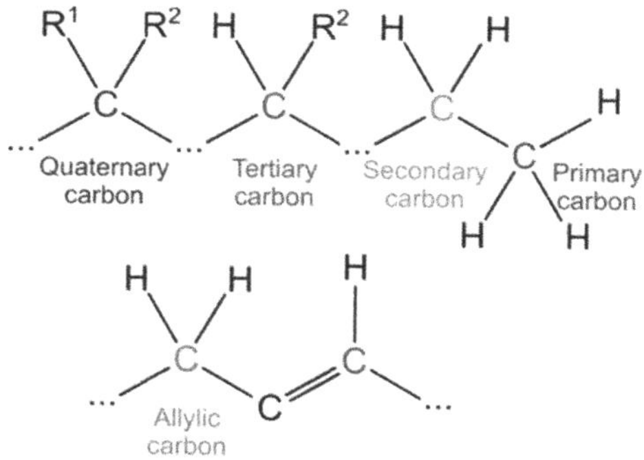

Figure 3.14. *Classification of carbon atoms. For a color version of this figure, see www.iste.co.uk/richaud/industrialpolymers1.zip*

	Condition	Groups in RU	Polymer example
Predominant cross-linkings	Secondary carbon	$-CH_2-$	Polyethylene
	One hydrogen atom on the carbon in α[15] of a CH_2 group	$-C(R)H-CH_2-$	Polypropylene*
	Unsaturations >C=C<	$-CH_2-CH=CH-CH_2-$	Polybutadiene
Chain scissions	Quaternary carbon	$-CH_2-C(CH_3)(R)-CH_2-$	Poly(methyl methacrylate) 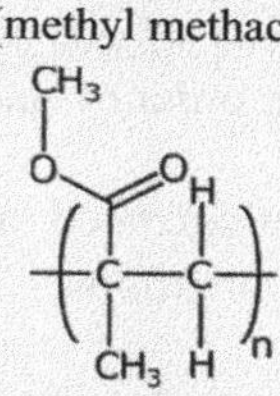
	Presence of halogen atom	$-CF_2-CF_2-$	PTFE
	Ester groups	RC(O)O-	Polyesters

Table 3.4. *Empirical rules for the classification of polymers depending on their tendency to chain scission or cross-linking at ambient temperature. The case of tertiary carbon atoms is slightly more complicated. However, applying the rule on the presence or absence of a hydrogen atom on the carbon in α a $-CH_2-$ makes it possible to decide. *Radiochemical yields depend on tacticity*

15 Carbon located next to the considered group.

The polymer crystallinity rate can also influence the cross-linking/scission rate to the extent that cross-linkings essentially occur in the amorphous phase.

3.4.4. *Parameters influencing the stability of polymers under radiation*

As mentioned at the beginning of this chapter, polymers are sensitive to radiolysis. Nevertheless, they do not have the same stability under ionizing radiation[16,17]. This depends on the groups constituting the RU, the irradiation conditions (type of ionizing radiation, dose rate and dose) and the irradiation environment (oxygen, temperature and humidity).

3.4.4.1. *Groups constituting the repeating unit*

The intrinsic behavior of polymers under ionizing radiation is normally studied under anoxia (inert atmosphere or vacuum) to bypass the highly complex oxidation chain reaction at room temperature and at low LETs. The defects formed and the efficiency of their formation depend on the chemical structure of the RU, their organization in the chain and the organization of chains with respect to one another (crystalline ratio).

A strict classification of polymers according to their resistance to ionizing radiation is a complex task. However, if gases are decorrelated from macromolecular defects, it is possible to provide some useful indications.

In *terms of gas emission*, the groups composing the RU can be classified in ascending order of stability according to the following:

Halogenated aliphatic groups < Aliphatic groups < Unsaturated groups < Aromatic groups

In *terms of macromolecular defects*, the RUs carrying exclusively aliphatic carbon atoms (in the absence of heteroatoms) are classified according to their tendency to establish cross-linkings or chain scissions. They can be classified in ascending order of stability (from less stable to more stable) according to the following:

Quaternary carbon < Tertiary carbon < Secondary carbon

---

16 This is also referred to as "radiostability".

17 Depending on the use, the stability of polymers under ionizing radiation can be defined by considering radiochemical yields of defect formation or by following the evolution of a specific property.

The RU containing heteroatoms other than halogens can be classified in ascending order of stability according to the following:

Esters < Ethers < Amines

The radiation stability of polymers containing benzene rings under an inert atmosphere has been known and used for several decades. Consequently, it is natural to think that stability increases when the number of benzene rings in the RU increases, and this is what happens. However, all RU with the same number of benzene rings do not have the same radiation stability. In the end, the RU stability will be defined by that of the groups between cycles, as shown by the order in Figure 3.16 for RU containing two benzene rings. The stability of the RU decreases from left to right.

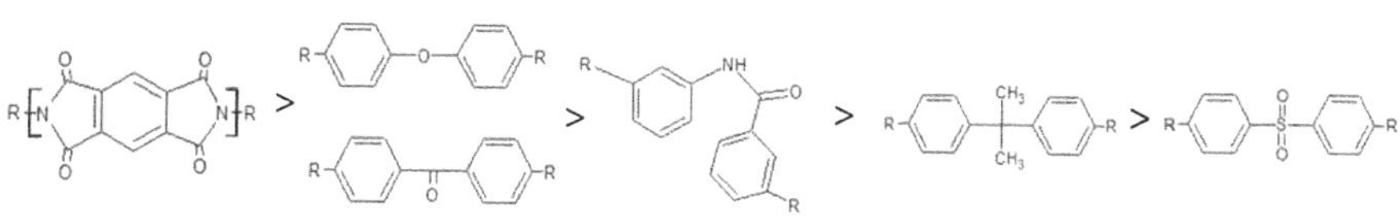

Figure 3.15. *Radiation stability of several RU containing benzene rings obtained from their resistance to traction after irradiation under electron beams at room temperature (Sasuga et al. 1985)*

Schematically, in the absence of carbon–halogen bonds, most synthetic polymers can be considered a combination of simple alkyl groups (–CH_2–CH_2– or –CH_2–) and energy-trapping groups. Consequently, the stability of a given polymer can be seen as the result of energy transfer between the alkyl groups composing the RU and the other groups of the RU. The trapping group will then represent a major modification point, and the formed defects and their radiochemical yields indicate the chemical structure and stability of the energy-absorbing group following these energy transfers. This also applies to benzene rings, for which it was shown several years ago (Ferry 2008) that they stabilized the aliphatic phase to their detriment.

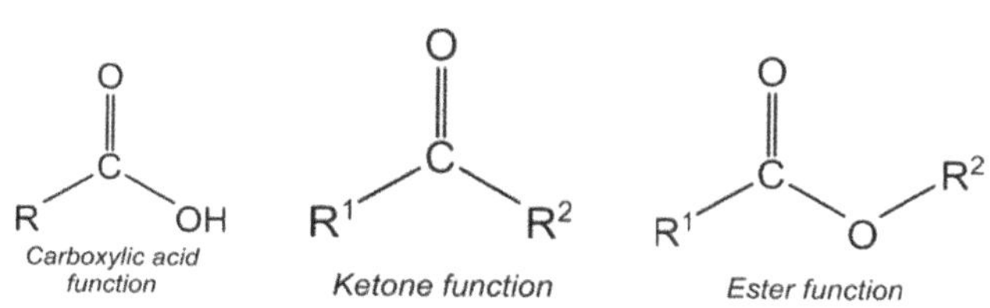

Figure 3.16. *Chemical functions involving the carbonyl group*

Groups other than benzene rings act as energy traps. These are alkene-type unsaturations (–CH=CH–), amines, ethers or functions containing the carbonyl group (C=O), such as ketones and aldehydes, carboxylic acids or esters (Figure 3.16).

3.4.4.2. *Influence of irradiation parameters: dose and dose rate*

Irrespective of the considered polymer, the evolution of the concentration of a given radiation-induced macromolecular defect as a function of the dose depends on its initial concentration in the virgin material.

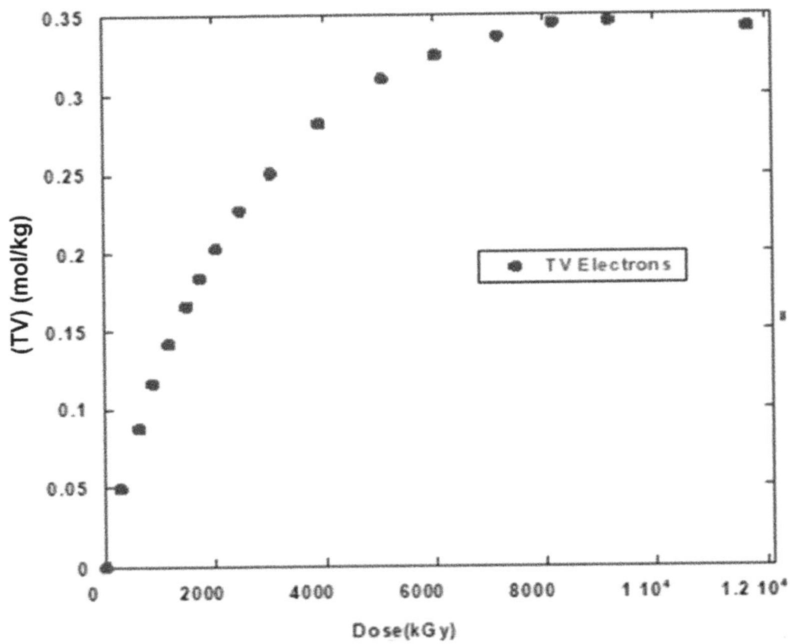

Figure 3.17. *Evolution of the concentration of trans-vinylene (TV), the predominant macroscopic defects in a pure PE irradiated under electron beams in an inert atmosphere, depending on the dose (Ventura et al. 2016)*

When the radiation-induced group is absent from the virgin material, its concentration increases when the *dose* increases. Under inert condition, the concentration of some defects peaks at a concentration known as the saturation concentration, Csat (Figure 3.17). The value of Csat for a given defect depends on the polymer. Under ionizing radiation, the creation of defects is concomitant with the destruction of defects created at lower doses. The concentration continues to increase as long as the creation remains more efficient than destruction (also known as consumption). For a given molecular defect, saturation is reached when its destruction yield equals its creation yield: the apparent radiochemical yield of creation is then zero.

In case the group considered is initially present in the polymer, its evolution will depend on its concentration relative to Csat. When the concentration prior to irradiation of the defect group is below the saturation concentration, its concentration will increase with the dose and get stabilized around Csat. On the other hand, if the initial concentration of the defect group was above its saturation concentration, the group would be destroyed, and its concentration would stabilize at Csat. This behavior was recently confirmed at CIMAP in PEXs containing controlled concentrations of TV groups (Ventura 2013) and explains the evolution of vinyl unsaturation in PEX samples subjected to γ or β radiation in early studies. These groups were reported to be consumed under irradiation in early studies (Dole et al. 1958) but were created in studies performed decades later. The main difference lies in the concentration of vinyl groups in these PEX films: high in the PE studied in the former and lower in the latter, as the PE synthesis was better controlled.

TVs are the major unsaturated macromolecular defects formed in PE irradiated under an inert atmosphere. Their concentration increases when the dose increases and saturates at very high doses (Figure 3.17): their creation radiochemical yield decreases when the dose increases and ends at zero.

TV formation is concomitant with the dihydrogen emission. The radiochemical yield of the dihydrogen emission also decreases when the dose increases, starting from low doses, and stabilizes at a value of approximately half the one at low doses (Figure 3.17).

The decrease in $G(H_2)$ with increasing dose could be intuitively assigned to the rarefaction of the hydrogen reservoir, but this is not the case. In fact, calculations have shown that enough hydrogen atoms remain in the material and enough $-CH_2-CH_2-$ groups exist for macromolecular defects resulting from hydrogen emission to be formed. The phenomenon at play here is energy transfers to radiation-induced defects (mentioned in section 3.4.2.1). The excitations created in the unmodified alkyl segments migrate in the material and are trapped by TV defects created at lower doses that act as energy traps and thus reduce the efficacy of H_2 emission (Ventura et al. 2016). In parallel, this energy trapping leads to the destruction of TV groups (in addition to that induced by radiation).

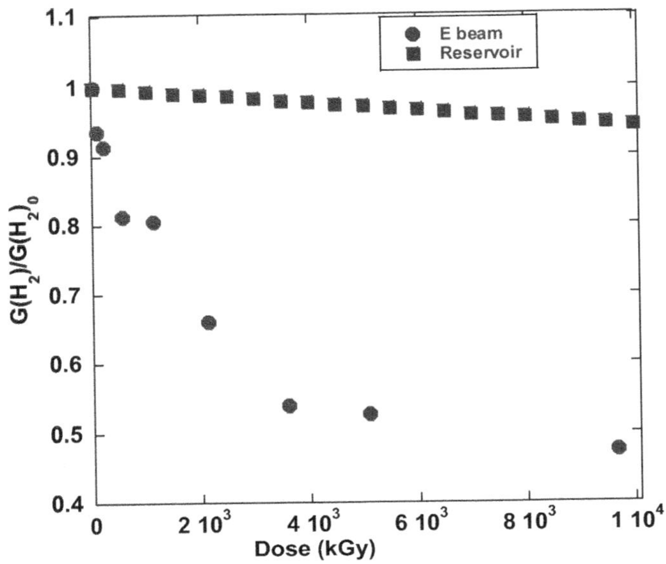

Figure 3.18. *Evolution of the radiochemical yield of emission of H_2, as a function of the dose, in a pure PE irradiated under an inert atmosphere with a 1 MeV electron beam. $G(H_2)_0$ represents the radiochemical yield at a very low dose (25 kGy). The reservoir represents the evolution of $G(H_2)$ if it were proportional to the H reservoir in the polymer*

As already stated, polymers have several levels of organization. The above-mentioned defects are formed at the molecular level. The accumulation of these defects influences the other levels of organization, but the consequences of such an accumulation may be visible only at significant doses. This is the case for the "amorphization" observed in the PE irradiated with electron beams (Balanzat et al. 1994). Indeed, the gradual increase in the in-fine defect concentration destroys the long-distance organization required for the formation of crystallites. This results in a decrease in the crystalline ratio and thus an increase in the amorphous component.

The dose rate has very little influence on the creation of defects in polymers irradiated with β and γ radiation under an inert atmosphere at low doses. On the other hand, at high doses, the higher the dose rate is, the higher the defect concentration at saturation.

3.4.4.3. *Influence of the irradiation environment*

The environment in which the material is irradiated can also influence its evolution. The oxygen concentration in the material or in its environment, the irradiation temperature and the relative humidity in the irradiation environment are the factors that are extrinsic to the material or to the radiation and are susceptible to modifying the polymer behavior under ionizing radiation.

3.4.4.3.1. Temperature

The influence of temperature is essential for chemical evolution after energy deposition[18]. In particular, secondary processes occur after radical formation, that is, their migration, recombination and molecular rearrangements lead to the formation of stable defects. Temperature affects both the type of defects created and their creation efficacy and the associated kinetics (Mélot et al. 2003a, 2003b).

As mentioned in the introduction to this chapter, polymers have various levels of organization, and these levels are associated with specific transitions or relaxations and therefore with various transition and relaxation temperatures. The various levels of mobility in the polymers evolve with temperature, particularly near relaxation or transition temperatures. One of the most important temperatures is the glass transition temperature (Tg), the temperature at which long-distance motions and collaborative phenomena become possible in the amorphous phase. This mobility of chains goes hand in hand with that of reactive species. The probability of an encounter between a reactive species and the subsequent reactions increases when the temperature increases.

The influence of temperature on the behavior of polymers under ionizing radiation should not be considered in a vacuum. Rather, to understand the influence of temperature on polymer behavior under ionizing radiation or when comparing the effects of temperature on two different polymers, it is important to consider the gap between the irradiation temperature and the various transition or relaxation temperatures of these materials.

Under low-permeability conditions, radiation-induced gases can remain trapped in the amorphous matrix in its vitreous state and diffuse only at a temperature near the Tg. Depending on the radiochemical yield of the gas emission, bubbles can then form in the material and threaten its integrity.

18 In the presence of charged species transfer, the effect of temperature can be observed upstream of the chemical stage. As for the transfers of excitation, they occur at temperatures as low as 8 K.

3.4.4.3.2. Hygrometry

During irradiation under immersion, the action of ionizing radiation on the polymer will either be direct (stemming from the energy deposition in the material) or indirect through the radiation-induced reactive species from water. The subsequent reactions of these species with the polymer, as well as the potential hydrolysis of the radiation-induced defects formed in the polymer, should then be considered.

During irradiation under a humid atmosphere, the hydration level of the materials and the interaction configurations of H_2O molecules (isolated or associated with other molecules) should be considered (Ngono-Ravache and Maréchal 2000, 2001). The behavior of a polymer under such an atmosphere will then depend on its hydration capacity, which in turn depends on the presence or lack of accessible hydrophilic groups in the material and on the available free volume.

3.4.4.3.3. Oxygen supply: radiation-induced oxidation

Under the usage conditions, oxidation is the main process at the origin of the degradation of polymer functional properties. Polymers are permeable to oxygen, even though the solubility of this gas, like that of other gases, differs from one polymer to another.

When oxygen molecules are present in the irradiation atmosphere, the radicals issued from the de-excitation of the material react instantaneously, with the latter following a chain mechanism: the stable defects yield is greater than the primary species creation yield. The chain mechanism used for the study of radiation-induced oxidation in polymers resulted from the works of Bolland and Gee (1946) on the thermal-induced oxidation of elastomers, whose simplified version is presented below. It proceeds in three main steps (equations [3.12]–[3.17]):

– initiation induced by irradiation;

– propagation that leads to the formation of new radicals by hydrogen atom abstraction from the virgin polymer;

– termination, during which stable defects are formed. Branching reactions are also present, for example, in the presence of antioxidants or following the decomposition of hydroperoxides.

Initiation:

$$PH \rightsquigarrow P^\circ + H^\circ \qquad [3.12]$$

Propagation:

$$P^\circ + O_2 \rightarrow PO_2^\circ \quad [3.13]$$

$$PO_2^\circ + PH \rightarrow PO_2H + P^\circ \quad [3.14]$$

Termination:

$$P^\circ + P^\circ \rightarrow \text{stable defects} \quad [3.15]$$

$$P^\circ + PO_2^\circ \rightarrow \text{stable defects} \quad [3.16]$$

$$PO_2^\circ + PO_2^\circ \rightarrow \text{stable defects} \quad [3.17]$$

Similar to those formed under anoxia, there are two types of radiation-induced defects under an oxidizing atmosphere: macromolecular defects and gases. The oxidation defects are quite different from those formed under an inert atmosphere: they have the characteristic of carrying at least one oxygen atom. The commonly encountered macromolecular defects are oxidized chemical groups (ketones, esters, amides in the presence of amine functions in the RU, alcohols and carboxylic acids), chain scissions and cross-linkings (Rivaton et al. 2005).

The ratio of radiochemical yields of chain scissions and cross-linking differs from that observed under an inert atmosphere and depends on the oxygen pressure in the environment and on the dose rate through the predominant termination reaction. Indeed, when the oxygen concentration in a material is significant (oxidation is not limited by oxygen diffusion from the environment), terminations preferentially occur via reaction [3.17], and the scission/cross-linking ratio increases with increasing irradiation.

In addition to macromolecular defects, gases, such as dihydrogen, H_2O and carbon oxides, are emitted by common polymers. When nitrogen or sulfur atoms are present in a material, the emitted gases can contain these heteroatoms.

The factors influencing polymer behavior under radiation-induced oxidation can be intrinsic or extrinsic to the material. At identical initiation efficacies, the tendency of a polymer to be oxidized is related to the propagation and termination stages. Owing to the stability of RO_2° radicals, the limiting stage in the oxidation process is the reaction of propagation by hydrogen atom abstraction (equation [3.14]). It is widely influenced by the ease with which the hydrogen atom can be removed. This is related to the stability of the resulting radical (P°): the more stable the radical is, the easier hydrogen abstraction.

In the absence of heteroatoms in the RU, the hydrogen atom abstraction ease would increase from hydrogen atoms carried by carbon atoms, forming an aromatic cycle ($CH_{aromatic}$) to the number of allylic carbon atoms (Figure 3.14).

$$CH_{aromatic} << CH_{primary} < CH_{secondary} < CH_{tertiary} < CH_{allylic}$$

The allylic proton is one of the easiest to abstract, which explains the sensitivity to oxidation of elastomers that contain C=C double bonds in their RU, such as polyisoprene or EPDM. In the presence of tertiary carbon atoms, the reactivity depends on the groups forming it. For example, in PS, the presence of the benzene ring stabilizes the radical created on the tertiary carbon by mesomerism, which facilitates the abstraction of the hydrogen carried by this carbon during the propagation stage.

Importantly, these classifications are indicative, and the specificity of the oxidation site is not straightforward. In PP, oxidation occurs primarily from tertiary carbons, but 20–25% of the oxidation sites are secondary carbon atoms (Billingham 1998).

In the presence of heteroatoms, hydrogen atoms carried by the same atom as the heteroatom are activated by the difference in electronegativity caused by the heteroatom and become sensitive to oxidation. Because of this phenomenon, the oxidized groups created under radiation-induced oxidation are very sensitive to oxidation (Rabek 1996) and may lead to oxidation acceleration when the material aging level increases (Figure 3.19). Moreover, the tendency toward saturation observed for some defects under an inert atmosphere may not exist under oxidation.

Oxidation can be maintained only if the material is constantly supplied with oxygen. Several factors can limit the oxygen concentration in the material. It may be the presence of crystallites, in which oxygen solubility is very low (if any), or a higher consumption rate of the oxygen molecules dissolved in the material compared to its diffusion rate. The two important factors in this latter case are the oxygen pressure in the environment and the dose rate.

The oxygen pressure first influences the quantity of oxygen dissolved in a material via Henry's law. The oxidation rate increases when the oxygen pressure increases in the environment and stabilizes at high pressures. In an EPDMh irradiated with 1 MeV electron beams, the variation in the oxidation rate beyond 20 mbar is low: stabilization is already reached at oxygen pressures corresponding to atmospheric pressure (Dély 2005; Dély et al. 2005).

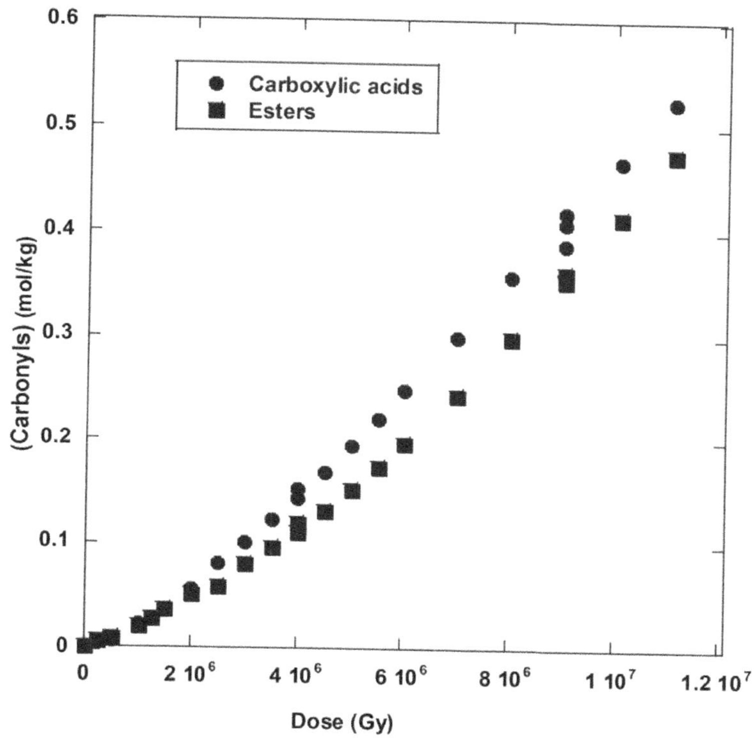

Figure 3.19. *Evolution of the concentration of carboxylic acids and esters as a function of dose in an irradiated EPDMh*[19] *under 1,000 mbar of dioxygen, with Ne^{9+} beams*[20] *of 12 MeV A^{-1}*

To achieve homogeneous oxidation throughout the irradiated thickness, the conditions should be chosen to prevent oxidation from being limited by oxygen diffusion. This is achieved by adapting the polymer thickness to the dose rate and to the oxygen pressure in the environment. For this purpose, it is important to choose a sample thickness significantly below the critical thickness. These data represent the thickness of the polymer through which there is no (or at least a quite reduced one) oxidation profile. It is related to the radiochemical yield of oxygen consumption,

19 Ethylene/propylene-1,4 hexadiene terpolymer (elastomer containing C=C unsaturations in pendant chain).

the oxygen pressure and the dose rate according to equation [3.18] (Guillen and Clough 1991):

$$L_c = \sqrt{\frac{P(O_2)}{[G(-O_2)].I}.p.X} \qquad [3.18]$$

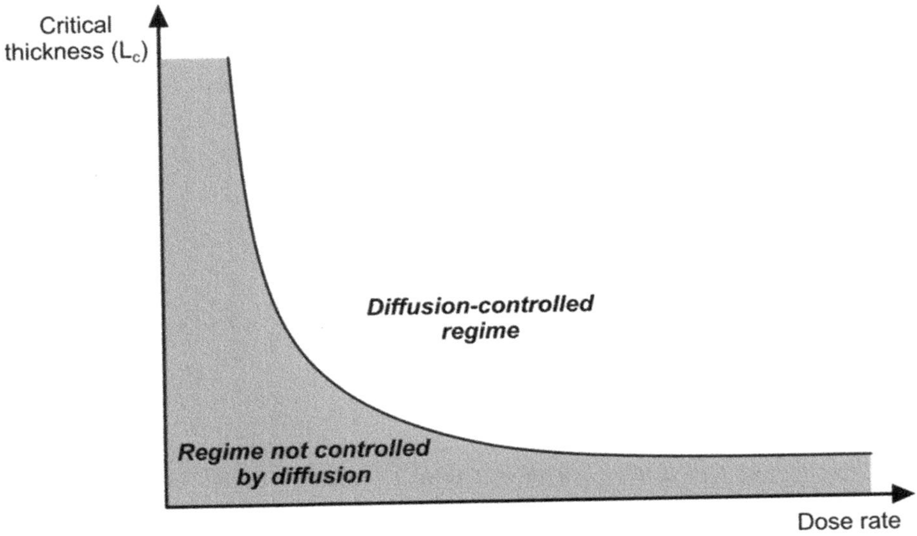

Figure 3.20. *Domains of homogeneous or heterogeneous oxidation as a function of the oxygen pressure in the environment and the dose rate*

In this equation:

– $P(O_2)$, expressed in mbar, represents the oxygen pressure in the irradiation environment;

– P is the oxygen coefficient of permeation in the polymer, expressed in $mol.cm^2.g^{-1}.atm^{-1}.s^{-1}$;

– I is the dose rate, expressed in $Gy.s^{-1}$. $G(-O_2)$ is the radiochemical yield of oxygen consumption, expressed in $mol.J^{-1}$, and X is a reduced parameter associated with the oxygen pressure.

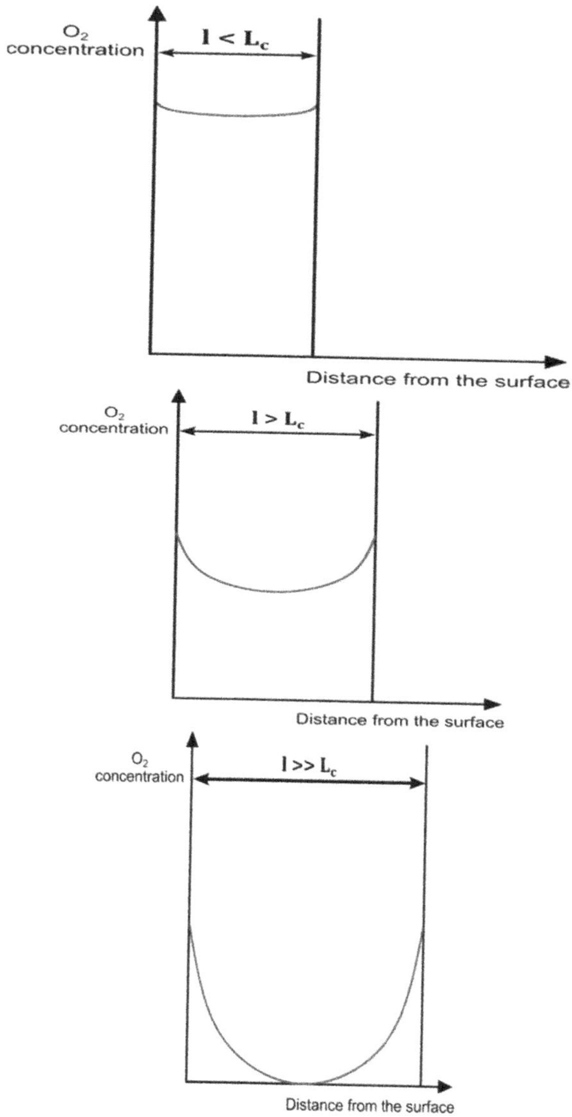

Figure 3.21. *Schematic representation of oxidation profiles as a function of sample thickness compared with the critical thickness*

Under usage conditions, it is not always possible to comply with the critical thickness condition. An oxidation profile then appears in the sample thickness

(Figure 3.21): the oxidation degree decreases while getting farther away from the sample surface. In a material with a thickness significantly above the critical thickness, L_c, oxidation essentially occurs near the surface.

Under conditions where oxidation is not limited by oxygen diffusion, once the stationary regime is reached, the oxidation rate, $V(-O_2)$, can be linked to the dose rate (I) according to [3.19] (Decker and Mayo 1973; Decker et al. 1973):

$$V(-O_2) = \alpha I + \beta I^{0.5} \qquad [3.19]$$

The radiochemical yield of oxygen consumption (G = V/I) is then expressed as follows:

$$G(-O_2) = \alpha + \frac{\beta}{\sqrt{I}} \qquad [3.20]$$

The values of α and β depend on the polymer and the type of ionizing radiation applied (low LET or high LET).

3.4.5. *Specificity of swift heavy ions*

There are two types of defects created in polymers subjected to SHI beams: defects common to all ionizing radiation (presented in the previous sections) and defects specific to accelerated heavy ions.

As previously noted, heavy ions deposit their energy along their path in a cylinder of several nanometers known as "the track". This track is composed of a core that is very dense in excitations and ionizations and of a less dense halo. These high densities of excitation and ionization involve a high probability of de-excitation of primary species, neutralization, bond scission or recombination of the radicals created. Therefore, within the track, there are collaborative phenomena and complex rearrangements leading to specific modifications in polymers irradiated with accelerated ions (Gaté 1997; Ngono-Ravache et al. 2007, 2015).

The *macromolecular defects* specific to heavy ions are highly unsaturated. These are groups carrying triple bonds, such as alkynes (Ngono-Ravache et al. 2007) or nitriles in the presence of a nitrogen atom; groups carrying cumulative double bonds, such as allenes ($-CH_2-CH=C=CH-CH_2-$), or isocyanates (Stekenreiter et al. 1999a, 1999b; Sun et al. 2002); and vinyl groups. The latter are also created in polymers subjected to γ and β radiation but with very low radiochemical yields.

Under heavy ions, the radiochemical yield at low doses of vinyl groups is greater and increases when the LET increases (Mélot 2003b).

The characteristic of defects specific to heavy ions is that they are created in the track core (Le Bouëdec et al. 1999), above an LET threshold that depends on the considered defect and on the chemical structure of RU. As an example, the threshold LET for alkyne creation in saturated polyolefines without heteroatoms is on the order of 6 $MeV.mg^{-1}.cm^2$, which decreases in the presence of alkene unsaturations (Ngono-Ravache et al. 2007).

The influence of LET on the defects created under SHI irradiation differs depending on whether these defects are specific to heavy ions or common defects. At low doses, the radiochemical yields of the creation of specific defects increase when LET increases, that is, when the density of excitation and ionization increases. Logically, at high doses, the concentration at saturation increases when LET increases. The effect of LET on the radiochemical yield of the creation of common defects (TV, *trans-trans*-diene in the PE) is very limited, if any, at low doses. On the other hand, the saturation concentrations of these defects increase when LET increases. The efficiency of creation at low doses is independent of LET. This behavior is explained by a lesser destruction at high doses and at high LET (Ventura et al. 2016).

As mentioned above, under γ and β radiation, there is a gradual "amorphization" of polymers when the concentration of molecular defects increases, meaning when the dose increases. Under SHIs, amorphization occurs directly in the track, starting with the lowest doses (Balanzat et al. 1994, 1995).

Another characteristic of heavy ions is their effect on *radiation-induced gases*. The nature of the formed gases depends on LET. At low LET, similar to β and γ irradiations, the emitted radiation-induced gases reflect the structure of the RU. Under high LET, the emitted gases are no longer the footprint of the RU. The gases produced contain a relatively high percentage of unsaturated hydrocarbons (Pick et al. 1998).

The specific structure of the energy deposition with SHI influences *energy transfers* in polymers. These transfers remain effective, even at very low temperatures, but only at LET, at which the fragmentation is very limited (LaVerne et al. 2001; Ferry et al. 2008). Transfers become ineffective when fragmentation becomes important. Hence, at LET above 20 $MGy.mg^{-1}.cm^2$, the radiochemical yields of hydrocarbon emission in the PE and in the PS are identical (Picq et al. 1998; Picq and Balanzat 1999).

	β, γ beams	Swift heavy ions
Energy deposition	More or less homogeneous	Highly heterogeneous. Track structure
Dose distribution	More or less homogeneous	Radial Very high local dose
Influence of RU	Radiolysis gases reflect the chemical structure of the RU	Depends on LET At high LET: radiolysis gases no longer reflect the chemical structure of RU Radiochemical yields of emitted gaseous hydrocarbons ↗ when LET ↗ Unsaturation degree of radiolysis gases ↗ when LET ↗
Efficiency of energy transfers	Transfers of species when mobility is allowed Transfers of excitation even very far below Tg	Effective energy transfers in the absence of fragmentation
Effect of dose rate	Effective under oxidation conditions	Present under oxidation conditions, but reduced effect
Oxidation	No creation of unsaturated groups, except at very high dose rates	Creation of unsaturated groups
Influence of the dose in the absence of O_2	Tendency to saturation of macromolecular defects Tendency to the saturation of radiochemical yields of some gases	Tendency to saturation of macromolecular defects. C[20]$_{sat}$ (ions) > C_{sat} (β, γ) Tendency to saturation of radiochemical yields of some gases
Influence of the dose under O_2	No tendency to saturation of oxidized products	No tendency to the saturation of oxidized products Saturation of defects containing >C=C< bonds

Table 3.5. *Characteristics of swift heavy ions: parameters influencing the polymer behavior under ionizing radiation*

20 Saturation concentration.

The influence of the strong heterogeneity of the energy deposition of SHI has three main consequences *in the presence of oxygen*:

– The first one logically concerns the very heterogeneous distribution of POO° radicals in the material (Gervais et al. 2021).

– The second concerns the persistence of P°+P° recombinations in the track, irrespective of the oxygen pressure in the environment. As a result, unsaturated groups are formed during irradiation under ion beams, independently of the dose rate and the oxygen pressure, whereas under electron beams, such defects are created only in the presence of low oxygen pressures and/or high dose rates (Ngono-Ravache et al. 2015).

– The third consequence is related to the high local dose rate in the track, which leads to a lower radiochemical yield of oxidation under ion beams than under electron beams at an identical average dose rate. The [21]$G(-O_2)_{ions}/G(-O_2)_{electrons}$ ratio in an EPDMh is approximately 1/7 at 100 kGy h^{-1} under 1 000 mbar of O_2 (Dély 2005).

3.5. Conclusion

Owing to their simple chemical structure, which is formed by the repetition of one or several monomers, studying the behavior of polymers under ionizing radiation is an easy task. This involves overlooking the fact that the RU, solely, is not sufficient to characterize a given polymer. The various levels of organization and their consequences in terms of transition temperatures or crystallinity ratio should also be considered.

Radiation-induced defects are formed at multiple scales of polymer organization. There may be defects at the level of the RU or chain scission (which leads to chain length reduction) and cross-linking between chains at the macromolecular level. The accumulation of these defects may eventually lead to the modification of the crystalline ratio, that is, the supra-macromolecular level.

As already noted, several intrinsic or extrinsic parameters are involved in the radiation-induced evolution of polymers. For a pure polymer, it is first of all a matter of the chemical structure of its RU, and then of the irradiation parameters, such as the ionizing radiation type (γ, β and ions), the dose, the dose rate and, finally, the irradiation environment (temperature, oxygen pressure and humidity).

21 ^{13}C ion beams of 13 MeV/A.

In the absence of oxygen, radiochemical yields of defect formation decrease when the dose increases. This evolution is assigned to the implementation of energy transfers toward defects created at lower doses. In fact, these defects bring to the polymeric chain "alien groups" that differ from the RU either by their excitation energy or by their energy of first ionization, which may play the role of energy trap, stabilizing the material. An increase in the concentration of defects at the molecular level leads to the modification of the macroscopic properties or usage properties. The lifetime of the considered materials is defined based on the evolution of these properties: the dose by itself does not define the lifetime.

Concerning the influence of the ionizing radiation type, at a first approximation, I.R. can be divided into two large groups: those inducing low LET (β, γ) and those inducing high LET (α, SHI). For the former, radiation-induced defects are distributed more or less homogeneously in the sample volume, whereas the defects created under irradiation with SHI are distributed in a highly heterogeneous way, at the nanometric scale, along the passage of ions (in the track). Moreover, at high LET, highly unsaturated specific defects are created above an LET threshold that depends both on the considered defect and on the polymer.

Concerning the influence of the irradiation environment, the most important factor is the presence or absence of oxygen molecules. When they are dissolved in the polymer, oxygen molecules react instantaneously with the radiation-induced radicals to lead, according to a chain oxidation mechanism, to oxidation defects. Under oxidation conditions, the dose rate effect is the most significant: the lower the dose rate is, the greater the degree of oxidation. Radiation-oxidative aging is the most damaging process for materials usage properties.

Taking into account the influence of temperature on the behavior of a given polymer subjected to ionizing radiation should not be done ex nihilo, simply by comparing the irradiation temperature with room temperature. It is important to consider all mobility levels (side groups, side chains, chain segments, long-distance mobility or mobility of chains with respect to one another) and the corresponding transition temperatures.

One of the latter is the glass transition temperature. The number of defects for which long-distance or long-range mobility is mandatory increases with increasing temperature, especially near the Tg. Under ion beams, radiochemical yields of specific defects are slightly influenced by the irradiation temperature, as these defects are created at very short times in regions with a high density of excitations and ionization.

Polymers are rarely pure. To elaborate a material, various organic additives should be added to the resin (pure polymer). These include antioxidants, plasticizers and/or mineral fillers. These compounds produce "alien" groups that may present excitation potentials, first ionization energies or electronic affinities that differ from those of the groups forming the resin. Compared with those of pure polymers, energy transfer (excitations, radicals and charges) can then take place, and these are susceptible to drastically modify the behavior under radiation of the industrial material compared to that of pure polymer.

Energy transfer phenomena play an important role in polymer evolution under IR. In the presence of accepting groups, primary species (ions, electrons, and electronic excitations) will migrate over significant distances until they are trapped by these groups. Depending on their stability, the energy-accepting groups can be the starting point for material degradation.

It is possible to push this energy transfer principle by considering most RUs of common synthetic polymers as a more or less substituted version of the RU of PEX. The substitution groups are considered potential energy traps toward which energy transfer from alkyl residues can occur. The efficiency of these energy transfers will set the behavior of the material, as compared to polyethylene. Notably, this efficiency depends on the excitation potential for electronic excitation transfers and the first ionization energy for charge transfers.

The values of radiochemical yields for gas emission or for macromolecular defect creation are sometimes quite scattered in the literature. A detailed study of the polymers at stake often reveals the presence of unsaturations resulting from synthesis, and as mentioned above, these unsaturations play the role of radiation stabilizers. Therefore, depending on their concentration in the material, defects creation yields will vary. The most astonishing example is that of vinyl groups in PEX and the diversity of radiochemical yields of H_2 emission.

The effect of energy transfer is presented in this chapter under positive light, with radiation protection of the initial polymer. These transfers may also be negative in the presence of groups that transfer the absorbed energy to the rest of the polymer.

3.6. Acknowledgments

The illustrations presented in this chapter are the work of Florian Aubrit, a postdoctoral researcher at CIMAP, whom we warmly thank for his contribution.

3.7. References

Aubrit, F. (2017). Films nanocomposites plasmoniques auto-assemblés. PhD Thesis, Université de Bordeaux, Bordeaux.

Balanzat, E., Bouffard, S., Le Moel, A., Betz, N. (1994). Physico-chemical modifications induced in polymers by swift heavy ion. *Nuclear Instruments and Methods in Physics Research B*, 91(1), 140–145.

Balanzat, E., Betz, N., Bouffard, S. (1995). Swift heavy ion modification of polymers. *Nuclear Instruments and Methods in Physics Research B*, 105(1/4), P46.

Billingham, N.C. (1998). Some notes on oxidative degradation of polymers. In *Post-Doctoral Course on Degradation, Stabilisation and Reactive Modification of Polymers*. COREP, Turin.

Blanc, D. (1997). *Les rayonnements ionisants, détection, dosimétrie, spectrométrie*. Masson, Paris.

Bolland, J.L. and Gee, G. (1946). Kinetics studies in the chemistry of rubber and related materials: The kinetics of oxidation of conjugated olefins. *Trans. Faraday Soc.*, 42, 236.

Bovey, F.A. (1958). *The Effects of Ionizing Radiation on Natural and Synthetic High Polymers*. John Wiley & Sons, New York.

Bowmer, T.N. and O'Donnell, J.H. (1977). Nature of the side chain branches in low-density polyethylene: Volatile products from gamma radiolysis. *Polymer*, 18(10), 1032–1040.

Brede, O., Wojnarovits, L., Stephan, L., Taplick, T. (1989). Pulse-radiolytic investigation of reactions of phenolic stabilizers in molten polyethylene. *Polymer Degradation and Stability*, 24(4), 257–268.

Calcagno, L. and Foti, G. (1991). Ion irradiation of polymers. *Nuclear Instruments and Methods in Physics Research B*, 59(60), 1153–1158.

Chang, Z. and LaVerne, J.A. (2000). Hydrogen production in the heavy ion radiolysis of polymers. 1. Polyethylene, polypropylene, poly(methyl methacrylate), and polystyrene. *J. Phys. Chem. B*, 104(45), 10557–10562.

Chapiro, A. (1962). *Radiation Chemistry of Polymeric Systems*. John Wiley & Sons, New York.

Chapiro, A. (1988). Chemical modifications in irradiated polymers. *Nuclear Instruments and Methods in Physics Research B*, 32(111), 114.

Corbin, D. (2001). Étude de l'oxydation et de la tenue d'élastomères irradiés : conséquences sur l'intégrité des câbles électriques lors d'une situation accidentelle d'un réacteur à eau pressurisée. PhD Thesis, Université de Caen Basse Normandie, Caen.

Decker, C. and Mayo, F.R. (1973). Ageing and degradation of polyolefins. II: G-initiated oxidation of atactic polypropylene. *Journal of Polymer Science: Polymer Chemistry*, 11(11), 2847–2877.

Decker, C., Mayo, F.R., Richardson, H. (1973). Ageing and degradation of polyolefins. III. Polyethylene and ethylene-propylene copolymers. *Journal of Polymer Science: Polymer Chemistry*, 11(11), 2879–2898.

Dély, N. (2005). Radio-oxydation d'un élastomère de type EPDM lors d'irradiations faiblement ou fortement ionisantes : mesure et modélisation de la consommation d'oxygène. PhD Thesis, Université de Caen Basse Normandie, Caen.

Dély, N., Ngono-Ravache, Y., Ramillon, J.M., Balanzat, E. (2005). Oxygen consumption in EPDM irradiated under different oxygen pressures and at different LET. *Nuclear Instruments & Methods in Physics Research Section B*, 236(1–4), 145–152.

Dole, M. (1972). *The Radiation Chemistry of Macromolecules*. Academic Press, New York.

Dole, M., Milner, D.C., Williams, T.F. (1958). Irradiation of polyethylene. 2. Kinetics of unsaturation effects. *J. Am. Chem. Soc.*, 80(7), 1580–1588.

Ferry, M. (2008). Comportement des verres cyclohexane/benzène et des copolymères éthylène/styrène sous rayonnements ionisants : transferts d'énergie et d'espèces entre les groupements alipathiques et aromatiques. PhD Thesis, Université de Caen Basse Normandie, Caen.

Ferry, M. and Ngono, Y. (2020). Energy transfer in polymers submitted to ionizing radiation: A review. *Radiation Physics and Chemistry*, 180, 109320. doi: 10.1016/j.radphyschem.2020.109320.

Ferry, M., Ngono-Ravache, Y., Picq, V., Balanzat, E. (2008). Irradiation of atactic polystyrene: Linear energy transfer effects. *Journal of Physical Chemistry B*, 112(35), 10879–10889.

Ferry, M., Bessy, E., Harris, H., Lutz, P.J., Ramillon, J.M., Ngono-Ravache, Y., Balanzat, E. (2012). Irradiation of ethylene/styrene copolymers: Evidence of sensitization of the aromatic moiety as counterpart of the radiation protection effect. *Journal of Physical Chemistry B*, 116(6), 1772–1776.

Ferry, M., Ngono-Ravache, Y., Aymes-Chodour, C., Clochard, M.C, Coqueret, X., Cortella, L., Pellizzi, E., Rouif, S., Esnouf, S. (2016). Ionizing radiation effects in polymers. In *Reference Module in Materials Science and Materials Engineering*, Holm, D.D. (ed.). Elsevier, Berlin [Online]. Available at: http://www.sciencedirect.com/science/article/pii/B9780128035818020956.

Ganil (2022). Ganil [Online]. Available at: https://www.ganil-spiral2.eu/scientists/ganil-spiral-2-facilities/available-beams/ [Accessed 20 March 2022].

Gardner, J.B. and Harper, B.G. (1965). Radiation protection of polyethylene. *Journal of Applied Polymer Science*, 9, 1585–1591.

Gaté, C. (1997). Polymères aliphatiques sous irradiation pas des ions lourds. PhD Thesis, Université de Caen Basse Normandie, Caen.

Gervais, B. (1993). Étude du dépôt d'énergie des ions lourds rapides dans les solides par simulation Monte Carlo. PhD Thesis, Université de Caen Basse Normandie, Caen.

Gervais, B. and Bouffard, S. (1994). Simulation of the primary stage of interaction of swift heavy ions with condensed matter. *Nuclear Instruments and Methods in Physics Research B: Beam Interactions with Materials Atoms*, 88(4), P355.

Gervais, B., Ngono, Y., Balanzat, E. (2021). Kinetic Monte Carlo simulation of heterogeneous and homogeneous radio-oxidation of a polymer. *Polymer Degradation and Stability*, 185, 109493.

Gillen, K.T. and Clough, R.L. (1991). *Irradiation Effects in Polymers*. Elsevier Applied Science, Amsterdam.

Hedvig, P. (1972). Theory of free radicals. In *The Radiation Chemistry of Macromolecules I*, Dole, M. (ed.). Academic Press, New York.

Hickel, B. (2000). La radiolyse de l'eau. In *Comprendre et utiliser les irradiations*, La Londes-les-Maures.

Kang, H.Y., Saito, O., Dole, M. (1967). The radiation chemistry of polyethylene. IX. Temperature coefficient of cross-linking and other effects. *Journal of the American Chemical Society*, 89(9), 1980–1986.

Laverne, J.A., Chang, Z., Araos, M.S. (2001). Heavy ion radiolysis of organic materials. *Radiation Physics and Chemistry*, 60(4/5), 253–257.

Mélot, M., Ngono-Ravache, Y., Balanzat, E. (2003a). Very low temperature irradiation of aliphatic polymers: Role of radical migration on the creation of stable groups. *Nuclear Instruments and Methods in Physics Research B*, 208, 345–352.

Mélot, M., Ngono-Ravache, Y., Balanzat, E. (2003b). Role of the irradiation temperature on the modifications of swift-heavy-ion irradiated polyethylene. *Nuclear Instruments and Methods in Physics Research B*, 209, 205–211.

Miller, A., Lawton, E., Balwit, J. (1954). Effect of chemical structure of vinyl polymers on crosslinking and degradation by ionizing radiation. *Journal of Polymer Science*, 14(77), 503.

Mozumder, A. and Magee, J.L. (1966). Model of tracks of ionizing radiations for radical reaction mechanisms. *Radiation Research*, 28(2) 203–214.

Ngono-Ravache, Y. (2015). Spectroscopic study of chemical modifications induced by swift heavy ions on polymers: The contribution of the CIRIL platform and the CIMAP laboratory. *Journal of Physics: Conference Series*, 629, 012006.

Ngono-Ravache, Y. and Maréchal, Y. (2000). Epoxy-amine reticulates observed by infra-red spectrometry. II. Modifications of the structure and hydration abilities after irradiation in a dry atmosphere. *Journal of Polymer Science Part B: Polymer Physics*, 38(2), 329–340.

Ngono-Ravache, Y. and Maréchal, Y. (2001). Epoxy-amine reticulates observed by infra-red spectrometry. III. Modifications of the structure and hydration abilities after irradiation in a humid atmosphere. *Journal of Polymer Science Part B: Polymer Physics*, 39(11), 1129–1136.

Ngono-Ravache, Y., Corbin, D., Mélot, M., Gaté, C., Balanzat, E. (2007). Alkyne creation in aliphatic polymers: Influence of side groups. *Journal of Physical Chemistry B*, 111(11), 2813–2819.

Ngono-Ravache, Y., Damaj, Z., Dannoux-Papin, A., Ferry, M., Esnouf, S., Cochin, F., De Combarieu, G., Balanzat, E. (2015). Effect of swift heavy ions on an EPDM elastomer in the presence of oxygen: LET effect on radiation-induced chemical ageing. *Polymer Degradation and Stability*, 111, 89–101.

Ngono-Ravache, Y., Ferry, M., Esnouf, S., Balanzat, E. (2016) Polymers under ionizing radiations: The specificity of swift heavy ions. *EPJ Web of Conferences*, 115, 02003.

O'Donnell, J.H. (1989). *The Effects of Radiation on High-Technology Polymers.* American Chemical Society, Washington, D.C.

Picq, V. and Balanzat, E. (1999). Ion-induced molecular emission of polymers: Analytical potentialities of FTIR and mass spectroscopy. *Nuclear Instruments and Methods in Physics Research B*, 151(1–4), 76–83.

Picq, V., Ramillon, J.M., Balanzat, E. (1998). Swift heavy ions on polymers: Hydrocarbon gas release. *Nuclear Instruments and Methods in Physics Research B*, 146(1–4), 496–503.

Rabek, J.F. (1996). *Photodegradation of Polymers. Physical Characteristics and Applications.* Springer, Berlin.

Rivaton, A., Cambon, S., Gardette, J.L. (2005). Radiochemical ageing of EPDM elastomers. 2. Identification and quantification of chemical changes in EPDM and EPR films γ-irradiated under oxygen atmosphere. *Nuclear Instruments and Methods in Physics Research B*, 227(3), 343–368.

Rullier-Albenque, F. (1993). Vieillissement des polymères sous irradiation. Technical Note, SESI No. 126.

Sasuga, T., Hayakawa, N., Yoshida, K., Hagiwara, M. (1985). Degradation in tensile properties of aromatic polymers by electron beam irradiation. *Polymer*, 26(7), 1039–1045.

Seguchi, T. (2001). Mechanisms and kinetics of hydrogen yield from polymers by irradiation. *Nuclear Instruments and Methods in Physics Research Section B*, 185(1/4), 43–49.

Steckenreiter, T., Balanzat, E., Fuess, H., Trautmann, C. (1999a). Chemical degradation of polyimide and polysulfone films under the irradiation with heavy ions of several hundred meV. *Journal of Polymer Science: Part A: Polymer Chemistry*, 37(23), 4318–4329.

Steckenreiter, T., Balanzat, E., Fuess, Trautmann, C. (1999b). Pyrolytic effects induced by energetic ions in polymers. *Nuclear Instruments and Methods in Physics Research B*, 151(1–4), 161–168.

Sun, Y., Zhu, Z., Jin, Y., Liu, C., Wang, Z., Liu, J., Hou, M., Zhang, Q. (2002). The effects of high electronic energy loss on the chemical modification of polyimide. *Nuclear Instruments and Methods in Physics Research B*, 193(1–4), 214–220.

Thomas, J.K. (2007). Fundamental aspects of the radiolysis of solid polymers, crosslinking and degradation. *Nucl. Instrum. Methods Phys. Res. Sect. B-Beam Interact. Mater. Atoms*, 265(1), 1–7.

Toimil-Molares, M.E. (2012). Characterisation and properties of micro- and nanowires of controlled size, composition, and geometry fabricated by electrodeposition and ion-track technology. *Belstein Journal of Nanotechnology*, (2)3, 860–883.

Ungar, G. (1981). Radiation effects in polyethylene and normal-alkanes. *J. Mater. Sci.*, 16(10), 2635–2656.

Ventura, A. (2013). Polymères sous rayonnements ionisants : étude des transferts d'énergie vers les défauts d'irradiation. PhD Thesis, Université de Caen Basse Normandie, Caen.

Ventura, A., Ngono-Ravache, Y., Marie, H., Levavasseur-Marie, D., Legay, R., Dauvois, V., Chenal, T., Visseaux, M., Balanzat, E. (2016). Hydrogen emission and macromolecular radiation-induced defects in polyethylene irradiated under inert atmosphere: The role of energy transfers towards trans-vinylene unsaturations. *J. Phys. Chem. B.*, 120(39), 10367–10380.

Waterman, D.C. and Dole, M. (1971). Infrared spectrum of polyethylene irradiated at 4.deg.K. *The Journal of Physical Chemistry*, 75(26), 3988–3992.

Wündrich, K. (1974). Effect of temperature and physical state on the inhibition by additives of radiation-induced degradation of poly(methyl methacrylate). *J. Polym. Sci.*, 12(1), 201.

Zhang, G.H. and Thomas, J.K. (1996). Energy transfer via ionic processes in polymer films irradiated by 0.4 MeV electrons. *Journal of Physical Chemistry*, 100(27), 11438–11455.

Ziegler, J.F. (1985). Particle interactions with matter [Online]. Available at: www.srim.org.

Ziegler, J.F., Ziegler, M.D., Biersack, J.P. (2010). SRIM – The stopping and range of ions in matter. *Nucl. Instrum. Methods Phys. Res. Sect. B-Beam Interact. Mater. Atoms*, 268(11/12), 1818–1823.

4

Stabilization

Emmanuel RICHAUD
PIMM, CNRS, CNAM, ENSAM, Paris, France

4.1. Introduction

Some chapters of this book describe the degradation mechanisms, focusing on the molecular scale transformations that induce alterations in physical properties. While these modifications are nearly always irreversible, they can also be significantly slowed, which contributes to the lifetime extension of the polymer. Stabilization is the action through which the polymer structure or formulation is modified to extend its lifetime. Broadly speaking, there are two paths:

– Internal stabilization involves the modification of the polymer structure to increase its resistance to external aggressions. This is, for example, the case when the nature of the diol used to synthesize an unsaturated polyester is changed so that its water absorption is reduced or when a comonomer is employed to slow the thermal depolymerization of POM or PMMA.

– External stabilization involves the addition of additives that slow the aging mechanism. This chapter focuses essentially on this second aspect.

4.2. Chemical reactions of stabilization against thermo-oxidative aging

4.2.1. *Strategy of stabilization against oxidative aging*

There are two main paths for stabilization against thermo-oxidative aging:

Aging of Industrial Polymers 1,

– reducing the initiation rate, which is the rate at which species, such as hydroperoxides, create radicals;

– increasing the termination rate, that is, the rate at which radicals are converted into stable products. For this purpose, sacrificial or regenerative antioxidants are added.

4.2.1.1. *Hydroperoxide reducers*

Organophosphites and organosulfides can react with hydroperoxides, which reduce into alcohols (Figure 4.1).

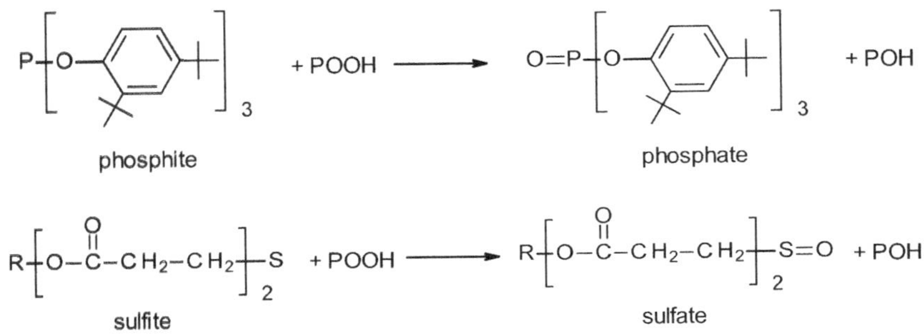

Figure 4.1. *Structure of hydroperoxide reducers*

These are considered primarily process antioxidants since they would be efficient mainly at high temperatures (in contrast to other stabilizers, such as phenols, which will be described in the next section) (Bauer et al. 1995). This seems partly justified by the value of the activation energy of the reaction between a hydroperoxide and a reducer (100 kJ.mol^{-1}), which is effectively above that of the reaction of hydrogen removal from a phenol by a radical.

In other words, phosphites and/or sulfides are the first to react under process conditions, thus limiting the consumption of other antioxidants. A limitation, however, is that the molecules in Figure 4.1, which are of quite low molar mass, can evaporate quite rapidly under processing conditions.

4.2.1.2. *Sacrificial antioxidants*

The most widespread antioxidant family is that of hindered phenols, most of which are derivatives of di-tert-butylphenol. These react by giving their hydrogen atom to a peroxyl radical. The resulting phenoxyl radical can then be isomerized, and the isomerized form reacts with another peroxyl radical to yield an inactive form (Figure 4.2).

Notably, in the absence of an antioxidant, the balance is as follows:

POO° + PH → POOH + P°

POOH → PO° + °OH

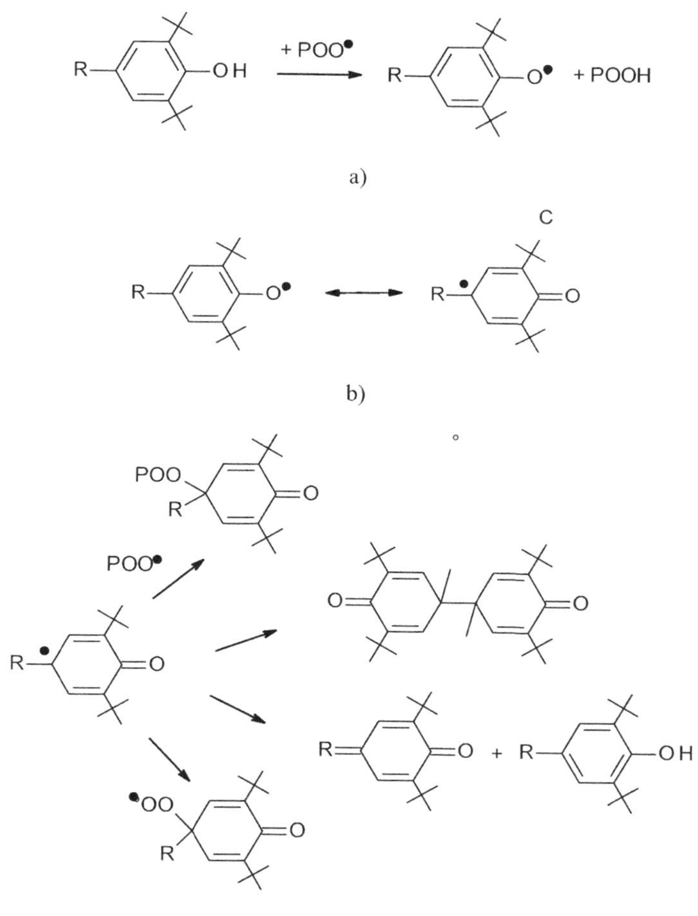

Figure 4.2. *Action of phenolic antioxidants: (a) removal of phenolic hydrogen; (b) isomerization; (c) secondary reactions*

This means that a (POO°) radical leads to the formation of three radicals.

On the other hand, in the presence of antioxidant, we have:

$$POO° + AH \rightarrow POOH + A°$$

$$POO° + A° \rightarrow POO - A$$

$$POOH \rightarrow PO° + °OH$$

Two radicals disappear, and only two are formed.

In reality, the fate of radical A° depends on many factors related to the structure of the phenol substituents and, undoubtedly, also on the aging conditions. Many secondary reactions have been proposed and lead to structures (quinone methide, cinnamates, etc.) responsible for polymer yellowing (Pospíšil et al. 2002).

Aromatic amines, which act as sacrificial antioxidants, can also be employed. They lead to highly conjugated and blackening compounds and are generally reserved for elastomer matrices mixed with carbon black (Cibulková et al. 2005; Zhao et al. 2023). They react as illustrated in Figure 4.3 (Cibulková et al. 2005; Zhao et al. 2023).

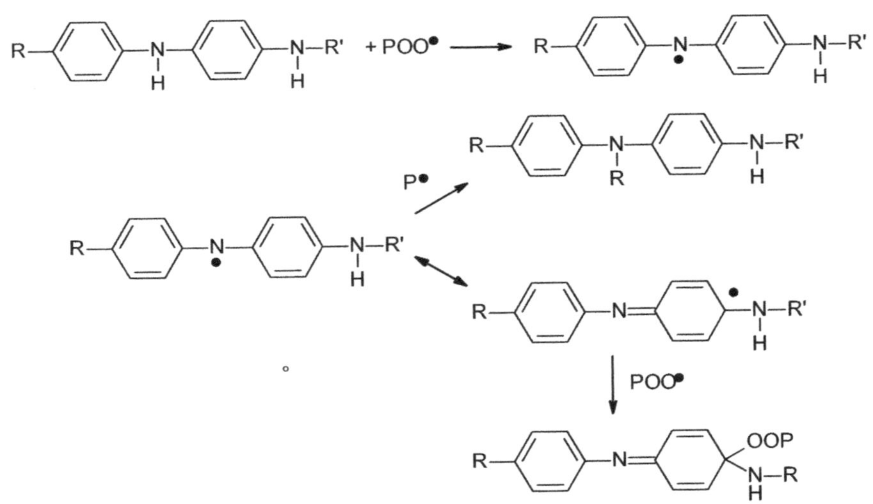

Figure 4.3. *Action of aromatic amines*

4.2.1.3. *Regenerative antioxidants*

This family essentially consists of hindered amine stabilizers (HAS), also known as hindered amine light stabilizers (HALS), which are derived from tetramethylpiperidine (Figure 4.4(a)).

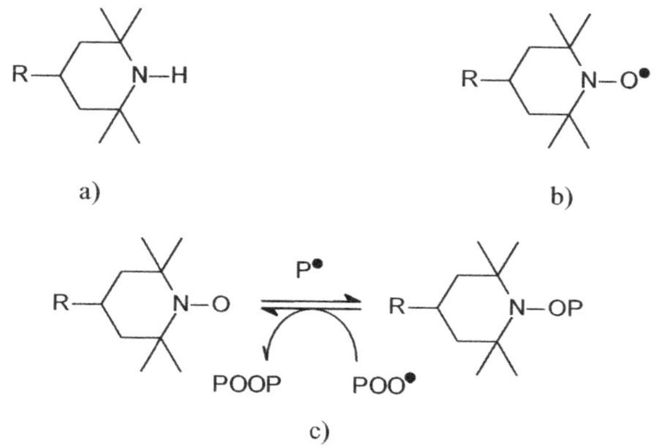

Figure 4.4. *(a) Structure of the tetramethylpiperidine group; (b) nitroxy; (c) cyclic stabilization mechanism*

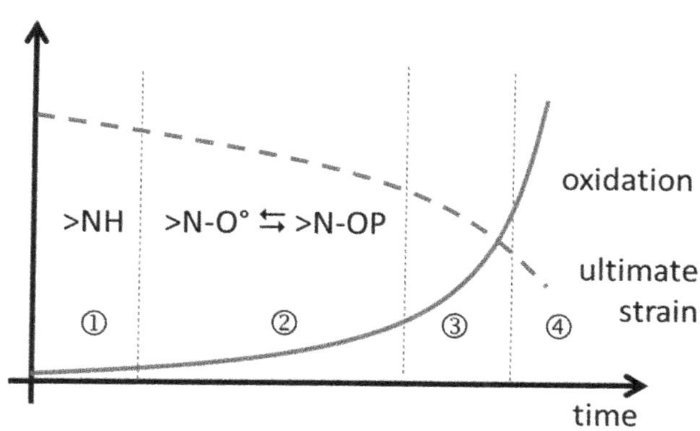

Figure 4.5. *The four phases of stabilization using HALS according to Sedlář and Zahradníčková. 1: HALS conversion into active forms; 2: regeneration of stabilizing forms; 3: transformation of active forms into non-stabilizing derivatives; 4: degradation (Sedlář and Zahradníčková 1989)*

It is generally admitted that their intrinsic power of stabilizing >N–OH bonds by reactions, such as those presented for phenolic antioxidants, is average, but they are precursors of nitroxyl radicals (Figure 4.4(b)).

The high performances of these stabilizers were rapidly acknowledged to be related to the existence of a cyclic process of >N–O° regeneration based on their reaction products (Figure 4.4(c)). The HALS stabilization process can generally be described as a sequence of four phases (Figure 4.5).

In summary:

– During phase 1, amine gradually transforms into oxidation products, >N–O° and >N–OP (and sometimes >N–OH). At the end of this phase, the initial amine has practically disappeared.

– Phase 2 is the most important. During this phase, the stabilization cycle reaches its maximum efficiency. However, the secondary oxidation products build up, some of them (peracids, peresters) react with the >N–O° to yield less efficient stable species.

– Phase 3 corresponds to a gradual loss of stabilizing efficiency. Until the end of this phase, the piperidine group is preserved, and the structural modifications impact only the nitrogen atom.

– However, in the long term (Phase 4), this cycle (and the other parts of the molecule) ends up by being itself oxidized and the stabilizing effect completely disappears.

4.2.1.4. *Copper salts*

Finally, the case of copper salts ($CuCl_2$, $CuBr_2$, etc.) can be mentioned. While it is an established fact that copper derivatives have a negative action on polyolefins, they are seemingly efficient against the oxidative stabilizer in the case of polyamides (Cerruti et al. 2004):

$$P° + Cu^{2+} \rightarrow P^+ + Cu^+$$

$$POO° + Cu^+ \rightarrow POO^- + Cu^{2+}$$

$$P^+ + POO^- \rightarrow POOP$$

4.2.2. *Kinetic aspects*

There are two operating modes of the stabilization system:

– the "sacrificial" system, in which the stabilizer is destroyed during stabilization;

– the "regenerative" system, in which the stabilizing species are involved in a cycle in which one regenerates the other, as in the case of HALS, when the cycle is renewed in principle indefinitely (with effects close to the chemical fatigue of the stabilizer).

For the "sacrificial" systems, oxidation resumes the course it would have in the non-stabilized polymer when the entire stabilizer was consumed. The end of the induction period is, in general, well marked. Hindered phenols are typical for this behavior. In the case of HALS "regenerative" systems, the stabilizer efficiency gradually decreases, and the end of the induction period is not strongly marked; it even disappears completely (Zeynalov and Allen 2006).

The difference between the two types of systems is illustrated by many works.

Figure 4.6 represents two possible cases (Gensler et al. 2000; Zeynalov and Allen 2006; Richaud et al. 2009).

In the case of purely sacrificial antioxidants, the kinetic oxidation curves are shifted to long periods of time, and the induction period varies linearly with the antioxidant concentration, at least in the absence of an overlapping physical phenomenon, such as antioxidant evaporation, which will be addressed further (Richaud et al. 2011).

On the other hand, at the end of induction, the maximal oxidation rate is, in a first approximation, independent of the initial stabilizer concentration, which means that once the antioxidants have been destroyed, the polymer reactivity is very close to that of the polymer without an antioxidant.

In the case of regenerative antioxidants, the main modification induced by an increase in the antioxidant concentration is that the maximal oxidation rate decreases. In other words, even if oxidation is detected, there are still active forms of antioxidant that control the degradation rate.

It seems that the differences in reactivity (sacrificial versus regenerative) are reflected in the rate of reduction in the mechanical properties, as shown, for example, by the results of Gensler et al. (2000), and as illustrated in Figure 4.6(b).

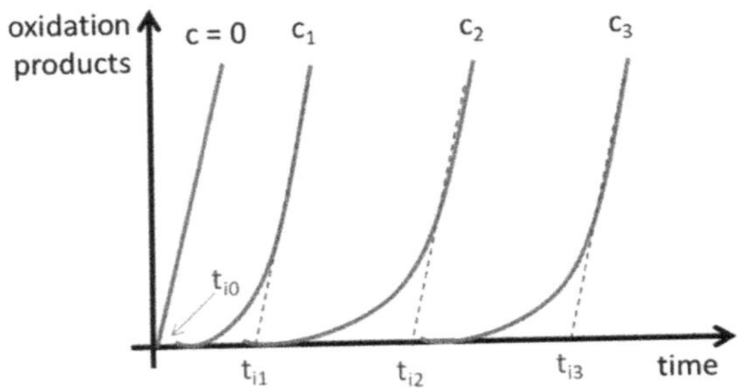

a)

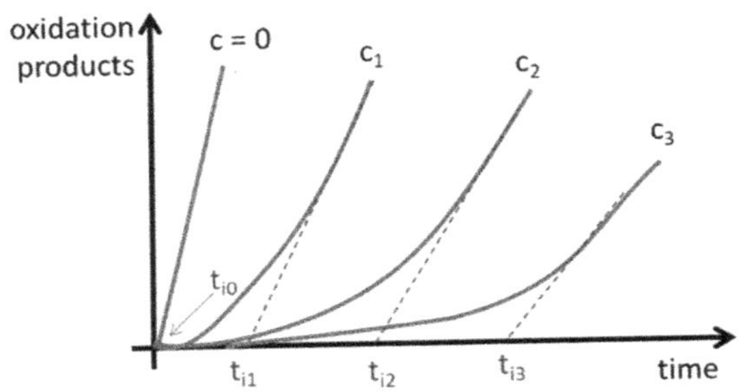

b)

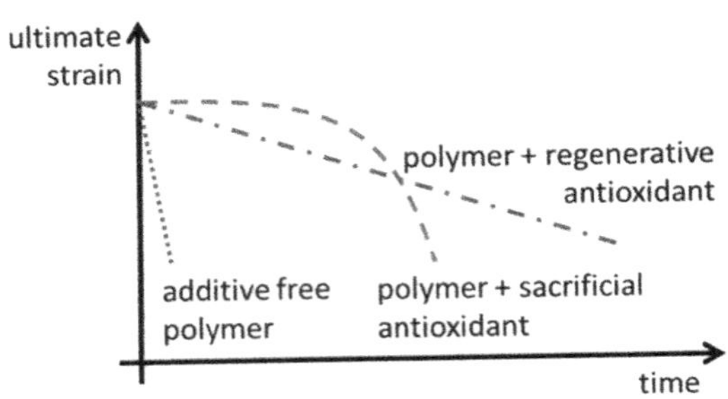

c)

Figure 4.6. *(a) Effects of sacrificial antioxidants and (b) regenerative antioxidants on the degradation rate (the induction time characteristic of the pure polymer, t_{i0}, is not necessarily zero but is often very low compared with t_{i1}, t_{i2}, etc.), and embrittlement kinetics of phenol- or HALS-stabilized PP at moderate temperature (c). For a color version of this figure, see www.iste.co.uk/richaud/industrialpolymers1.zip*

Figure 4.7 shows a remarkable result obtained in the case of polyamides stabilized by a phenolic antioxidant. The oxidation of these polymers may lead to chain scissions, but a polycondensation of the ends of the amine and carboxylic acid chains can also lead to an increase in the molar mass. When the polymer is not stabilized, oxidation prevails, and the molar mass decreases (Okamba-Diogo et al. 2020).

When phenolic antioxidants are added, there is a period during which the oxidation effects are not visible, while the main phenomenon is postpolycondensation in the solid state, which may extend the period of time after which the average molar mass drops below a threshold, leading to embrittlement.

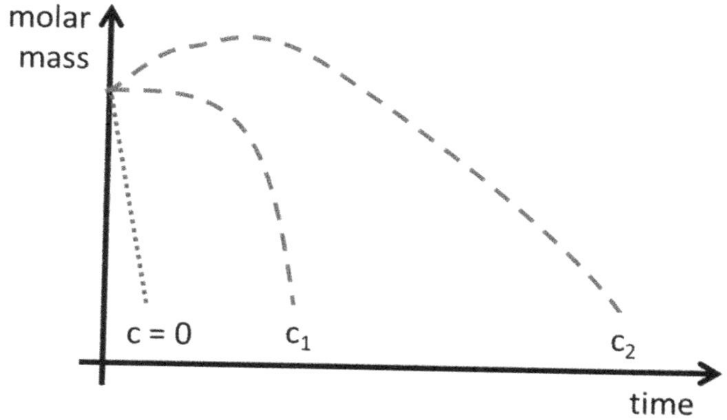

Figure 4.7. *PA11 molar mass variation with various quantities of antioxidants*

In terms of reaction rates, the antioxidant is efficient if its reactivity with the radicals involved in the aging process is greater than that in the second reaction and if it competes with the POO° + PH → POOH + P° reaction.

In terms of reaction rates:

$POO° + PH \rightarrow POOH + P°$ $\qquad r_3 = k_3.[POO°][PH]$

$POO° + St \rightarrow \text{inactive products}$ $\qquad r_{St} = k_{St}.[POO°][St]$

The stabilizing effect occurs if the rate ratio is on the order of the unit or above 1:

$$\frac{r_{St}}{r_3} = \frac{k_{St}.[St]}{k_3.[PH]} \geq 1$$

4.3. Stabilization against photochemical aging

UV aging is initiated by the interaction between a chromophore group and a photon of "convenient energy" that converts this group to an excited photochemical state.

The chromophore groups can be polymer-specific groups for example, the carbonate group in the polycarbonate, which leads to photo-Fries rearrangement (Lemaire et al. 1986), or impurities resulting from polymer synthesis or implementation for example, ketone or peroxide groups in polyolefins. The objective of UV stabilization is to limit these interactions via UV radiation "filtering" molecules.

4.3.1. *UV absorbers*

UV absorbers are molecules that systematically involve internal hydrogen bonds. They capture UV photons with high molar absorptivity (typically $\varepsilon \sim 10{,}000$ L.mol^{-1}.cm^{-1}), and under their action, they pass from a standard state to an excited photochemical state (singlet state S_1 and then are subjected to an intersystem crossing to the excited triple state T_1*, as illustrated in the Jablonski diagram; Figure 4.8).

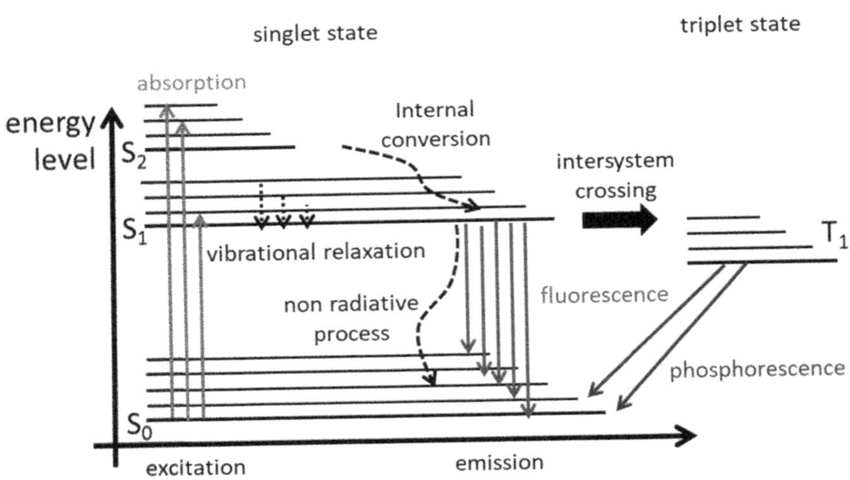

Figure 4.8. *Simplified Jablonski diagram. For a color version of this figure, see www.iste.co.uk/richaud/industrialpolymers1.zip*

Then, they are deactivated to the initial state (the quantum yield of the cycle is very close to 1, which makes it possible for the molecule to regenerate). In all cases, an excited state capable of reliably producing radicals is converted into an emission of heat or of a less energetic photon than the initial one, the set having less negative consequences on the polymer properties.

The molecules that are able to act as UV stabilizers include benzophenones, benzotriazoles and triazines (Figure 4.9).

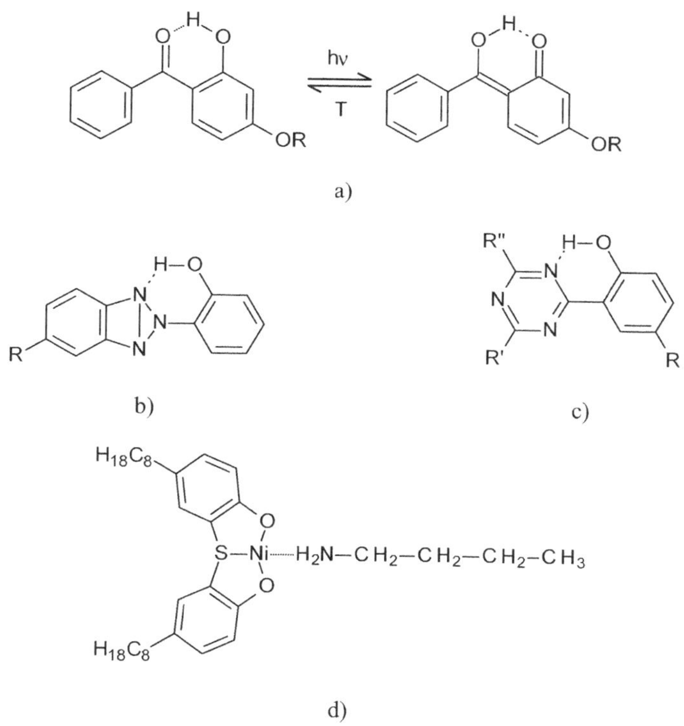

Figure 4.9. *Photostabilization mechanism by (a) hydroxybenzophenones; (b) hydroxybenzotriazoles; (c) triazine; and (d) quenchers*

4.3.2. *Quencher*

Quenchers (deactivators) (Figure 4.9(d)) differ from UV absorbers by the fact that they are brought to an excited state by deactivating the polymer group, which was in an excited state and returns to its standard state via a nondestructive process:

$$\text{polymer} + h\nu \rightarrow (\text{polymer})^*$$

$$(\text{polymer})^* + Q \rightarrow \text{polymer} + Q^*$$

$$Q^* \rightarrow Q + h\nu' \text{ or heat}$$

Among these molecules, there are, for example, various complexes of transition metals (Figure 4.9).

4.3.3. *Pigments*

Pigments are a major class of additives used to modify color. These are (most often) insoluble mineral particles. The size of the particles usually ranges between 0.1 and 1 µm.

These pigments include:

- carbon black;
- titanium dioxide (white);
- iron oxide (red);
- phthalocyanine (blue–green).

4.4. Synergies/antagonism

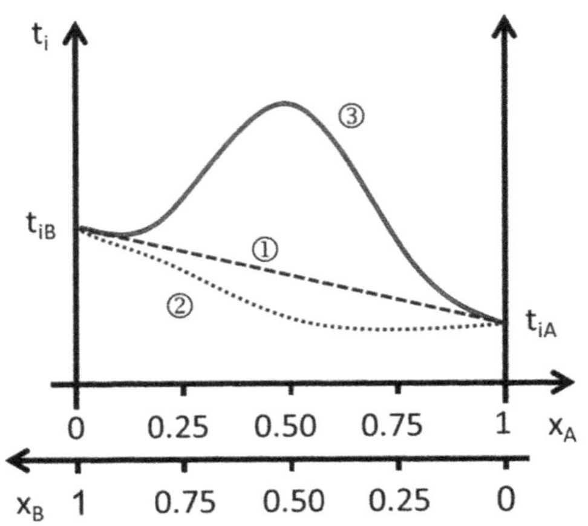

Figure 4.10. *Diagram summarizing the synergies/antagonisms between two antioxidants A/B*

Single antioxidants are rarely used as such. Combinations of antioxidants are generally used, particularly because each has an optimal activity within a given range of temperature. These combinations lead to antagonisms or synergies, as described in Figure 4.10.

Antioxidant A and antioxidant B at a concentration of x_0 lead to induction times t_{iA} and t_{iB}, respectively. Importantly, the interaction between two stabilizers A and B (respecting the condition $x_0 = x_A + x_B$) can be characterized as follows:

– synergistic if $t_i > x_A.t_{iA} + x_B.t_{iB}$;

– without interaction if $t_i = x_A.t_{iA} + x_B.t_{iB}$;

– antagonistic if $t_i < x_A.t_{iA} + x_B.t_{iB}$.

Among the mixtures susceptible to synergistic effects, the phenol + phosphite mixture is commonly used in polyolefins (Chirinos-Padrón et al. 1987). In the case of polyolefins, antagonisms have been reported between HALS and phosphites during photooxidative aging (Bauer et al. 1997) and, sometimes, between HALS and phenols during thermal aging (Allen et al. 1981). Pigments can also play complex roles, particularly carbon black. In unstabilized matrices, this seems to play an antioxidant role (Klemchuk 1983). In stabilized matrices, there is an antioxidant adsorption phenomenon at the surface of carbon black, which strongly impacts their performances (Peña et al. 2001). Finally, antagonisms have been reported between antifire derivatives and some antioxidants (Sinturel et al. 2000).

4.5. Example of specific stabilizations

4.5.1. *PVC stabilization against dehydrochlorination*

PVC features an oxidizing degradation mechanism comparable to that of other polymers (Lemaire et al. 1991), to which a dehydrochlorination mechanism is superimposed (Figure 4.11(a)). Degradation by sequential elimination of HCl is recognizable by the strong coloration it induces on the samples.

There are many strategies to cope with this, the following two being the main ones:

1) Substitution of labile chlorines by stable groups through carboxylates (Figure 4.11(b)). It is well established that the elimination is initiated by the decomposition of carbon–chlorine bonds having a low dissociation energy, such as allylic C–Cl and tertiary C–Cl. Metallic soaps, such as zinc stearates, which replace the labile chlorine with an ester during extrusion, can be used. Zinc stearates (StZn)

are used in synergy with calcium stearates (StCa), which regenerate zinc stearates in a double decomposition reaction:

$$PVC–Cl + StZn \rightarrow PVC – St + ZnCl_2$$

$$ZnCl_2 + StCa \rightarrow StZn + CaCl_2$$

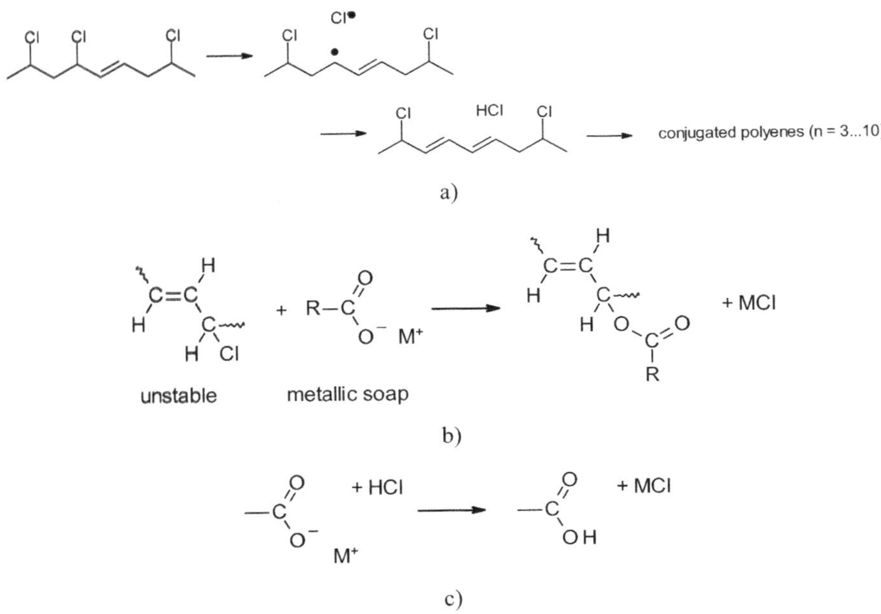

Figure 4.11. *PVC degradation mechanism*

Calcium stearate does not play the role of direct stabilizer; however, in addition to the fact that it regenerates zinc stearate, it eliminates zinc chloride, which is a powerful catalyst of the elimination reaction.

Dialkyl tin thioglycolates and other derivatives of the same type are also highly efficient stabilizers (Lequan 1980):

$$PVC–Cl + Alk_2SnY_2 \rightarrow PVC–Y + Alk_2SnYCl$$

$$PVC–Cl + Alk_2SnYCl \rightarrow PVC–Y + Alk_2SnCl_2$$

$$Alk_2SnY_2 + Alk_2SnCl_2 \rightarrow 2\ Alk_2SnYCl$$

where $Y = –S–CH_2–COO–C_nH_{2n+1}$.

Here, there is autosynergy, and the stabilizer itself eliminates dichloride, whose catalytic effect on dehydrochloruration is detrimental.

2) Neutralization of HCl: its catalytic effect is also detrimental. All of the stabilizers able to give rise to the above substitutions can react with HCl.

3) Repair of damaged sequences: this is the case for dienophiles, capable to add up to conjugated dienes at the core of polyenes and thus break the conjugation and the resulting coloration.

4.5.2. *Stabilization against depolymerization*

Thermal depolymerization can be initiated by unstable chain ends, which are head-to-head chaining type of structural impurities (Kashiwagi et al. 1986). These defects result in the formation of radicals, leading to scissions of monomer–monomer bonds, which are favored by low monomer–monomer energy:

$$-AAAAAAAAAAA^{\circ} \rightarrow -AAAAAAAAAA^{\circ} + A$$

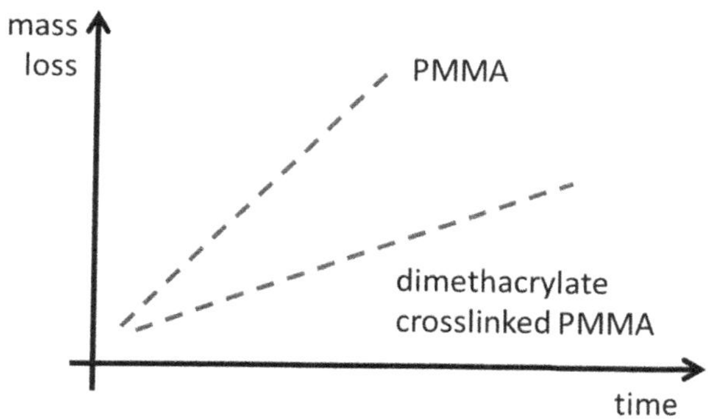

Figure 4.12. *Effect of EGDMA on the PMMA depolymerization rate (Bate and Lehrle 1998)*

The principle of stabilization against depolymerization consists of the following steps:

– adding a monomer that blocks initiation by end capping, which occurs in POM (Kubico and MacDonald 1959);

– adding a "depolymerization stopper" that is a comonomer leading to a stable chain end radical, which is illustrated by the effect of adding a comonomer (ethylene glycol dimethacrylate) on slowing down the thermal decomposition of PMMA (Figure 4.12):

$$-AAAAA\text{–}B\text{–}(A)_n\text{–}A^\circ \rightarrow -AAAAAB^\circ \text{ (stable)} + (n+1)A$$

4.6. Physical phenomena involved in the external stabilization of polymers

An adjuvant molecule has stabilizing properties provided that it performs the above-described functions, which is a necessary but not sufficient condition, as this molecule should also be soluble in the polymer so that its concentration reaches the level required for the stabilizing effect to be fully manifested. Many polymers among the most important polymers in the industrial field (polyolefins, hydrocarbon elastomers) are not very polar and are therefore not highly compatible with moderately or strongly polar molecules, which is the case for most small molecules with only the antioxidant function. Solubility problems in polymers can therefore be critical and require thorough investigations.

Let us now consider a polymer sample stabilized under standard conditions. The antioxidant molecules near the surface are more or less prone to "evading" in the environment (depending on their volatility in the atmosphere) or on their extractability (in a liquid medium). The departure of the first molecules depletes the surface zone and creates a concentration gradient that will drive the diffusion of the molecules present in the core to the surface. This phenomenon will only end (in an unconfined medium) upon complete disappearance of the antioxidant. Evaporation–diffusion (or extraction–diffusion) coupling is susceptible to controlling the material lifetime, if it is more rapid than the chemical stabilization reaction. This explains why relatively significant research activity has focused on this topic.

4.6.1. *Solubility of stabilizers*

In the case of adjuvants, solubility is characterized by the concentration [S]sat of the stabilizer at saturation. Thermodynamic approaches, particularly the Flory–Huggins theory, consider the volume fraction v_S in the polymer at saturation (Billingham et al. 1981):

$$-\ln v_S = \frac{\Delta H_{fS}}{RT} \cdot \left(1 - \frac{T}{T_{fS}}\right) + v_P + \chi . v_P^2$$

where v_S and v_P are the volume fractions of the stabilizer and polymer, ΔH_{fS} and T_{fS} are the enthalpy and the melting temperature of the stabilizer, respectively, and χ is the coefficient of interaction of the polymer–stabilizer couple depending on the molar volume of the stabilizer and on the solubility parameter difference between the polymer and the stabilizer. The following three factors are important in this context:

– Stabilizer cohesion (ΔH_{fS} and T_{fS}): the more cohesive the stabilizer is, the more difficult it is to dissociate molecules.

– The size of the stabilizer molecule (its molar volume): the larger it is, the more difficult the dissolution.

– The polarity of the molecule and its capacity to bind to the polymer via weak bonds: the miscibility of a stabilizer and of a polymer is greater, as their structures are more similar and are able to establish the same types of interactions.

For the stabilizers of polyolefins, for example, a general tendency to graft long alkyl chains with stabilizing functions can be noted, as is the case for these three commercial stabilizers, which belong to three different families (Figure 4.13).

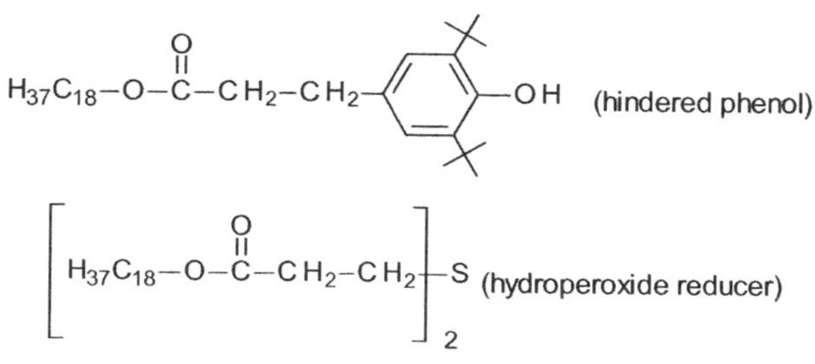

Figure 4.13. *Optimized stabilizing structures for good solubility in polyolefins*

By grafting an aliphatic chain (low polarity) with a stabilizing function (rather polar), the overall polarity of the molecule is reduced, and it is rendered compatible with polymers whose polarity is also low.

4.6.2. *Volatility/surface loss of stabilizers*

The kinetic analysis of evaporation relies on two hypotheses:

1) In an arbitrary sample, the evaporation rate, characterized by the mass loss per unit surface, is proportional to the stabilizer concentration in the surface layer.

2) In a sample saturated with a stabilizer, the evaporation rate or the surface loss rate is equal to its value for a pure stabilizer.

Therefore, the following law can be written (Calvert and Billingham 1979):

$$\frac{dm}{dt} = v_0 \cdot \frac{[S]_{surf}}{[S]_{sat}}$$

Or, in terms of the number of evaporated stabilizer moles:

$$\frac{dn}{dt} = \frac{v_0}{M_S} \cdot \frac{[S]_{surf}}{[S]_{sat}} = H.[S]_{surf}$$

where v_0 is the evaporation rate of the pure stabilizer, whose molar mass is M_S and the saturation concentration in the polymer is $[S]_{sat}$. $[S]_{surf}$ is the stabilizer concentration in the vicinity of the surface.

Here, $H = \frac{v_0}{M_S[S]_{sat}}$ is a coefficient specific to the considered polymer–stabilizer mixture.

Let us now consider the case when the thickness L of the sample is quite small and the diffusivity of the antioxidant is sufficiently high so that at any instant, the stabilizer is homogeneously redistributed throughout the sample. The following can be written as:

$$\frac{d[S]_{surf}}{dt} = -\frac{H}{L}.[S]_{surf} = -K_V.[S]_{surf}$$

where K_V (s^{-1}) is a first-order rate constant. The characteristic evaporation time t_V can be deduced as follows:

$$t_V = \frac{1}{K_V} = \frac{L}{H}$$

The main factors governing volatility are cohesion and the molar mass M_S of the stabilizer. t_V increases with these two factors. An evaporation phenomenon can be generally identified by the fact that, for a quite small sample, it follows first-order kinetics, which is as follows:

$$[S] = [S]_0.\exp(-K_V.t)$$

In a structural series written as A-$(CH_2)_{n-1}$-CH_3, where A is a stabilizing function with a molar mass on the order of 200 g.mol^{-1}, K_V might decrease by a factor of approximately 1,000 when n increases from 1–12 (Durmis et al. 1975).

Within the HALS-stabilized polyolefins in contact with water, the migration rate V_0 is given in Table 4.1, with their saturation concentration in water (Malík et al. 1992a).

	V_0 (g.cm^{-2}.s^{-1})	$S_{20°C}$ (%)
M = 396–423 g.mol^{-1}	$(3–8) \times 10^{-7}$	0.169
M = 481 g.mol^{-1}	$(1–3) \times 10^{-7}$	0.05
M = 2,300 g.mol^{-1}	8×10^{-8}	0.011

Table 4.1. *Dissolution rate (from plates) and solubility in water at 20°C of several HALS*

Thus, the loss rate decreases as the molar mass increases. However, this happens to the detriment of the diffusion coefficient.

4.6.3. *Diffusivity of stabilizers*

In the above studied case, the sample diffusivity is quite high, and the sample thickness is quite small for the stabilizer loss to be fully controlled by evaporation.

Let us now consider the opposite case of very rapid evaporation in relation to diffusion. Then, a high concentration gradient will be established in the thickness (Figure 4.14).

Let us first consider the extreme case of very rapid evaporation and zero surface concentration. The kinetics of stabilizer loss are exclusively governed by diffusion. At (standard) low concentrations, Fickian diffusion occurs:

$$\frac{\partial[S]}{\partial t} = D_S \cdot \frac{\partial^2[S]}{\partial x^2}$$

where x is the coordinate of the elementary layer considered in thickness and D_S is the diffusion coefficient of the stabilizer in the amorphous phase of the polymer. The common values of D_S are generally below 10^{-10} m^2.s^{-1} and significantly vary in the same direction as the volatility constant K_V (see above) with the molar mass and flexibility of the stabilizer molecule (Földes and Turcsányi 1992; Földes 1993, 1994).

The apparent activation energy of diffusion is generally an increasing function of these factors (Limm and Hollifield 1996). In the case of polyolefin-HALS binaries, values can be found in the work of Malík et al. (1992b) (Table 4.2).

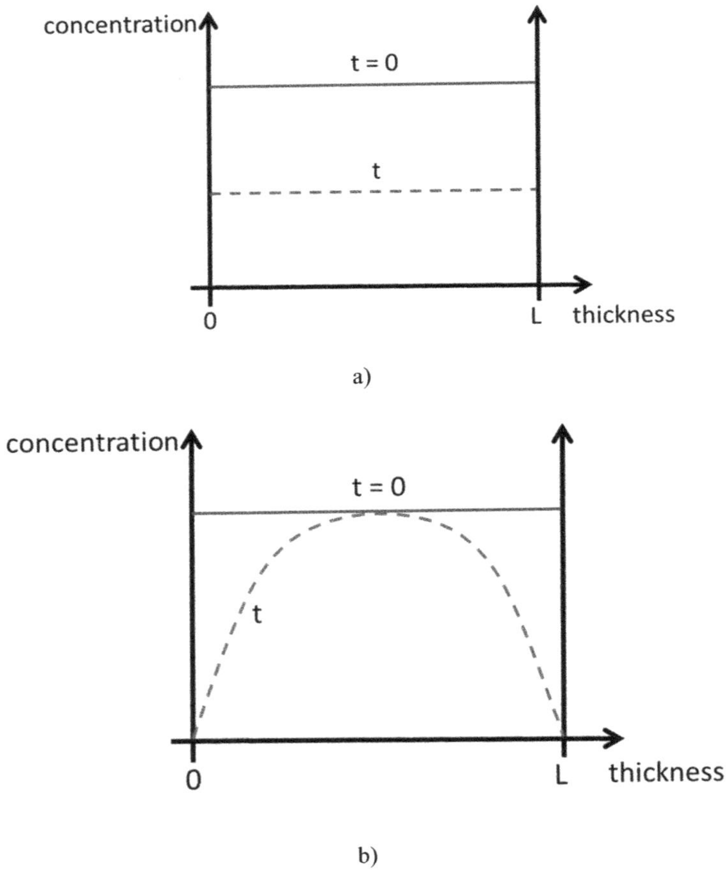

Figure 4.14. *Schematic representation of the evolution of the profile of the stabilizer concentration with respect to the thickness of the initial sample (t = 0) and aged sample (t), with kinetics governed by (a) evaporation and (b) diffusion*

	E_D (kJ.mol^{-1})	log D_0	Temperature range (°C)
M = 396-423 g.mol^{-1}	57.2	1.55	50–75
M = 481 g.mol^{-1}	83.1	5.3	23–75
M = 367 g.mol^{-1}	38.3	–1.6	50–80

Table 4.2. *Arrhenius diffusivity parameters (D_0 in cm².s^{-1}) for HALS in LDPE*

4.6.4. *Kinetic analysis of migration phenomena*

The characteristic time of diffusion t_D can be defined as follows:

$$t_D = \frac{L^2}{D_S}$$

The characteristic times of evaporation and diffusion can be compared via the ratio r:

$$r = \frac{t_D}{t_V} = \frac{L.H}{D_S}$$

Notably, the process is governed by diffusion if $r >> 1$ and by evaporation if $r << 1$.

Calvert and Billingham (1979) reached the same conclusions using another path and reported that loss is governed by diffusion if $r > 10$ and by evaporation if $r < 0.6$. Obviously, the intermediate case can be considered, in which the stabilizer concentration is lower at the surface than at the core, but it is not zero. The migration in the internal layers is governed only by diffusion, but in the kinetic analysis, specific boundary limits should be applied, taking into account the evaporation rate when describing the behavior in the surface layers.

4.7. Methods for antioxidant detection and evaluation

There are several methods for the detection of antioxidants.

– Spectroscopic methods: phenolic antioxidants can be, in principle, identified owing to specific absorptions (hydroxyl groups of phenols, ester groups of non-functional groups, etc.).

– Chromatographic methods require most often a preliminary stage of extraction, except for gas chromatography. In general, such techniques reveal especially fragments of the molecule.

– One of the most classical methods used to estimate the efficiency of stabilizers is the measurement of the oxidation induction time, as schematically represented in Figure 4.15 (Schmid and Affolter 2003).

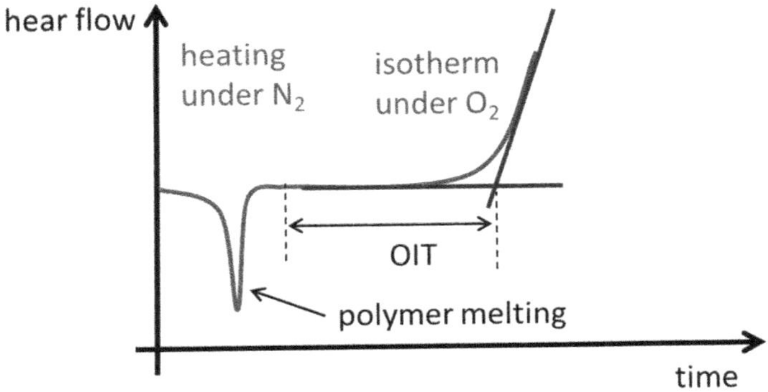

a)

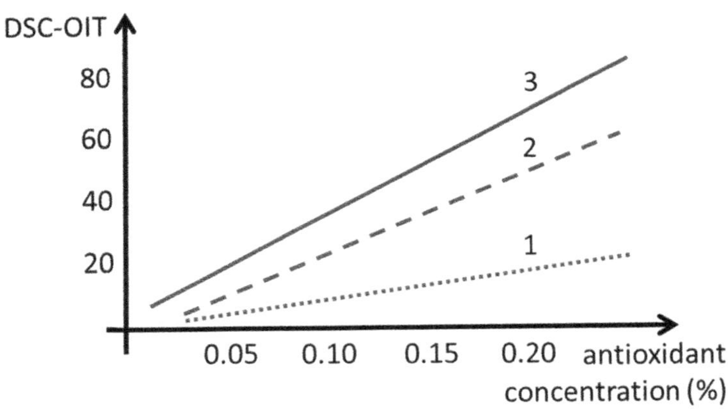

b)

Figure 4.15. *(a) Principle of OIT measurement; (b) OIT variations with the antioxidant concentration for PP (1), HDPE obtained with a titanium catalyst (2) and HDPE obtained with a chromium catalyst (3) at 200°C under O_2 (Pospíšil et al. 2003). For a color version of this figure, see www.iste.co.uk/richaud/industrialpolymers1.zip*

In general, OIT increases quasilinearly with the stabilizer concentration unless the physical loss is not negligible at the time scale of the experiment. The selection of the test temperature is thus crucial. Importantly, this method, which can be readily implemented, can be at best used to compare two samples, but there are many arguments against its use as a dosage method:

– antioxidant evaporation during the measurement strongly disturbs the measurement;

– in general, in situ oxidation of a polymer sample during measurement can be controlled by oxygen diffusion, which renders the quantitative interpretation of the curves in Figure 4.15 a complicated task;

– in the case of aged samples, unstable species created during aging (radicals, peroxides, etc.) impact the value of OIT and reduce it, even if antioxidant is still present. In other words, the correlation of Figure 4.15 is only valid for an unaged polymer but does not apply as this polymer has been previously aged (which includes the case of polymers containing recycled mixtures).

Other tests, which are based on the measurement of the oxidation onset temperature, are also employed (Okamba-Diogo et al. 2016). In terms of simplicity, they are comparable to the measurement of OIT, and the limitations mentioned in the previous section are still applicable.

4.8. Conclusion

This chapter summarizes the main paths for polymer stabilization against various aging types. The mechanisms are generally known, which opens the way for their integration into kinetic models for the prediction of degradation. The first impediment for the full validation of these models is the ability to fully quantify the stabilizers and their (active or inactive) derivatives in the polymer matrix, where their concentration may be very low. The second is the complexity related to industrial materials, in which "cocktail" effects (synergies and antagonisms) are numerous and difficult to predict.

Furthermore, polymer stabilization seems to be a subject bound for renewal given the many societal constraints, namely, the gradual replacement of petrochemical products by stabilizers derived from biomass and the elimination of products considered potentially dangerous or harmful to health (for example, the withdrawal of 6PPD antioxidants used in tires). Finally, the implementation of ambitious public policies in recycling and plastic packaging reuse will inevitably stimulate new formulations of polymer materials that meet societal expectations.

4.9. References

Allen, N.S., Gardette, J.L., Lemaire, J. (1981). Stabiliser interactions with hindered piperidine compounds in polypropylene: Influence of processing. *Polymer Degradation and Stability*, 3(3), 199–208.

Bate, D.M. and Lehrle, R.S. (1998). A new approach for measuring the rate of pyrolysis of cross-linked polymers: Evaluation of degradation rate constants for cross-linked PMMA. *Polymer Degradation and Stability*, 62, 67–71.

Bauer, I., Habicher, W.D., Rautenberg, C., Al-Malaika, S. (1995). Antioxidant interaction between organic phosphites and hindered amine light stabilisers during processing and thermoxidation of polypropylene. *Polymer Degradation and Stability*, 48, 427–440.

Bauer, I., Habicher, W.D., Korner, S., Al-Malaika, S. (1997). Antioxidant interaction between organic phosphites and hindered amine light stabilizers: Effects during photoxidation of polypropylene – II. *Polymer Degradation and Stability*, 55(2), 217–224.

Billingham, N.C., Calvert, P.D., Manke, A.S. (1981). Solubility of phenolic antioxidants in polyolefins. *Journal of Applied Polymer Science*, 26, 3543–3555.

Calvert, P.D. and Billingham, N.C. (1979). Loss of additives from polymers: A theoretical model. *Journal of Applied Polymer Science*, 24(2), 357–370.

Cerruti, P., Rychly, J., Matisová Rychlá, L., Carfagna, C. (2004). Chemiluminescence from oxidation of polyamide 6,6: II. The effect of metal salts. *Polymer Degradation and Stability*, 84, 199–206.

Chirinos-Padrón, A.J., Hernández, P.H., Allen, N.S., Vasilion, C., Marshall, G.P., de Poortere, M. (1987). Synergism of antioxidants in high density polyethylene. *Polymer Degradation and Stability*, 19(2), 177–189.

Cibulková, Z., Šimon, P., Lehocký, P., Balko, J. (2005). Antioxidant activity of p-phenylenediamines studied by DSC. *Polymer Degradation and Stability*, 87(3), 479–486.

Durmis, J., Karvaš, M., Čaučik, P., Holčik, J. (1975). Loss of light stabilizers from polypropylene. *European Polymer Journal*, 11, 219–222.

Földes, E. (1993). Transport of small molecules in polyolefins. II. Diffusion and solubility of irganox 1076 in ethylene polymers. *Journal of Applied Polymer Science*, 48, 1905–1913.

Földes, E. (1994). Transport of small molecules in polyolefins. III. Diffusion of topanol CA in ethylene polymers. *Journal of Applied Polymer Science*, 51, 1581–1589.

Földes, E. and Turcsányi, B. (1992). Transport of small molecules in polyolefins. I. Diffusion of irganox 1010 in polyethylene. *Journal of Applied Polymer Science*, 46, 507–515.

Gensler, R., Plummer, C.J.G., Kausch, H.H., Kramer, E., Pauquet, J.R., Zweifel, H. (2000). Thermo-oxidative degradation of isotactic polypropylene at high temperatures: Phenolic antioxidants versus HAS. *Polymer Degradation and Stability*, 67(2), 195–208.

Kashiwagi, T., Inaba, A., Brown, J.E., Hatada, K., Kitayama, T., Masuda, E. (1986). Effects of weak linkages on the thermal and oxidative degradation of poly(methyl methacrylates). *Macromolecules*, 19, 2160–2168.

Klemchuk, P.P. (1983). Influence of pigments on the light stability of polymers: A critical review. *Polymer Photochemistry*, 3(1), 1–27.

Kubico, M.A. and MacDonald, R.N. (1959). Stabilization of polyoxymethylene with ureas and thioureas. U.S. Patent, 2,893,972A.

Lemaire, J., Gardette, J.L., Rivaton, A., Roger, A. (1986). Dual photo-chemistries in aliphatic polyamides, bisphenol A polycarbonate and aromatic polyurethanes – A short review. *Polymer Degradation and Stability*, 15(1), 1–13.

Lemaire, J., Arnaud, R., Gardette, J.L. (1991). Low temperature thermo-oxidation of thermoplastics in the solid state. *Polymer Degradation and Stability*, 33(2), 277–294.

Lequan, M., Besace, Y., Poller, R.C. (1980). Stabilisation thermique du PVC : relation structure-efficacite mode d'action des stabilisants organostanniques. *European Polymer Journal*, 16(11), 1109–1113.

Limm, W. and Hollifield, H.C. (1996). Modelling of additive diffusion in polyolefins. *Food Additives and Contaminants: Part A, Chemistry, Analysis, Control, Exposure & Risk Assessment*, 13, 949–967.

Malík, J., Hrivík, A., Alexyová, D. (1992a). Physical loss of hindered amine light stabilizers from polyethylene. *Polymer Degradation and Stability*, 35, 125–130.

Malík, J., Hrivík, A., Tomová, E. (1992b). Diffusion of hindered amine light stabilizers in low density polyethylene and isotactic polypropylene. *Polymer Degradation and Stability*, 35, 61–66.

Okamba-Diogo, O., Richaud, E., Verdu, J., Fernagut, F., Guilment, J., Pery, F., Fayolle, B. (2016). Quantification of hindered phenols in polyamide 11 during thermal aging. *Polymer Testing*, 52, 63–70.

Okamba-Diogo, O., Fernagut, F., Guilment, J., Pery, F., Fayolle, B., Richaud, E. (2020). Thermal stabilization of polyamide 11 by phenolic antioxidants. *Polymer Degradation and Stability*, 179, 109206.

Peña, J.M., Allen, N.S., Edge, M., Liauw, C.M., Valange, B. (2001). Interactions between carbon black and stabilisers in LDPE thermal oxidation. *Polymer Degradation and Stability*, 72(1), 163–174.

Pospíšil, J., Habicher, W.D., Pilař, J., Nešpůek, S., Kuthan, J., Piringer, G.O., Zweifel, H. (2002). Discoloration of polymers by phenolic antioxidants. *Polymer Degradation and Stability*, 77, 531–538.

Pospíšil, J., Horák, Z., Pilař, J., Billingham, N.C., Zweifel, H., Nešpůrek, S. (2003). Influence of testing conditions on the performance and durability of polymer stabilisers in thermal oxidation. *Polymer Degradation and Stability*, 82(2), 145–162.

Richaud, E., Colin, X., Monchy-Leroy, C., Audouin, L., Verdu, J. (2009). Polyethylene stabilization against thermal oxidation by a trimethylquinoleine oligomer. *Polymer Degradation and Stability*, 94(3), 410–420.

Richaud, E., Fayolle, B., Verdu, J. (2011). Polypropylene stabilization by hindered phenols – Kinetic aspects. *Polymer Degradation and Stability*, 96(1), 1–11.

Schmid, M. and Affolter, S. (2003). Interlaboratory tests on polymers by differential scanning calorimetry (DSC): Determination and comparison of oxidation induction time (OIT) and oxidation induction temperature (OIT*). *Polymer Testing*, 22(4), 419–428.

Sedlář, J. and Zahradníčková, A. (1989). HALS – The mechanistic aspects of their action. In *Advances in the Stabilization and Controlled Degradation of Polymers*, Patsis, A.V (ed.). Technomic Publishing Company, Lancaster.

Sinturel, C., Lemaire, J., Gardette, J.L. (2000). Photooxidation of fire retarded polypropylene: III. Mechanism of HAS inactivation. *European Polymer Journal*, 36(7), 1431–1443.

Zeynalov, E.B. and Allen, N.S. (2006). Modelling light stabilizers as thermal antioxidants. *Polymer Degradation and Stability*, 91(12), 3390–3396.

Zhao, W., He, J., Yu, P., Jiang, X., Zhang, L. (2023). Recent progress in the rubber antioxidants: A review. *Polymer Degradation and Stability*, 207, 110223.

5

Effect of Chemical Aging on Mechanical Properties

Emmanuel RICHAUD
PIMM, CNRS, CNAM, ENSAM, Paris, France

5.1. Introduction

In studies on durability, it is essential to have structure–property relationships to be able to evaluate the link between a structural modification (e.g. chain scission or cross-linking) and changes in mechanical properties. These will be presented in this chapter, focusing on the mechanical properties (elastic modulus, fracture properties, etc.).

This chapter focuses on the properties of polymer materials (thermoplastics, elastomers or thermosets), in which degradation can be considered spatially homogeneous, meaning that there is no degradation gradient. In such samples, embrittlement is described by direct relationships between structure and properties and not by finite element simulation methods specific to heterogeneously degraded thick materials.

This chapter starts with thermoplastic polymers, which can be considered the easiest to approach, particularly owing to the architecture of these (inherently soluble) materials, which can be readily analyzed (particularly by measurements of average molar masses), unlike elastomer and thermosetting networks, for which the macromolecular architecture is described by "indirect" approaches.

5.2. Thermoplastics

This section will first address the changes in average molar masses, which are descriptors of the macromolecular architecture. These are related to many macroscopic properties (viscosity of the molten state, glass transition temperature and fracture properties), which are addressed second.

5.2.1. *Change in the average molar mass*

Let us first consider the simple case of thermal aging leading to depolymerization, with a concentration of chain scission(s). In this case, the number-average molar mass varies as follows:

$$M_n = M_{n0} - s.M_{n0}.M_0 \quad [5.1]$$

where M_{n0} is the initial value of the number molar mass and M_0 is the molar mass of the unit released during a depolymerization act.

Let us now consider the case of statistical chain scissions, which take place on a monomer unit randomly located on the polymer chain. Such scissions are generally expected for all chemical aging modes (radio-, photo- and thermo-oxidative). Their effect on the molar mass is more significant than that of depolymerization, as indicated by the following relationship:

$$\frac{1}{M_n} - \frac{1}{M_{n0}} = s \quad [5.2]$$

In some polymers, statistical scissions may coexist with a depolymerization process initiated by chain end scissions (whose nature and reactivity result from the polymerization process). Let us consider, for example, the case of a PMMA (molar mass of monomer unit equal to 100 g.mol^{-1}) and the initial average molar mass M_n = 100 kg.mol^{-1}. In the case of a chain scission concentration of 0.01 mol.kg^{-1} (which is one scission per chain), if depolymerization occurs (in the case of thermal aging under an inert atmosphere at very high temperature), the average molar mass will decrease from 100 to 99.9 kg.mol^{-1}. On the other hand, in the case of statistical scissions, the molar mass will drop from 100 to 50 kg.mol^{-1}.

Finally, in the case of aging, during which there are simultaneous statistical (tetrafunctional) chain scissions and cross-linking, the number of molar masses (M_n)

and weight molar masses (M_w) vary according to the following relationships (Saito 1958):

$$\frac{1}{M_n} - \frac{1}{M_{n0}} = s - x \qquad [5.3]$$

$$\frac{1}{M_w} - \frac{1}{M_{w0}} = \frac{s}{2} - 2x \qquad [5.4]$$

When the aging mechanism leads to trifunctional nodes, we have:

$$\frac{1}{M_w} - \frac{1}{M_{w0}} = \frac{s}{2} - x \qquad [5.5]$$

It is therefore difficult in practice to identify s and x based on the systematic follow-up of number- and weight-average molar mass such that it can be achieved with a steric exclusion chromatography device (Reano et al. 2018).

5.2.2. *Glass transition temperature*

The glass transition temperature defines the range of operating temperatures of an amorphous thermoplastic material, as it separates its glassy domain (glass-like behavior, i.e., in general, fragile and brittle) from the rubbery domain (the latter is sometimes very narrow for certain amorphous thermoplastics of low molar mass). In linear polymers, the glass transition temperature (T_g) is given by the Flory–Fox equation (Fox and Flory 1950):

$$T_g = T_{g\infty} - \frac{k_{FF}}{M_n} \qquad [5.6]$$

in which $T_{g\infty}$ is the glass transition temperature of an "infinite" chain, k_{FF} is a constant specific to the chain related to its stiffness and M_n is the number-average molar mass. Notably, k_{FF} and $T_{g\infty}$ are strongly correlated (Bicerano 2002; Richaud et al. 2010). It can then be shown that:

$$T_g = T_{g0} - k_{FF}.(s - x) \qquad [5.7]$$

In other words, the increase in T_g is direct proof of the existence of a (predominant) chain cross-linking mechanism, while its decrease indicates the predominance of chain scissions. The glass transition temperature is particularly easy to measure via DSC for amorphous polymers, for which monitoring the T_g facilitates an understanding of the degradation mechanism.

5.2.3. *Viscosity of the molten state*

It is known that in polymers in the molten state, viscosity changes in the same way as the average molar mass in weight, since both values are related by the following relationship (Mark et al. 1984):

$$\eta_0 = k.M_w{}^{3.4} \quad [5.8]$$

where η_0 is the Newtonian viscosity, that is, the value of viscosity when pulsation tends to 0, and k is a constant that depends on temperature and decreases with temperature according to the Arrhenius law.

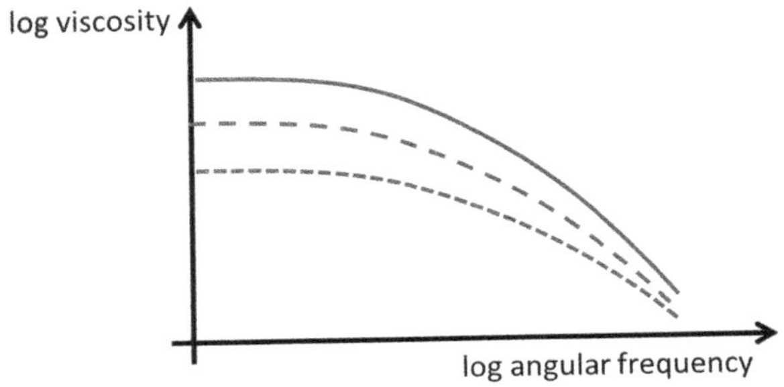

Figure 5.1. *Schematic representation of the rheogram showing the viscosity of polypropylene in the molten state (—: new; - - -: degraded; ....: highly degraded)*

It is therefore possible to identify the predominance of chain scissions or cross-linking based on viscosity variation. Here are some examples:

– the case of polypropylene, in which chain scissions are exclusive (Fayolle et al. 2004). This type of analysis is, however, complicated for two reasons:

- from a theoretical point of view, it cannot always be a priori known whether scissions are exclusive. For example, it is obvious that polyamide aging in a wet environment will undergo a hydrolysis process (and that aging will lead to 100% chain scissions). On the other hand, the situation is far more complicated, for example, for thermally aged polyethylene, in which scissions and cross-linking coexist (Fayolle et al. 2007),

- a rheological analysis is conducted in the molten state, and the increase in temperature required to conduct the experiment induces additional in-situ degradations;

– the case of initially non-cross-linked elastomers, but aging leads to predominant cross-linking. An evolution of rheological modules G' and G'' can be noted, and they are used to identify the "freezing point", that is, the formation of an insoluble network (Winter and Chambon 1986). More precisely, for polymers in the molten state, we have the following:

$$G' \sim \omega^{n'} \quad [5.9]$$

$$G'' \sim \omega^{n''} \quad [5.10]$$

with $n' = 2$ and $n'' = 1$ for new polymers. During aging, n' and n'' gradually decrease with a critical point, or $n' = n''$ is associated with the freezing point, that is, at cross-linking (Kumar et al. 2004).

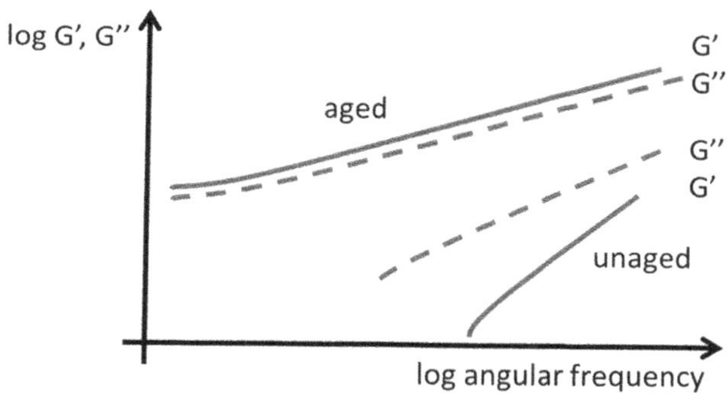

Figure 5.2. *Schematic representation of rheograms of a virgin and aged polymer undergoing major cross-linking*

5.2.4. *Fracture properties*

The crack propagation equation can be used to predict the behavior of a material under fracture:

$$\sqrt{E.G_{IC}} = \beta.\sigma.\sqrt{\pi a} \quad [5.11]$$

where E is the elastic modulus, β is a criterion that depends on the geometry, σ is the stress applied to the material and a is the size of the defect. The energy dissipation rate G_{IC} corresponds to the sum of the energy required for surface creation (2.γ), to which plastic dissipation is added:

$$G_{IC} = 2\gamma + G_P \quad [5.12]$$

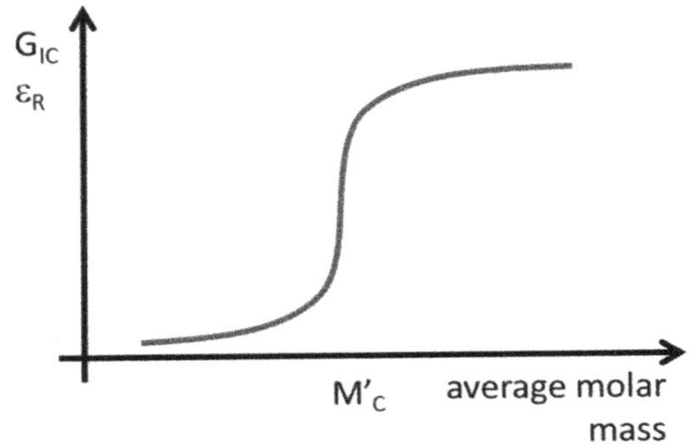

Figure 5.3. *Relationships between ductility and molar mass*

In brittle materials, G_{IC} is close to 2γ and increases when materials are ductile. According to this equation, the crack propagates:

– if the stress or size of the defect is above a limit value;

– if the toughness K_{IC} or the energy release rate G_{IC} is below a limit threshold, it can be considered a criterion for the polymer end-of-life.

	Symbol	M'_C (kg.mol^{-1})	M'_E (kg.mol^{-1})	M'_C/M'_E
Bisphenol A polycarbonate	PC	28	1.78	16
Polymethyl methacrylate	PMMA	40	9.2	4
Polystyrene	PS	60	18.7	3
Polyethylene	PE	70	1.4	50
Polytetrafluoroethylene	PTFE	200	3.7	54
Polypropylene	PP	200	3.5	57
Polyoxymethylene	POM	70	2.5	28

Table 5.1. *Critical molar masses at embrittlement (M'_C) and between entanglements (M'_E) for several amorphous polymers (PC, PMMA, PS) and semicrystalline polymers (PE, PP, PTFE, POM)*

The polymer loses its plastic properties when the average molar mass decreases below a critical threshold M'_C. Let us start by comparing the ratio of the critical molar mass M'_C and the average molar mass between the entanglements M_E for amorphous and semicrystalline thermoplastics (Table 5.1):

– In amorphous polymers, the ratio $q = M'_C/M_E$ is close to 5, which means that the polymer becomes brittle when the network of entanglements has been significantly damaged.

– In semicrystalline polymers (in any case, those whose amorphous phase is in a rubbery state), q is close to 50, suggesting the existence of another mechanism driving embrittlement. This is known as chemicrystallization: the chain segments released by scissions in the amorphous phase have enough mobility to integrate into the crystalline phase. The latter is less deformable than the amorphous phase, and the phagocytation of the amorphous phase by the crystalline phase often leads to embrittlement. Chemicrystallization is characterized by the following yield:

$$y_c = \frac{1}{M_m} \cdot \frac{dx_C}{ds} \quad [5.13]$$

This is an end-of-life criterion illustrated in Figure 5.4.

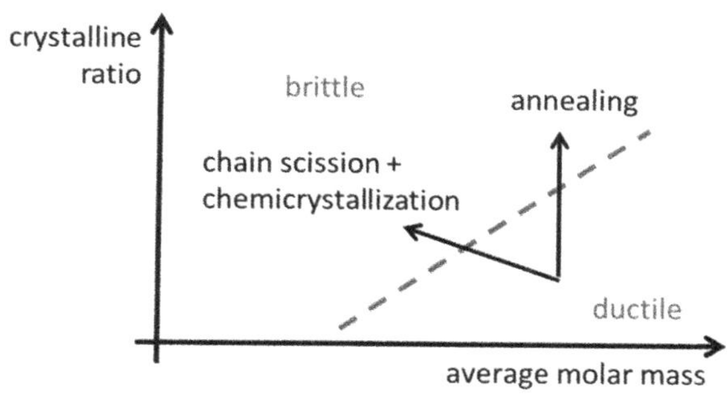

Figure 5.4. *Schematic representation of the "embrittlement map" of semicrystalline polymers*

Figure 5.4 illustrates two scenarios: embrittlement induced by annealing and embrittlement induced by chain scissions accompanied by chemicrystallization. Obviously, the two phenomena can coexist. For example, in the case of PA11 thermo-oxidized at 110°C, we have observed that the crystallinity rate (initially at 25%) increases from 1 to 2% in the first hours of aging and then increases by approximately 5% under the effect of chain scissions, as described by equation [5.13], annealing not explaining by itself embrittlement (Okamba-Diogo et al. 2016).

5.3. Networks

5.3.1. *Networks in the rubbery state: elastomers*

Elastomers are networks in which elasticity has an entropic origin. According to the hypothesis of an affine network (where the elastically active chain deformation is the same as the macroscopic deformation), it can be shown that:

$$E = 2(1 + \nu) = 3.n_A.RT \quad [5.14]$$

n_A is the elastically active chain concentration (in mol.m^{-3}), and ν is the Poisson's ratio (0.5 for the elastomers that are, in a first approximation, considered incompressible). In a network supposed to be ideal (without dangling chains), this equation is reformulated as follows:

$$E = 3G = \frac{3\rho RT}{M_C} \quad [5.15]$$

M_C is the mass between cross-linking nodes (kg.mol^{-1}) and ρ its density. Let us add that $E = 3G$ for incompressible materials.

An elastomer network can be schematically represented before and after degradation, inducing cross-linking (which is, for example, the case of an EPDM (Delor-Jestin et al. 2000) or polycholoroprene (Le Gac et al. 2016) during thermal aging) as follows.

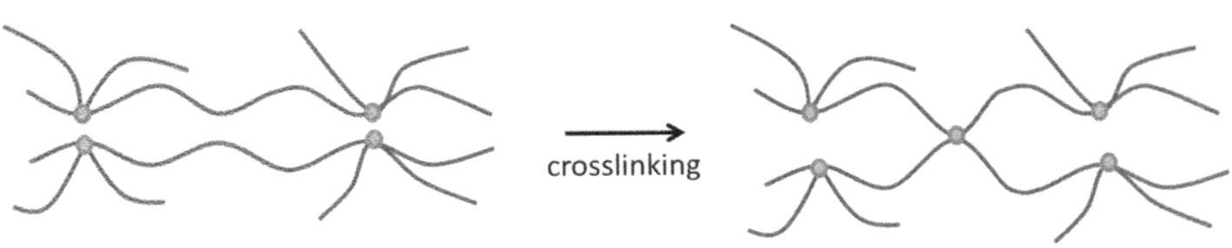

Figure 5.5. *Schematic representation of a network before and after cross-linking*

It can thus be predicted that:

$$n_A = n_{A0} + 2x \quad [5.16]$$

and, in the case of aging leading exclusively to the formation of tetrafunctional cross-linking nodes, regardless of the functionality of the initial network, that:

$$G = G_0 + 2xRT \quad [5.17]$$

For the chain scissions, it can be shown that:

– in a trifunctional network, $G = G_0 - 3.s.RT$ in the case of aging leading exclusively to chain scissions (e.g. hydrolysis of polyurethane networks obtained from polyesters terminated by alcohol functions and cross-linked with trifunctional isocyanates (Richaud et al. 2014));

– in a tetrafunctional network, $G = G_0 - s.RT$, in the case of aging leading exclusively to chain scissions (by the thermo-oxidation of vulcanized polyisoprene (Colin et al. 2007)).

This approach can be generalized as follows:

$$n_A = n_{A0} - \psi sRT + 2xRT \qquad [5.18]$$

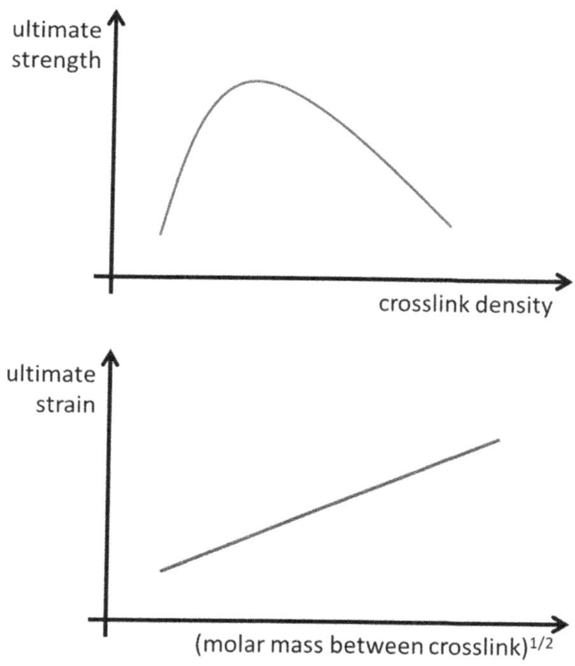

Figure 5.6. *Variations in fracture properties with cross-linking*

These relationships are, in reality, valid only at low conversion degrees. Given that the predominant consequence of aging (the occurrence of scissions or cross-linking) is well known, it is possible to predict the effect of aging on the elastic modulus under weak strains, which is a fundamental mechanical property of elastomers. On the other hand, module monitoring can be used to study aging architectural modifications.

The maximal extensibility ($\lambda_R = 1 + \varepsilon_R$) varies in principle linearly with the square root of the molar mass between the cross-linking nodes even if there may be gaps with this relationship. The fracture strength can be improved at a low cross-linking rate, but it inevitably decreases when the process is more advanced. Notably, it also decreases at a low cross-linking rate in the case of elastomers, which can, in principle, crystallize under stress. This phenomenon is inhibited by cross-linking, which is detrimental to fracture strength.

5.3.2. *Networks in the glassy state: thermosets*

The thermosets are used in the glassy state. The measurement of the concentration of elastically active chains via the theories of rubbery elasticity (as illustrated in the previous section) is complicated by the fact that the chains between cross-linking nodes are short and the material is naturally shifting away from the basic hypotheses of this theory.

In practice, measurements of the glass transition temperature may prove useful. In ideal networks (such as epoxies), they can be described by the following relationship (DiMarzio 1964):

$$T_g = \frac{T_{gl}}{1 - K_{DM}.F.X} \quad [5.19]$$

where T_{gl} is the glass transition temperature of the equivalent linear polymer. K_{DM} is a universal constant (2.91 for epoxies) (Bellenger et al. 1987). F is a parameter that describes network flexibility. X is the concentration of cross-linking nodes that is linked to the concentration of the elastically active chain by:

$$n_A = \frac{f.X}{2} \quad [5.20]$$

f is the functionality of cross-linking nodes. The concentration of chain scissions and cross-linking can be deduced via equations [5.17]–[5.19]. For example, in the case of a material undergoing only chain scissions, the following equation is obtained:

$$s = \frac{T_{gl}}{2K_{DM}F} \cdot \left(\frac{1}{T_g} - \frac{1}{T_{g0}}\right) \quad [5.21]$$

In practice, however, it seems that very rigid networks undergo exclusively chain scissions, but the presence of aliphatic segments (in the epoxies) or double bonds induces the appearance of cross-linking. In this case, we have:

$$3s - 2x = \frac{3T_{gl}}{2K_{DM}F} \cdot \left(\frac{1}{T_g} - \frac{1}{T_{g0}}\right) \quad [5.22]$$

In other words, the variations in T_g can only be used to draw conclusions when $s > 2x/3$ or $x > 3s/2$.

Measures of the soluble fraction (sol) can also be used. Charlesby and Pinner (1959) and Galant et al. (2010) proposed a relationship between them and the concentration in elastically active chains:

$$\frac{n_A}{n_{A0}} = \left(\frac{1 - sol^{1/2}}{1 - {sol_0}^{1/2}}\right)^2 \quad [5.23]$$

This can be simplified in the case of aging leading exclusively to chain scissions in a trifunctional network (equation [5.24]):

$$s = \frac{\nu_0}{3}\left(1 - \left(\frac{1 - sol^{1/2}}{1 - {sol_0}^{1/2}}\right)^2\right) \quad [5.24]$$

It is often considered that toughness in networks decreases with the concentration of cross-linking nodes (Figure 5.6). In the case of aging leading to cross-linking, for example, for epoxies cross-linked by an unsaturated anhydride (Le Huy et al. 1993), embrittlement can, in fact, be noted. A possible explanation was proposed based on Figure 5.7. At low temperature, the maximal stress is the brittle fracture stress σ_F, which does not depend on temperature. Above a certain temperature T_{DF}, plasticity can develop beyond a stress σ_y, which decreases with temperature, as indicated by the law of Kambour (1983):

$$\sigma_y = c.\left(T_g - T\right) \quad [5.25]$$

Cross-linking is reflected by an increase in T_g and leads to a shift of T_{DF} toward high temperatures, meaning that the material becomes brittle when the operating temperature T_u is equal to T_{DF}. Assuming that the coefficient of Kambour's law remains constant during aging, the end-of-life criterion can be expressed as follows:

$$T_u - T_{DF} = T_g - T_{g0} \quad [5.26]$$

The critical concentration in cross-linking nodes can be extracted, for example, with equations [5.19] and [5.20].

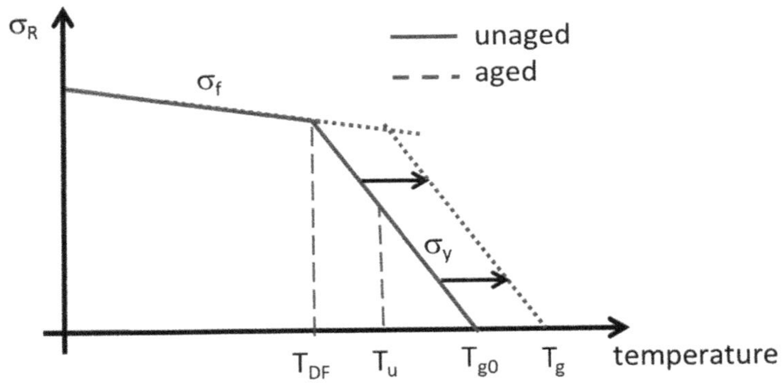

Figure 5.7. *Variations with temperature of the fracture properties of a virgin and aged polymer undergoing major cross-linking*

Other predictions of mechanical properties, particularly those of epoxies, during their aging rely on the disappearance of relaxation β, which seems to be due to hydroxypropyl ether functions (Heux et al. 1997). Figure 5.8 describes the main tendencies.

1) For a new material, the low-temperature module is given by the cohesive density and would be of the order of 5 GPa at 0 K. In general, its value can be approximated by measurements of the ultrasonic module (Rasoldier et al. 2008). It is accompanied by a drop in the real module around the relaxation β (ΔE_β, on the order of GPa), in addition to that induced by expansion. Notably, in the general case, we have (Gilbert et al. 1986):

$$E(T) = E_0 \cdot \left(1 - \alpha \frac{T}{T_g}\right) - \sum_i \Delta E_i \quad [5.27]$$

Therefore, in the case of epoxies between T_g and T_β:

$$E(T) = E_0 \cdot \left(1 - \alpha \frac{T}{T_g}\right) - \Delta E_\beta \quad [5.28]$$

where α is related to the thermal expansion coefficient. In the subglass domain ($T_\beta - T_g$), it gradually decreases in relation to the thermal expansion before dropping significantly at the glass transition passage.

Beyond the glass transition, the module in the rubbery state depends only on the cross-linking density.

For epoxies, a semiempirical relationship (Nielsen 1969) was proposed:

$$\log G' = 7 + 293 \cdot \frac{\rho}{M_C} \quad [5.29]$$

where G' is the shear module (10^{-1} Pa), ρ is the density (kg.L^{-1}) and M_C is the average molar mass between cross-linking nodes (mol.kg^{-1}).

2) In the case of an aged network (Figure 5.8 shows an example of chain scission aging), it can be noted that:

– the module at 0 K is slightly modified;

– under the effect of chemical processes, plasticization at the passage of relaxation β is gradually attenuated, leading to an increase in the sub-glass module;

– as predicted by the approach of DiMarzio, the T_g decreases, and also the rubbery module.

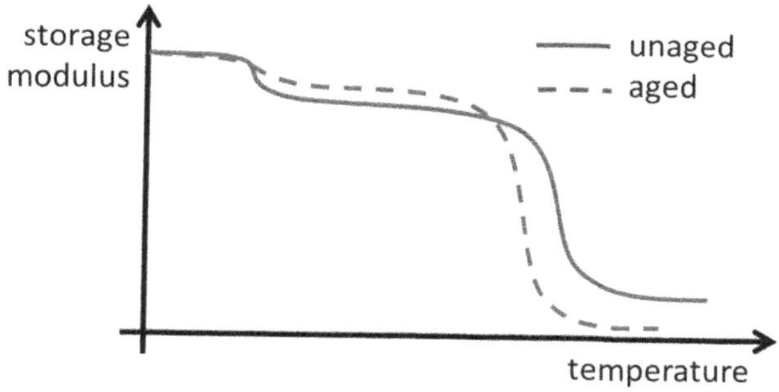

Figure 5.8. *Variation in the thermomechanical properties of an epoxy network during chemical aging (according to equations [5.19], [5.28] and [5.29])*

Besides the increase in the "subglass" module, the decrease in the amplitude of β relaxation is also accompanied by a decrease in toughness and by the ultimate elongation, in other words. Recent works suggest a linear relationship between the residual ultimate elongation and the amplitude of β relaxation, as measured by DMA (Ernault et al. 2017). This approach should, however, be generalized.

5.4. Conclusion

This chapter presents the consequences of aging (expressed on the basis of a concentration in chain scissions and/or cross-linking) on the mechanical properties (elastic modulus, fracture properties, etc.) of thermoplastic polymers and elastomer networks.

As the concentrations in scission and/or cross-linking can easily be predicted based on kinetic analyses of chemical aging mechanisms (Chapters 1 and 2), a procedure for predicting the loss rate of mechanical properties of polymer materials can thus be implemented, at least in the case of homogeneously degraded samples. More complex cases of samples, for which the degree of aging varies as a function of the depth of the sample (e.g. in the case of materials undergoing oxidation controlled by oxygen diffusion), involve the implementation of finite element methods, which is illustrated, for example, in the article of Ernault et al. (2019).

5.5. References

Bellenger, V., Verdu, J., Morel, E. (1987). Effect of structure on glass transition temperature of amine crosslinked epoxies. *Journal of Polymer Science Part B Polymer Physics*, 25(6), 1219–1234.

Bicerano, J. (2002). Transition and relaxation temperatures. In *Prediction of Polymer Properties*, Bicerano, J. (ed.). CRC Press, Boca Raton, New York. doi: 10.1201/9780203910115.

Charlesby, A. and Pinner, S.H. (1959). Analysis of the solubility behaviour of irradiated polyethylene and other polymers. *Proceedings of the Royal Society of London A*, 249, 367–386.

Colin, X., Audouin, L., Verdu, J. (2007). Kinetic modelling of the thermal oxidation of polyisoprene elastomers. Part 3: Oxidation induced changes of elastic properties. *Polymer Degradation and Stability*, 92(5), 906–914.

Delor-Jestin, F., Lacoste, J., Barrois-Oudin, N., Cardinet, C., Lemaire, J. (2000). Photo-, thermal and natural ageing of ethylene–propylene–diene monomer (EPDM) rubber used in automotive applications. Influence of carbon black, crosslinking and stabilizing agents. *Polymer Degradation and Stability*, 67(3), 469–477.

DiMarzio, E.A. (1964). On the second-order transition of a rubber. *Journal of Research of the National Bureau of Standards, Section A: Physics and Chemistry*, 68, 611–617.

Ernault, E., Richaud, E., Fayolle, B. (2017). Origin of epoxies embrittlement during oxidative ageing. *Polymer Testing*, 63, 448–454.

Ernault, E., Dirrenberger, J., Richaud, E., Fayolle, B. (2019). Prediction of stress induced by heterogeneous oxidation: Case of epoxy/amine networks. *Polymer Degradation and Stability*, 162, 112–121.

Fayolle, B., Audouin, L., Verdu, J. (2004). A critical molar mass separating the ductile and brittle regimes as revealed by thermal oxidation in polypropylene. *Polymer*, 45(12), 4323–4330.

Fayolle, B., Colin, X., Audouin, L., Verdu, J. (2007). Mechanism of degradation induced embrittlement in polyethylene. *Polymer Degradation and Stability*, 92(2), 231–238.

Fox, T.G. and Flory, P.J. (1950). Second-order transition temperatures and related properties of polystyrene. I. Influence of molecular weight. *Journal of Applied Physics*, 21(6), 581–591.

Galant, C., Fayolle, B., Kuntz, M., Verdu, J. (2010). Thermal and radio-oxidation of epoxy coatings. *Progress in Organic Coatings*, 69, 322–329.

Gilbert, D.G., Ashby, M.F., Beaumont, P.W.R. (1986). Modulus-maps for amorphous polymers. *Journal of Materials Science*, 21(9), 3194–3210.

Heux, L., Halary, J.L., Lauprêtre, F., Monnerie, L. (1997). Dynamic mechanical and 13C n.m.r. investigations of molecular motions involved in the β relaxation of epoxy networks based on DGEBA and aliphatic amines. *Polymer*, 38(8), 1767–1778.

Kambour, R.P. (1983). Correlations of the dry crazing resistance of glassy polymers with other physical properties. *Polymer Communications*, 24(10), 292–296.

Kumar, A., Commereuc, S., Verney, V. (2004). Ageing of elastomers: A molecular approach based on rheological characterization. *Polymer Degradation and Stability*, 85(2), 751–757.

Le Gac, P.Y., Celina, M., Roux, G., Verdu, J., Davies, P., Fayolle, B. (2016). Predictive ageing of elastomers: Oxidation driven modulus changes for polychloroprene. *Polymer Degradation and Stability*, 130, 348–355.

Le Huy, H.M., Bellenger, V., Verdu, J., Paris, M. (1993). Thermal oxidation of anhydride cured epoxies—III. Effect on mechanical properties. *Polymer Degradation and Stability*, 41(2), 149–156.

Mark, J.E., Eisenberg, A., Graessley, W.W., Mandelkern, L., Koenig, J.L. (1984). *Physical Properties of Polymers*. ACS, Washington, D.C.

Nielsen, L.E. (1969). Cross-linking–effect on physical properties of polymers. *Journal of Macromolecular Science, Part C Polymer Reviews*, 3(1), 69–103.

Okamba-Diogo, O., Richaud, E., Verdu, J., Fernagut, F., Guilment, J., Fayolle, B. (2016). Investigation of polyamide 11 embrittlement during oxidative degradation. *Polymer*, 82, 49–56.

Rasoldier, N., Colin, X., Verdu, J., Bocquet, M., Olivier, L., Chocinski-Arnault, L., Lafarie-Frenot, M.C. (2008). Model systems for thermo-oxidised epoxy composite matrices. *Composites Part A: Applied Science and Manufacturing*, 39(9), 1522–1529.

Reano, A.F., Guinault, A., Richaud, E., Fayolle, B. (2018). Polyethylene loss of ductility during oxidation: Effect of initial molar mass distribution. *Polymer Degradation and Stability*, 149, 78–84.

Richaud, E., Ferreira, P., Audouin, L., Colin, X., Verdu, J., Monchy-Leroy, C. (2010). Radiochemical ageing of poly(ether ether ketone). *European Polymer Journal*, 46(4), 731–743.

Richaud, E., Gilormini, P., Coquillat, M., Verdu, J. (2014). Crosslink density changes during the hydrolysis of tridimensional polyesters. *Macromolecular Theory and Simulations*, 23(5), 320–330.

Saito, O. (1958). On the effect of high energy radiation to polymers: I. Cross-linking and degradation. *Journal of the Physical Society of Japan*, 13(2), 198–206.

Winter, H.H. and Chambon, F. (1986). Analysis of linear viscoelasticity of a crosslinking polymer at the gel point. *Journal of Rheology*, 30(2), 367–382.

6

Physical Aging by Structural Relaxation in Polymers

Blandine QUÉLENNEC, Nicolas DELPOUVE and Laurent DELBREILH
GPM, CNRS, Université de Rouen Normandie, Saint-Étienne-du-Rouvray, France

6.1. Introduction: glass transition and the glassy state

The use of the term "physical aging" in relation to a material covers a wide range of behaviors in time, their common feature being the reversibility of the phenomenon. This reversibility is related to the fact that the chemical structure of a physically aged system is preserved. As an example, a polymer material involved in a physical aging phenomenon undergoes no transformation of the covalent bond, whereas intermolecular interactions are susceptible to being impacted.

The term "physical aging" most often refers to aging by structural relaxation. It is a spontaneous process characterized by the conformational rearrangement of polymer chains, and results in a modification of macroscopic properties that will affect the stability of polymer materials over a longer or shorter term. Considering, for example, the evolution of mechanical properties over time, it is possible to note an increase in stiffness or embrittlement of the polymer. As shown in this chapter, this undesired evolution of properties is particularly the consequence of decreases in molecular mobility and specific volume under the effect of aging.

Physical aging by structural relaxation, which is the focus of this chapter, is a property of "glasses" – a definition that will not be limited here to silicon dioxide-based materials, but cover all of the materials undergoing the glass transition phenomenon during heating. It is in the amorphous phase that aging by structural relaxation takes place. Therefore, this phenomenon concerns both amorphous and semicrystalline polymers. This phenomenon is reversible in the sense that to “rejuvenate” a polymer, it is sufficient to bring it above its glass transition temperature. Consequently, aging by structural relaxation only occurs when the amorphous phase is in a glassy state, in contrast with the rubbery state. The first step in approaching the glassy state is to provide several elements for understanding what a glass transition is.

6.1.1. *Glass transition*

6.1.1.1. *Kinetic character of glass transition*

6.1.1.1.1. Viscous slowing down

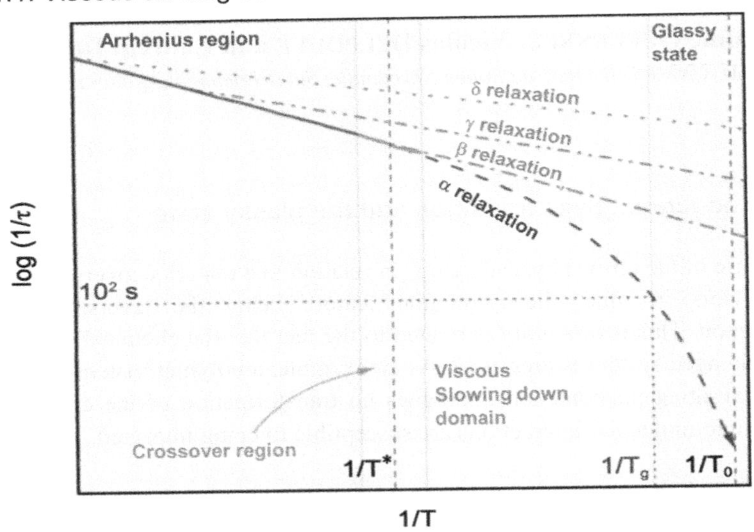

Figure 6.1. *Schematic illustration of the dependence of the processes of relaxation of a glass-forming liquid on temperature during cooling. For a color version of this figure, see www.iste.co.uk/richaud/industrialpolymers1.zip*

COMMENT ON FIGURE 6.1.– *T^* corresponds to the crossover temperature where viscous slowing down emerges. T_g is the glass transition temperature, where relaxation times reach hundreds of seconds. T_0 is the Vogel temperature, which is defined as the temperature at which the relaxation time tends to infinity (according to Quélennec (2022)).*

The cooling of a liquid induces an increase in viscosity η and in the relaxation time τ. At temperatures far beyond the glass transition temperature, this dependence is described by the Arrhenius law:

$$\tau = \tau_0 \exp\left(\frac{E_a}{RT}\right) \qquad [6.1]$$

where E_a is the activation energy associated with the relaxation process, T is the temperature, R is the universal constant of the ideal gas and τ_0 is a preexponential factor that corresponds to the anticipated relaxation time for a phonon, that is, the time needed for a molecule to vibrate at a frequency close to the Debye frequency.

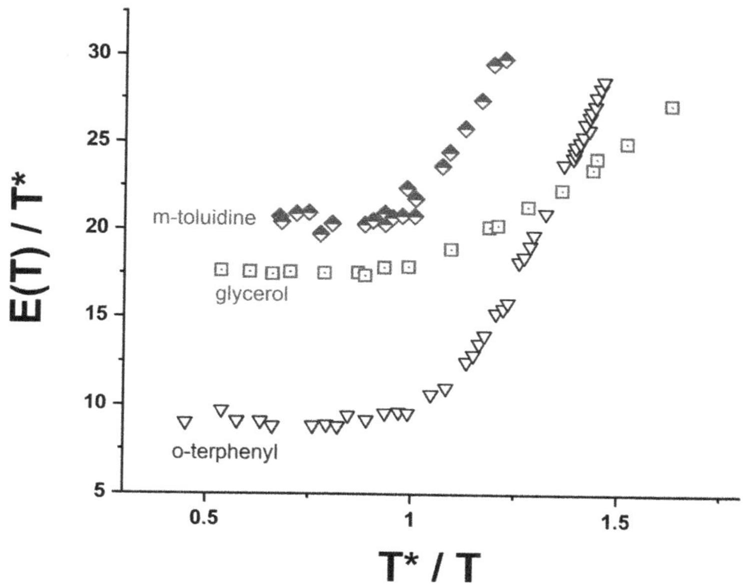

Figure 6.2. *Transition from Arrhenius to super-Arrhenius behavior for several glass-forming liquids: effective activation energy E(T) versus 1/T. E(T) and T are both normalized at crossover temperature T* (adapted from Tarjus et al. (2005)). For a color version of this figure, see www.iste.co.uk/richaud/industrialpolymers1.zip*

Nevertheless, this law can no longer be used to describe the dependence of viscosity and of the relaxation time near the glass transition; however, these values increase significantly. Figure 6.1 illustrates the dependence on temperature of the relaxation processes of a glass-forming liquid during cooling. Relaxations can be differentiated depending on their kinetics. α relaxation is the main relaxation.

The other relaxations are secondary relaxations. The dependence of the viscosity on temperature during α relaxation is referred to as viscous slowing down or the "super-Arrhenius regime". This is common in glass-forming liquids (Lunkenheimer et al. 2010).

As illustrated in Figure 6.2, in this regime, the activation energy is a function of the temperature, and the term E_a (T) replaces E_a in equation [6.1] (Aouadi et al. 2000).

6.1.1.1.2. Kinetic fragility

To compare glass-forming liquids, Angell (1985) proposed a representation of the dependence of the viscosity or relaxation time as a function of T_g/T, where T_g is the temperature at the relaxation time corresponding to the definition selected for the glass transition. An example of this representation is given in Figure 6.3.

NOTE.– T_g is most often defined as the temperature for which $\tau = 100$ s or $\eta = 10^{12}$ Pa s (Figure 6.1).

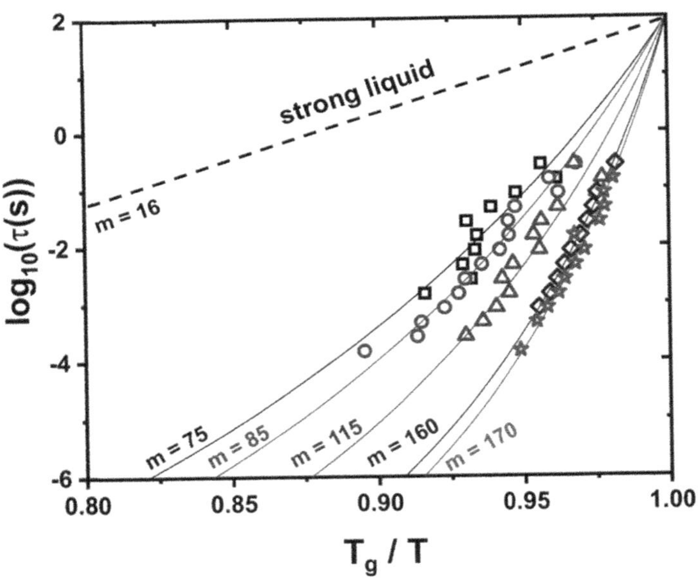

Figure 6.3. *Diagram of kinetic fragility or representation of viscous slowing down for various polymer systems, according to Angell's concept of fragility. For a color version of this figure, see www.iste.co.uk/richaud/industrialpolymers1.zip*

COMMENT ON FIGURE 6.3.– *The Arrhenius behavior is represented by the dotted lines. The solid lines are VTFH fits of the experimental data (equation [6.3]). The gap in the Arrhenius behavior is observed by the larger curvature of the VTFH fits and the increasingly steep slope near the T_g. This gap is reflected by the kinetic fragility index m that evolves here from 16 (the case of a glass-forming liquid referred to as "strong" when its main relaxation process obeys Arrhenius behavior) to 170 for the most fragile glass-forming liquid (adapted from Araujo et al. (2018)).*

Under constant pressure, the value of the slope near T_g constitutes the definition of the kinetic fragility index m:

$$m = \left(\frac{d \log_{10}(\tau)}{d\left(\frac{T}{T_g}\right)} \right) \Bigg|_{T=T_g} \qquad [6.2]$$

A glass-forming liquid for which $m = 16$ is referred to as a "strong liquid", whereas a high value of m reflects the behavior of a "fragile liquid", that is, a liquid for which the viscous slowing down is abrupt. Polymers are considered fragile liquids, as their kinetic fragility index ranges mainly between 50 and 200 (Kunal et al. 2008).

6.1.1.1.3. Behavior of fragile liquids

To adjust the data related to the dependence on the temperature of the relaxation time of a fragile liquid (Figure 6.3), the Vogel–Fulcher–Tamann–Hesse (VFTH) (Vogel 1921; Fulcher 1925; Tamann and Hesse 1926) relationship can be used:

$$\tau = \tau_0 \exp\left(\frac{DT_0}{T - T_0}\right) \qquad [6.3]$$

where D is a parameter specific to the stiffness of the viscous slowing down and T_0 is the Vogel temperature, which is defined as the temperature at which the relaxation time tends to infinity (Dudowicz et al. 2015) and where a divergent behavior is expected (Lunkenheimer et al. 2010). Considering the following conditions:

$$C_2 = T^* - T_0 \qquad [6.4]$$

$$C_1 C_2 \mathrm{Ln}(10) = B \qquad [6.5]$$

$$\tau = \tau_0 \exp\left(\frac{B}{T - T_0}\right) \qquad [6.6]$$

where B is an activation parameter related to the energy barrier in the Arrhenius case (Lunkenheimer et al. 2010), as highlighted by Kovacs (1964), an equivalence can be noted between the VFTH law and the empirical law of Williams–Landel–Ferry (WLF) (Williams et al. 1955) describing the viscoelastic properties of a polymer as a function of temperature (equation [6.7]) according to the time–temperature superposition (TTS) principle.

The application of the TTS principle consists of superposing the experimental responses obtained by applying various solicitation frequencies over a range of temperatures on a master curve (Figure 6.4).

To plot the master curve, the data obtained at various temperatures are simply translated with respect to the results obtained at a reference temperature T^* using a shift factor a_T (Figure 6.5). Williams et al. (1955) have shown that the values of a_T in the liquid state follow the following law:

$$\log a_T = \log\left(\frac{\tau(T)}{\tau(T^*)}\right) = -\frac{C_1(T - T^*)}{C_2 + T - T^*} \qquad [6.7]$$

where C_1 and C_2 are constants associated with T^*. This relationship coincides with the experimental data in a limited range of temperature near the T_g (Kausch et al. 2001) and diverges at higher temperature when the influence of thermal activation becomes too significant.

Williams et al. (1955) postulated that the constants C_1 and C_2 vary slightly depending on the polymer nature and are universal ($C_1 = 17.4$ and $C_2 = 51.6$ K) when the glass transition temperature is selected as the reference temperature.

The universal character of WLF constants was challenged by Angell (1997), who linked the parameter C_2/T_g to the liquid fragility according to the experimental findings that, if the value of C_1 is always approximately 16, C_2 is, in fact, susceptible to have significantly higher values for "strong liquids", up to 300 K for silicate glasses.

Dudowicz et al. (2015) have more recently highlighted that the universality of the constants C_1 and C_2, in the context of the equivalence between the WLF and VFTH relationships, requires kinetic fragility to be proportional to the T_g.

This was globally observed (Qin and McKenna 2006) for polymers and metal glass formers, but it proves to be no longer valid when this comparison is extended to other classes of materials.

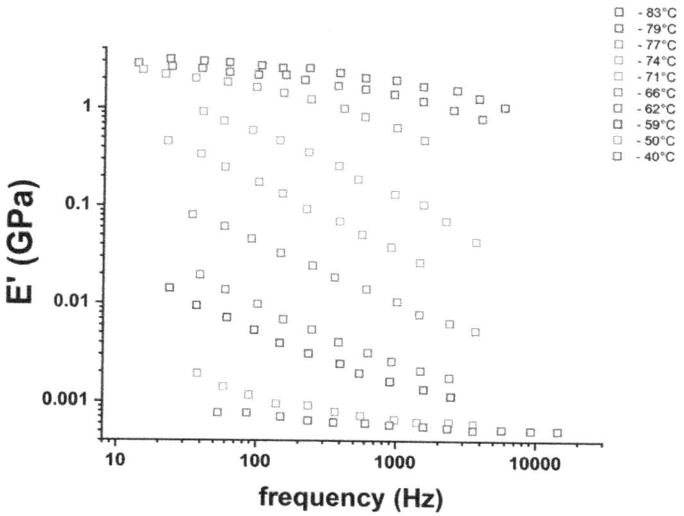

a)

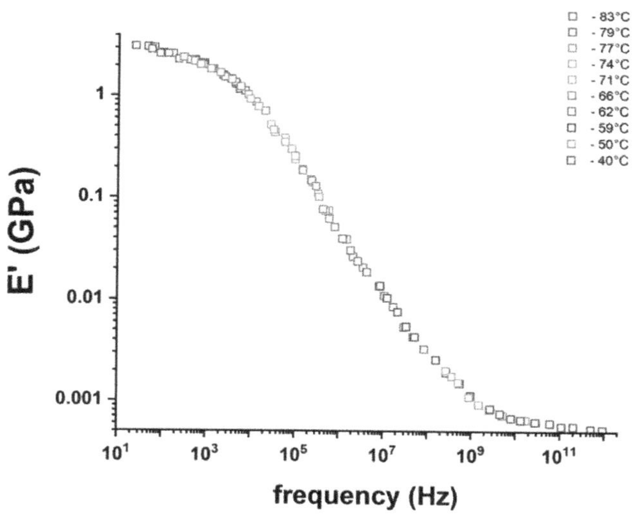

b)

Figure 6.4. *Time–temperature superposition principle applied to isotherms of the storage modulus E' obtained for an amorphous polymer at –83, –79, –77, –74, –71, –66, –62, –59, –50 and –40°C and at various frequencies (a). The reference isotherm is –83°C, and all other isotherms are shifted to form the master curve of the storage modulus over a wide range of frequencies (b) (adapted from Kaliappan (2007)). For a color version of this figure, see www.iste.co.uk/richaud/industrialpolymers1.zip*

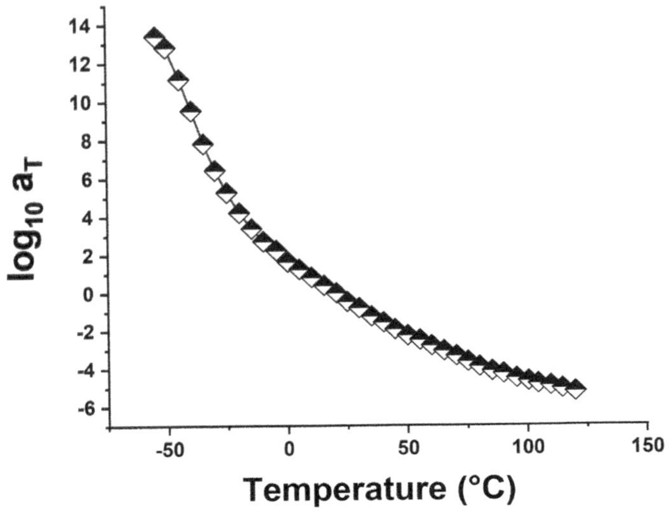

Figure 6.5. *Dependence of the a_T factor on temperature (adapted from Lorenz et al. (2013))*

6.1.1.2. *Dynamic character of glass transition*

6.1.1.2.1. Relaxation function

A dynamic perspective on the relaxation phenomenon governing the glass transition is obtained by exposing the system to a light perturbation and by studying its response in time. For example, the intermediate scattering function can be obtained via quasielastic neutron scattering (Figure 6.6). Frequency investigations are also possible; they are most often conducted using dielectric spectroscopy methods.

The intermediate scattering function can be used to highlight singularities in the behavior of a cooled liquid compared with the exponential relaxation observed at higher temperature (Dalle-Ferrier 2009). First, three regimes (two relaxations separated by a plateau) correspond to microscopic relaxation, the "cage" effect and the main relaxation (commonly known as "structural relaxation" or "alpha relaxation"). Notably, the main relaxation process cannot be described by a simple exponential. A stretched exponential function is most often used.

NOTE.– A clearer view of the three regimes can be obtained by numerical simulation, as was done for a binary mixture of Lennard–Jones particles (Kob and Andersen 1994).

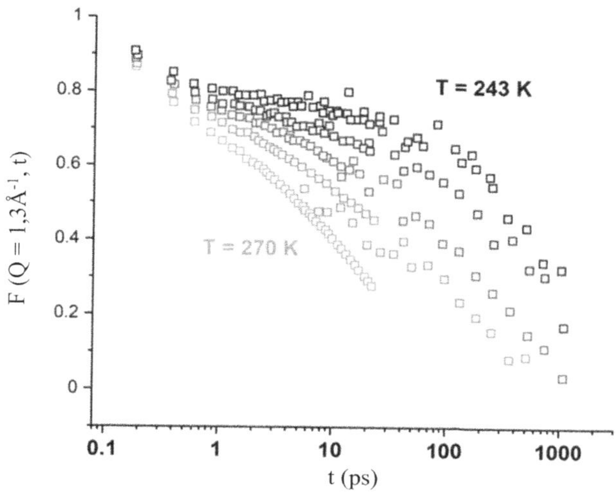

Figure 6.6. *Intermediate scattering function measured by combining two quasielastic neutron scattering techniques (adapted from Dalle-Ferrier (2009)). For a color version of this figure, see www.iste.co.uk/richaud/industrialpolymers1.zip*

6.1.1.2.2. Stretched exponential

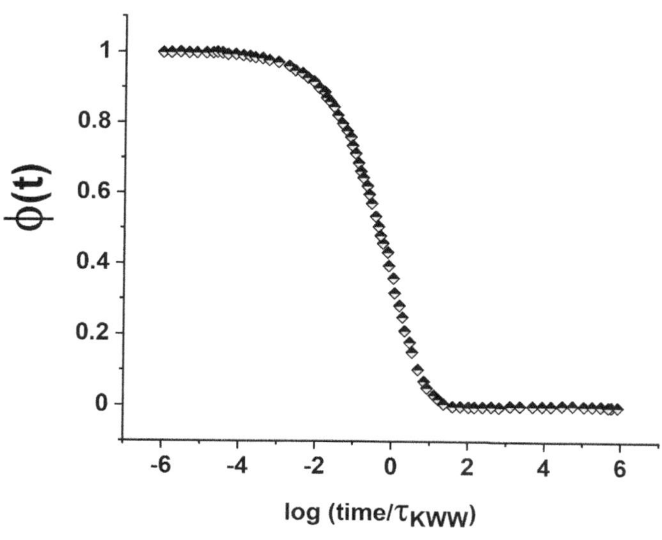

Figure 6.7. *Kohlrausch–Williams–Watts function (adapted from Andreozzi et al. (2008))*

The Kohlrausch–Williams–Watts (KWW) function is expressed as follows:

$$\Phi\,(t) = \Phi_0 \exp\left[-\left(\frac{t}{\tau_{KWW}}\right)^{\beta_{KWW}}\right] \qquad [6.8]$$

where Φ_0 is the preexponential factor corresponding to the response for time $t = 0$, τ_{KWW} represents the average characteristic relaxation time at the corresponding temperature and β_{KWW} is the stretching coefficient (Kohlrausch 1854; Williams and Watts 1970). Figure 6.7 illustrates the KWW function.

6.1.1.2.3. Dynamic heterogeneity

The stretched exponential character of the relaxation leads to two possible interpretations concerning the spatial distribution of dynamics (Brun 2011). According to the first method, relaxation is spatially homogeneous, and it is stretched at any point, identically with the same coefficient β_{KWW}. In the second case, the dynamics are heterogeneous, and the liquid involves several regions that relax exponentially but with various relaxation times. On the basis of many experimental and simulation results, Ediger (2000) concluded that relaxation dynamics were spatially heterogeneous.

6.1.1.3. *Thermodynamic character of glass transition*

6.1.1.3.1. Crystal, supercooled liquid, glass

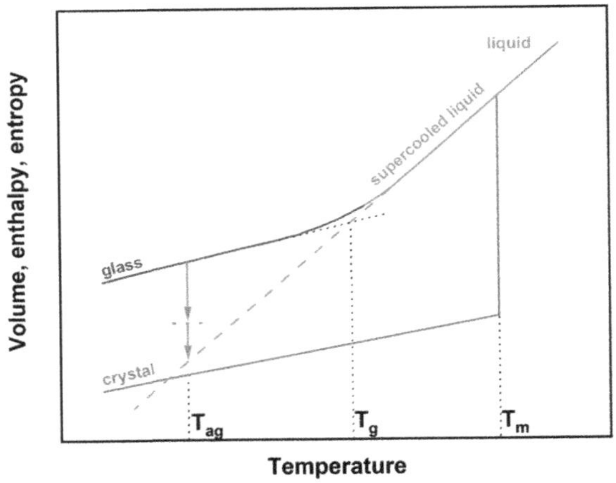

Figure 6.8. *Schematic illustration of the dependence of the thermodynamic properties (volume, enthalpy or entropy) of a glass-forming liquid on temperature. T_K: Kauzmann temperature; T_g: glass transition temperature; T_m: melting temperature (according to Quélennec (2022)). For a color version of this figure, see www.iste.co.uk/richaud/industrialpolymers1.zip*

Liquid cooling can lead to crystal formation by first-order phase transition, which induces a sudden change toward values that are far below the first derivatives of Gibbs free energy: entropy and specific volume (Figure 6.8). However, if the time allotted to system reorganization is insufficient, crystallization may be avoided. The system is then in a state of supercooled liquid. In the supercooled liquid, thermodynamic equilibrium is reached. Nevertheless, the enthalpy of the supercooled liquid is greater than that of the crystal, and the state is referred to as metastable (Brun 2011). Glass formation during the cooling of the supercooled liquid manifests as a change in the slope with respect to the level of temperature dependence of the entropy and of the specific volume; however, in contrast to crystallization, no discontinuity is observed. Therefore, it is not a first-order transition.

6.1.1.3.2. Influence of the cooling rate

Glass transition starts when the rate of molecular rearrangement has the same order of magnitude as the cooling rate imposed on the system. The chain movement rate becomes too low (too long a relaxation time) for the system to remain on the equilibrium line of the liquid. The system freezes and undergoes a transition to the glassy state. The thermodynamic characteristics of the glass formed in this way therefore depend on the cooling rate (Figure 6.9). The higher the cooling rate is, the further the system will thermodynamically be from equilibrium.

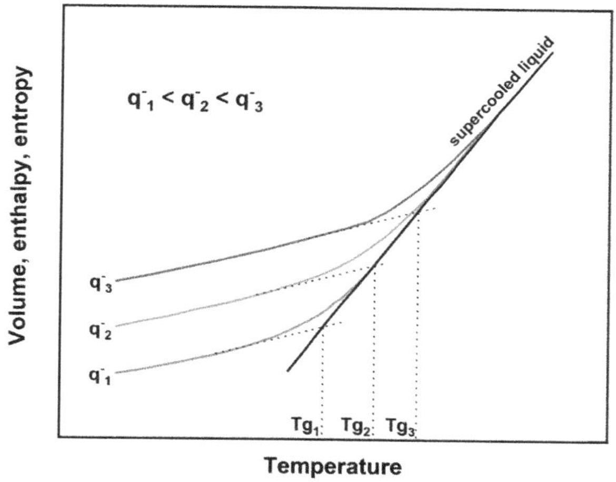

Figure 6.9. *Schematic illustration of the dependence of the thermodynamic properties (volume, enthalpy or entropy) of a glass-forming liquid on the cooling rate. T_{g1}, T_{g2}, T_{g3}: glass transition temperatures; q^{-}_{1}, q^{-}_{2}, q^{-}_{3}: cooling rates (according to Quélennec (2022)). For a color version of this figure, see www.iste.co.uk/richaud/industrialpolymers1.zip*

NOTE.– Glass transition manifests through a characteristic discontinuity of second derivatives of Gibbs free energy, such as the heat capacity. Nevertheless, the glass transition has many characteristics that led to it being qualified as a pseudo-second-order transition according to the Ehrenfest classification (van Ekeren 1998; Jaeger 1998). We can mention the gradual character of this change and the dependence on the cooling rate of the temperature at which it occurs, associated with the fact that the glassy state is not at thermodynamic equilibrium. Moreover, the Prigogine–Defay ratio (Prigogine and Defay 1954) is different from the unit ratio, which continues to be the subject of many discussions (Garden et al. 2012).

6.1.1.3.3. Thermodynamic critical temperature

The unique character of the glass transition from a thermodynamic point of view raises questions on the possibility that it may be a phenomenon governed only by kinetic and dynamic effects. Another hypothesis links the glass transition to a thermodynamic transition that would occur at a lower temperature (Brun 2011). This hypothesis resulted from the works of Kauzmann (1948), who reported that the entropy of the supercooled liquid decreases far more rapidly than that of the crystal during cooling. This leads to a temperature T_K, for which the configurational entropy, defined as the logarithm of the number of minima of potential energy accessible to the system, becomes zero, as this temperature indicates a situation for which the entropy of the crystal becomes higher than that of the liquid. This is often referred to as the Kauzmann paradox (Figure 6.8).

NOTE.– For the Kauzmann paradox, Berthier and Biroli (2011) mentioned that there is no general principle compelling the entropy of the liquid to be higher than that of the crystal. To illustrate this, Berthier and Biroli cited the example of the crystallization of hard spheres (Alder and Wainwright 1962). This occurs because at a sufficiently high density, the crystal becomes the state of higher entropy.

The Kauzmann temperature is roughly identical to the Vogel temperature, playing a predominant role in the kinetic description of the glass transition (Angell 1991; Roland 2011). Due to this fact, the hypothesis of a critical temperature sheltering a second-order transition has been considered (Brun 2011).

6.1.1.3.4. Observation of glass transition

Glass transition can be investigated by several techniques and methods by observing the evolution of several physical properties depending on the technique and operating conditions (Paroli and Penn 1994; Núñez et al. 2004; Hakme et al. 2005; Lixon et al. 2008; Menczel et al. 2009): differential scanning calorimetry

(DSC), modulated temperature differential scanning calorimetry (MT-DSC), dynamic mechanical analysis (DMA), dielectric relaxation spectroscopy (DRS) and thermomechanical analysis (TMA). In DSC, a jump in heat capacity is clearly observed, as illustrated in Figure 6.10. Calorimetric techniques have therefore been widely used and enhanced to evaluate the thermodynamic changes associated with the glass transition.

NOTE.– The glass transition signature can be observed upon heating, but the glass transition temperature (T_g) is defined upon cooling.

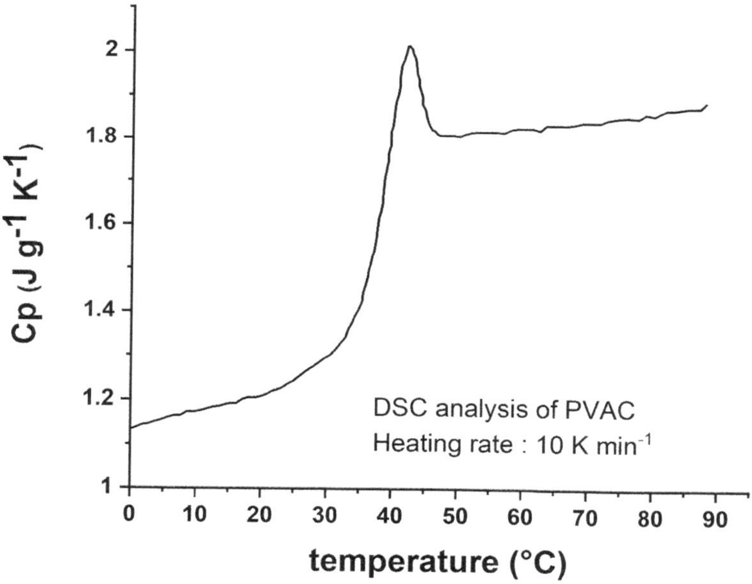

Figure 6.10. *Illustration of the jump in heat capacity of a PVAC sample (adapted from Donth (1996))*

MT-DSC can be used to observe the glass transition with greater clarity by separating kinetic events from those simply linked to the variation in heat capacity (Lacey et al. 2006). It is particularly efficient to investigate polymer systems with complex microstructure or morphology (Araujo et al. 2019).

The techniques offering high scanning rates, such as hyper DSC (Pijpers et al. 2002), or, more recently and more efficiently, chip-based fast scanning calorimetry (FSC) (Mathot et al. 2011; Minakov et al. 2019), can be used to form a completely amorphous polymer glass, even for systems with rapid crystallization kinetics, exceeding the cooling rate of reorganization of macromolecules. Considering the very high rates accessible by FSC (on the order of thousands of kelvin per second), it is possible to reach a sufficient level of sensitivity to observe the glass transition of samples of low mass (several dozen nanograms).

6.1.2. *Characteristics of the glassy state*

6.1.2.1. *Structure factor and static correlation length*

The results of the works of Morineau and Alba-Simionesco (1998), relating to the static structure factor of m-toluidine measured by inelastic neutron scattering, show that there is no major structural variation between liquid, supercooled liquid and glass. Below the glass transition, the order remains local, in contrast with the crystal, for which long-distance order is observed.

Nevertheless, theoretical arguments and numerical simulations support the existence of a static correlation length that increases during cooling (Biroli et al. 2008; Franz and Semerjian 2011). The nature of the link between static and dynamic correlation lengths has been studied (Berthier and Kob 2011), as has the experimental evidence of this length (Datin 2019).

6.1.2.2. *Physical aging by structural relaxation*

As previously mentioned, the glassy state is a state whose shift from thermodynamic equilibrium depends on the cooling rate applied to the glass-forming liquid. At equivalent temperatures in the glassy state, the enthalpy of the system will be higher for a high cooling rate compared to a low cooling rate (Kovacs 1981).

On the other hand, the glassy state has the possibility to evolve in time to a new state of structural equilibrium. The out-of-equilibrium glass has thermodynamic quantities in excess that will be reduced to reach a new state of metastable equilibrium (Hodge 1995), corresponding to the extrapolation of thermodynamic quantities of the supercooled liquid. For this evolution, structural relaxation is the origin of physical aging.

The phenomenon of structural relaxation is observed for many glassy systems, such as inorganic glass, metallic glass and polymer glass. This implies that glass formation and structural changes occurring in the glassy state are not directly associated with the chemical structure. This aspect should be taken into account in a physical interpretation of the structural relaxation.

The study of variations in properties such as the specific volume, enthalpy and dielectric and mechanical responses provides information on the nature of structural relaxation. The next sections will show that many scientific articles focus on the physical aging of amorphous polymers but do not always seem to agree with the interpretation of the results. Therefore, many questions remain to be answered.

6.2. Historical landmarks in the study of physical aging

This section briefly presents the main conclusions drawn from pioneering works that have significantly contributed to a better comprehension of physical aging. These results have already been addressed in many articles and comments. The authors particularly recommend the reviews authored by Hodge (1994) and Hutchinson (1995).

6.2.1. *Tool's observations and the concept of fictive temperature*

The first experiments dedicated to physical aging by structural relaxation have been conducted by Tool on silicate glass (Tool and Eichlin 1925, 1931; Tool et al. 1930), particularly via dilatometry measurements. On the basis of many results obtained at various temperatures and for various thermal histories, Tool showed that a glass return to equilibrium could not be described by an equation in which relaxation time was only a function of temperature (Hutchinson 1995). He suggested that the instantaneous glass structure during relaxation could act on the kinetics and translate this influence of the structure through the concept of fictive temperature T_f, which he defines as the temperature at which the glass in its instantaneous state would be at equilibrium if it was brought there by a very steep temperature ramp (Tool 1946).

6.2.2. *The limits stated by Ritland*

The limits of fictive temperature as a unique parameter for describing glass structure have been mentioned by Ritland (1956), who compared borosilicate glasses having different thermal histories. He noted that a cooled and aged sample, until reaching the same refractive index as its counterpart, having already reached the equilibrium at a higher aging temperature, had its refractive index vary when brought to this temperature.

At the same time, Ritland (1956) reported that two samples having the same refractive index and assumingly the same fictive temperature featured differences in terms of electrical resistivity and the coefficient of thermal expansion. Hodge (1994) estimates that these results illustrate that the structure of a glass must be defined with more than a fictive temperature and that the global fictive temperature is not necessarily identical from one physical property to another.

6.2.3. *Kovacs' experiments*

Kovacs highlighted three major properties of structural relaxation associated with physical aging: it is non-exponential, nonlinear and depends on the thermal history of the material (Kovacs 1964).

These results clarify the observations of Ritland (1956) and contribute to a re-examination of the concept of fictive temperature, particularly by Narayanaswamy and Moynihan (section 6.2.4.1).

6.2.3.1. *Intrinsic isotherm*

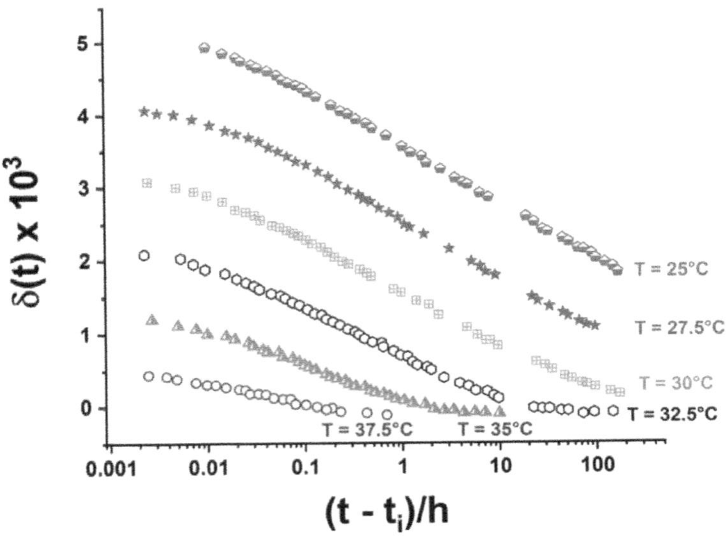

Figure 6.11. *Intrinsic isotherms for glassy polymers based on dilatometric measurements on polyvinyl acetate. The volume variation δ is represented as a function of the time passed after temperature quenching.* $\delta = \frac{V - V_\infty}{V_\infty}$, *where V is the volume at instant t,* V_∞ *is the volume at equilibrium and* t_i *corresponds to the quenching time (adapted from McKenna and Simon (2017)). For a color version of this figure, see www.iste.co.uk/richaud/industrialpolymers1.zip*

The intrinsic isotherm experiment is the most widely used method to investigate the evolution of the glassy response. The system is cooled from the liquid state to various temperatures at an identical rate.

The structural evolution of the glass was measured as a function of time at each aging temperature (Figure 6.11). Notably, the relaxation function has the stretched exponential character described in section 6.1.1.2.

6.2.3.2. *Asymmetric approach*

The asymmetric approach is highlighted by coupling two experiments: a positive temperature jump and a negative temperature jump.

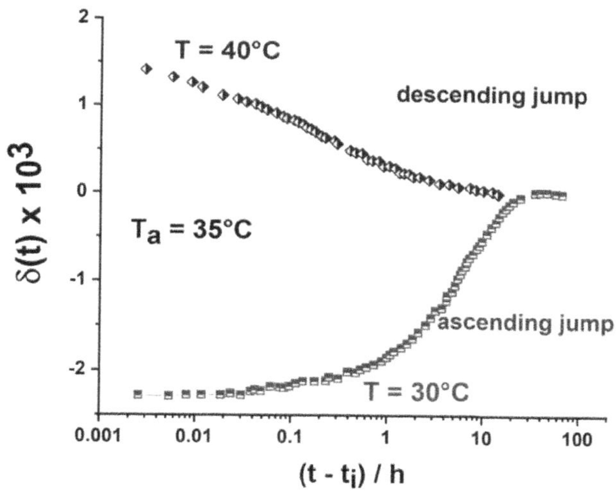

Figure 6.12. *Response to the asymmetric approach of polyvinyl acetate, comparing the behavior of a descending jump from T = 40°C to T_a = 35°C with the behavior of an ascending jump from 30°C to 35°C (adapted from McKenna and Simon (2017)). For a color version of this figure, see www.iste.co.uk/richaud/industrialpolymers1.zip*

In one case, the system is initially set at $T_1 = T_a + \Delta T$, and in the other case, the system is initially set at $T_2 = T_a - \Delta T$ for a sufficiently long time for the monitored physical property to be stabilized. Each system then undergoes an isotherm at T_a. In the case of Kovacs' experiment, it is a dilatometric measurement. During the isotherm at T_a, a logical observation is that one sample contracts, whereas the other expands. The originality of the result stems from the fact that the return to a new state of equilibrium is performed asymmetrically between the two samples

(Figure 6.12). For the system initially stored at T_1, the reaction of the system is rapid, followed by self-slowing near equilibrium as the glass density increases. For the system initially stored at T_2, the high density of the glass drives a slow evolution of the structure, followed by self-acceleration as it approaches equilibrium. According to Kovacs, this result provides additional information to the hypothesis of a distribution of relaxation times. Indeed, the latter does not predict the behavior observed for the system stored at T_2, namely, that slow processes will precede rapid processes. This is how Kovacs shows the nonlinear character of relaxation.

6.2.3.3. *Memory effect*

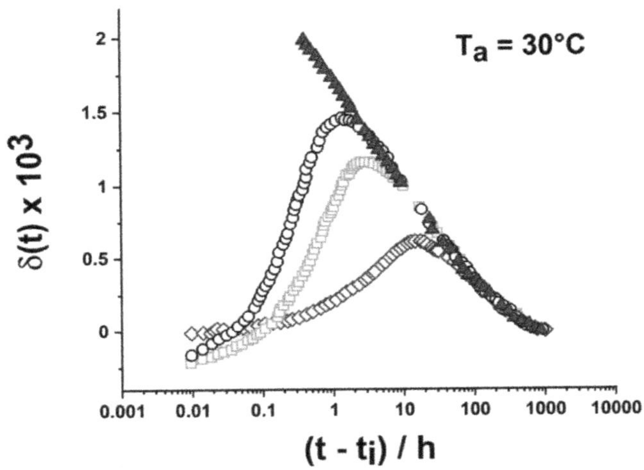

Figure 6.13. *Experiment of memory effect on a polyvinyl acetate at a testing temperature of 30°C. For a color version of this figure, see www.iste.co.uk/richaud/industrialpolymers1.zip*

COMMENT ON FIGURE 6.13.– *Red full triangles: direct experiment at 30°C following the descending jump from 40°C; black circles: partial structural recovery at 10°C for 160 h before the second jump at T_a = 30°C; pink squares: partial structural recovery at 15°C for 140 h before the second jump in T at T_a = 30°C; violet diamonds: partial structural recovery at 25°C for 90 h before the second jump in T at T_a = 30°C (adapted from McKenna and Simon (2017)).*

The memory effect applies to complex histories and can be observed via a two-step thermal treatment, reproducing one of Ritland's experiments globally. The sample is cooled from the liquid state and undergoes an isotherm at a temperature

T_1; then, a second jump in temperature is conducted at an aging temperature T_2 (slightly above the temperature of the first isotherm). The objective is that the aging at T_1 does not bring the sample to equilibrium but rather to induce a state, followed by dilatometry, corresponding to the equilibrium at T_2. It is then possible that aging at T_2 will not lead to any change in property. Instead, Kovacs reported very strong expansion of the sample until it reached quite high values (Figure 6.13). The behavior of the sample is subsequently aligned with the relaxation of a material that has directly reached T_2 from the liquid state. This result is consistent with the idea of a distribution of fictive temperatures and relaxation times.

6.2.4. *Theoretical contributions*

6.2.4.1. *The Tool–Narayanaswamy–Moynihan (TNM) model*

The TNM model resumes Tool's concept of fictive temperature and integrates the physical aging characteristics identified by Kovacs. This is how Narayanaswamy proposed a model in which nonlinearity can be transformed into complex thermal histories. This shows the possibility of describing the memory effect (Narayanaswamy 1971).

The most commonly cited form of the expression of TNM is the one proposed by Moynihan et al. (1976a). In this expression, the stretched exponential character of the relaxation function is described by the KWW relationship (equation [6.8]), and the relaxation time τ is:

$$\tau(T, T_{\mathrm{f}}) = \tau_{\mathrm{o}} \exp\left[\frac{x\Delta h^*}{RT} + \frac{(1-x)\Delta h^*}{RT_{\mathrm{f}}}\right] \qquad [6.9]$$

where τ_0 is the prefactor of the relaxation time, Δh^* is the activation energy of the Arrhenius process at equilibrium, that is, for $T = T_f$, and R is the ideal gas constant. In this equation, the dependences on temperature (first term of the exponential) and structure (second term of the exponential) are clearly differentiated. Their respective contribution to the relaxation process is given through the nonlinearity parameter x ($0 \leq x \leq 1$).

During physical aging, the fictive temperature evolves between two values, T_f' and T_{ag}. At zero aging time, the fictive temperature corresponds approximately to the T_g, except that it is determined during heating. This value is referred to as the limit fictive temperature T_f' (Gao and Simon 2015). This can be graphically determined by representing the enthalpy as a function of temperature (Figure 6.9) as the crossover of the extrapolated curves of the glass and liquid. When the aging time increases, the fictive temperature approaches the T_{ag} temperature at which aging occurs. Monitoring fictive temperature can therefore be used to characterize relaxation kinetics.

To translate the dependence of the limit fictive temperature (T_f') on the cooling rate q^-, Moynihan et al. (1976b) proposed the following equation:

$$\frac{d\ \ln|q^-|}{d\left(1/T_f'\right)} = -\frac{\Delta h^*}{R} \qquad [6.10]$$

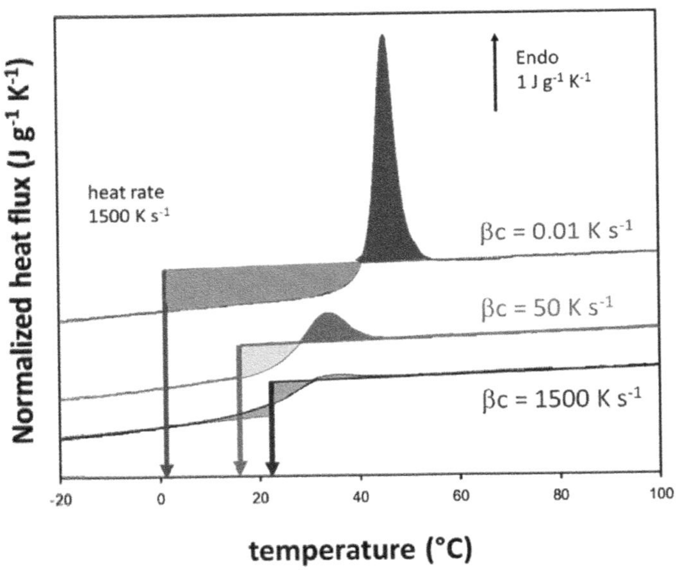

Figure 6.14. *Principle of the equivalent surface method used to determine the limit fictive temperature T_f' for various cooling rates on the basis of the results obtained after the analysis of lactic acid oligomers. The dotted lines and arrows mark T_f' (adapted from Moghaddam et al. (2021)). For a color version of this figure, see www.iste.co.uk/richaud/industrialpolymers1.zip*

Hutchinson (2009) noted that the concept of fictive temperature has two major advantages. It can be used to describe the glassy state after annealing, whereas the glass transition provides information only on its formation. Moreover, the fictive temperature is determined by a simple heating ramp in the DSC (Figure 6.14). For this, the method of equivalent areas is used (Moynihan et al. 1976b):

$$\int_{T_f}^{T_2} (Cp,l - Cp,g)dt = \int_{T_1}^{T_2} (Cp - Cp,g)dT \qquad [6.11]$$

where *Cp, l* is the heat capacity of the liquid, *Cp, g* is the heat capacity of the glass, and *Cp* is the heat capacity of the sample. T_1 and T_2 are two arbitrary temperatures chosen in the glassy state and in the liquid state, respectively, so that *Cp* = *Cp, g* at T_1 and *Cp* = *Cp*, l at T_2.

The works of Tool, Narayanaswami and Moynihan have strongly influenced contemporary studies on physical aging. The concept of fictive temperature and the TNM model are still being considered in recent studies (Gujrati 2010; Phalen et al. 2016; Grassia et al. 2018).

6.2.4.2. *Kovacs–Aklonis–Hutchinson–Ramos (KAHR) model*

The KAHR model was developed by Kovacs et al. (1979). It follows the same idea as the TNM model, with a dependence of the relaxation time on the temperature and glass structure according to a number of modes denoted by i. The relaxation time is given by equation [6.12]:

$$\tau_i(T, \delta_H) = \tau_{ir}\exp[-\theta(T - T_r)]\exp\left[-(1-x)\theta\delta_H/\Delta C_p\right] \qquad [6.12]$$

where δ_H represents the distance to equilibrium defined by Kovacs in terms of volume (Kovacs 1963), T_r is a reference temperature and ΔCp represents the gap between the heat capacities of the glass and liquid. θ is an analog of the activation energy in the TNM model (Hutchinson 1995) such that:

$$\theta \approx \frac{\Delta h^*}{\mathrm{R}T_g^2} \qquad [6.13]$$

In equation [6.12], each exponential represents the contribution to relaxation in terms of temperature and structure. They are partitioned by the *x* parameter. The major specificity of the KAHR model compared with the TNM model resides in the fact that the distribution of relaxation times is not represented by a KWW law but by a discrete function:

$$\delta = \sum_{i=1}^{N} \delta_i \qquad \text{with} \qquad \frac{\mathrm{d}\delta_i}{\mathrm{d}t} = -\frac{\delta_i}{\tau_{i(T,S)}} \qquad [6.14]$$

The authors' idea behind the use of this function is to facilitate the structural characterization of the glass at any step of aging (Hutchinson 1995), particularly to obtain a free volume distribution in the sample and to compare it with experimental results, such as those that can be obtained via positron annihilation lifetime spectroscopy (PALS).

As a continuation of the KAHR model, Hutchinson and Ruddy (1988) proposed a method known as the peak shift method for obtaining the partition parameter x through an independent experiment. This method relies on the observation of excess enthalpy via calorimetry following physical aging by structural relaxation. The signature of enthalpy recovery shifts linearly to higher temperatures with aging. The temperature of the main peak (T_p) is plotted as a function of the variation in the fictive temperature during structural relaxation. According to Hutchinson and Ruddy (1988), the slope of this line corresponds to the partition factor x, as illustrated in Figure 6.15 (Tool 1946; Ramos et al. 1984).

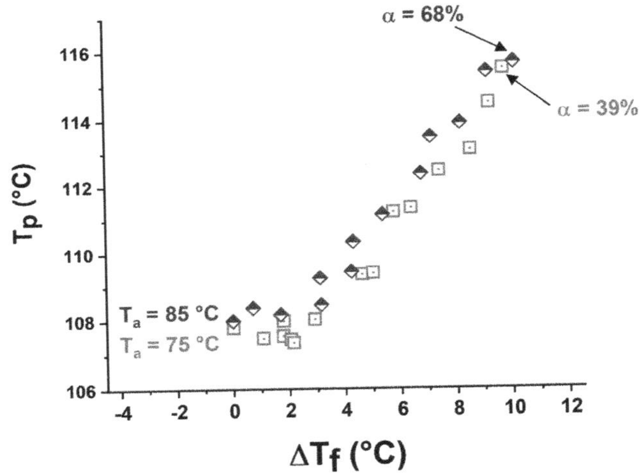

Figure 6.15. *Evolution of the temperature of the main peak (T_p) as a function of the variation in the fictive temperature $\Delta T_f = T_f' - T_f$, where T_f' is the limit fictive temperature and T_f is the fictive temperature (adapted from Zheng et al. (2002))*

In Hodge (1994)'s review mentioned that this was the first approach used to characterize and explain sub-T_g endotherms.

6.2.4.3. *Interpretation of physical aging in the context of Adam–Gibbs theory*

The Adam–Gibbs theory relies on the notions of the energy landscape and cooperativity. In the context of this theory, the kinetic phenomenon, that is, the increase in the relaxation time, can be connected with the thermodynamic phenomenon, that is, the decrease in configuration entropy.

The energy landscape can be used to explain the observed viscous slowing down. The general idea is to build an image of the hypersurface of the potential energy of a supercooled liquid in space using configurational coordinates. When the temperature decreases, the fraction of space statistically accessible by the system decreases. This happens during glass formation. The absolute minimum corresponding to the crystal is not reached, as it is kinetically inaccessible. Nevertheless, during the viscous slowing down, the system is trapped in increasingly deeper valleys, and exploration of the energy landscape involves breaking energy barriers (Dalle-Ferrier 2009).

The increase in relaxation time can be attributed to the decrease in the number of potentially accessible energy minima. Hence, the configurational entropy is defined as the logarithm of the number of these minima accessible for a given temperature. An increase in relaxation time is associated with a decrease in configurational entropy (Dalle-Ferrier 2009).

The decrease in configurational entropy is related to the increase in the height of the energy barriers to be crossed. In their theory, Adam and Gibbs have correlated the height of these barriers to the average number of molecules that must collectively relax to cross them.

Hence, according to this idea, the height of the energy barrier to be crossed for relaxation is proportional to the average volume of cooperatively rearranging regions (CRR) (Adam and Gibbs 1965). The relaxation time is expressed according to the following equation [6.15]:

$$\tau = Ae^{\left(\frac{N_a s_c^* \Delta\mu}{k_B T S_C}\right)} \qquad [6.15]$$

where N_a is Avogadro's number, s_c^* is the configurational entropy of the smallest region of cooperative rearrangements, $\Delta\mu$ is the elementary activation energy per segment, k_B is Boltzmann's constant and S_c is the macroscopic configurational entropy. The configurational entropy is expressed on the basis of the variation in heat capacity (ΔC_p) and temperature (T):

$$S_c = \int_{T_2}^{T} \frac{\Delta C_p}{T}\, dT \qquad [6.16]$$

T_2 is the temperature at which S_c is extrapolated at zero, that is, Kauzmann's temperature.

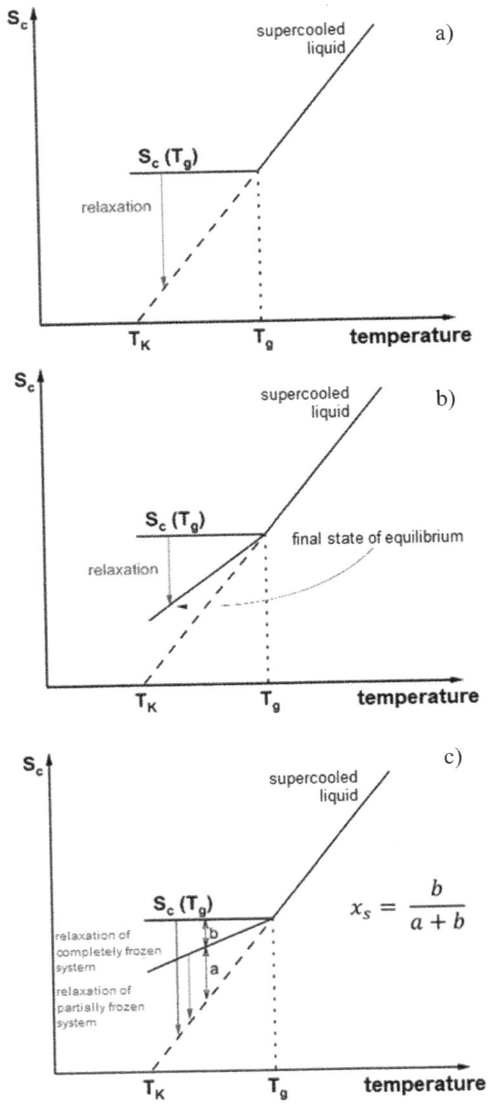

Figure 6.16. *Variation in the configurational entropy of the glass at the passage of the thermal glass transition temperature (T_g) and during structural relaxation: (a) Adam–Gibbs model; (b) model proposed by Gomez Ribelles; (c) model proposed by Huchinson (adapted from Saiter (2002)). For a color version of this figure, see www.iste.co.uk/richaud/industrialpolymers1.zip*

The application of Adam and Gibbs' theory to work on physical aging by structural relaxation led to the emergence of several models on glass-forming liquids (Saiter 2002). To summarize, Adam and Gibbs' model considers that the glass-forming liquid is fully frozen upon glass transition passage. The macroscopic configurational entropy of the glass formed is then constant under the glass transition for zero aging (Figure 6.16a). During aging, the configurational entropy decreases with the aging time until the value of the extrapolated line of the liquid is reached.

The Adam–Gibbs equation was extended from the glass transition to the glassy state by several researchers (Hodge 1994) by replacing T with T_f in the expression of configurational entropy Sc (equation [6.16]). The approximate relationships between the Narayanaswamy and Adam–Gibbs parameters are obtained by differentiating the relaxation time with respect to the temperature in the equilibrium ($T_f = T$) and glassy ($T_f = T_f'$) states.

The model proposed by Gomez Ribelles also considers that the glass-forming liquid is fully frozen at the passage of the glass transition. The glass formed in this way has a constant configurational entropy, irrespective of the temperature under the glass transition for zero aging (Figure 6.16(b)). However, in this model, it is assumed that during aging, the configurational entropy of the glass having already relaxed for an infinite time does not reach the extrapolated line of the liquid but rather a limit intermediate value (Saiter 2002).

Gomez Ribelles refers to an intermediate state of equilibrium justified by the existence of steric constraints and the cooperativity of molecular motions during structural relaxation (Gomez Ribelles et al. 1995; Gomez Ribelles and Monleon Pradas 1995).

The Hutchinson model takes into account both the effect of the structure and the temperature on the configurational entropy at the passage of the glass transition (Hutchinson et al. 2000). Hence, only one part of the glass is completely frozen below the glass transition (Figure 6.16(c)). Consequently, the configurational entropy of the formed glass varies depending on temperature. A parameter x_s, referred to as the "parameter of entropic nonlinearity", is introduced in this model. If $x_s = 0$, then the entire liquid is assumed to be frozen at the passage of the glass transition, whereas the opposite is true if $x_s \neq 0$, that is, only a part of the glass-forming liquid is frozen at the passage of the glass transition. This leads to two types of relaxation processes below the glass transition: rapid processes and slow processes.

NOTE.– There are arguments against the identification of configurational entropy as excess entropy (Dyre 2006). This hypothesis relies on the fact that at identical temperature, glass and crystal have the same contribution to the entropy. However, according to Goldstein, this scenario is not always realistic, as glass may have significant contributions to the specific heat released by anharmonic vibrations and secondary relaxations that are not present in the crystal (Goldstein 1976).

6.2.4.4. *Interpretation of physical aging according to the free volume theory*

Free volume theory was developed as a means to explain and quantify changes in dynamic and mechanical properties, such as viscosity (η) and relaxation time (τ), during a change in temperature. Roseveare et al. (1941) were the first to highlight the importance of volume in glass-forming liquids. They suggested that the lower the viscosity of the liquid, the greater the number of vacancies present in the liquid.

Special attention has been given to understanding the super-Arrhenius behavior, which is specific to many glasses, including polymers. Andrade's equation [6.17] for describing the Arrhenius dependence of viscosity on temperature has been the starting point of Doolittle's reasoning (Andrade 1934):

$$\eta = Ae^{B/T} \qquad [6.17]$$

where η is the viscosity, T is the temperature and A and B are constants. B is generally identified as the apparent activation energy and is associated with diffusion movement.

A link between this first equation and the free volume has been added. Doolittle's equation [6.18] introduces the notion of free volume:

$$\eta(T) = Ae^{B/V_f(T)} \qquad [6.18]$$

where A and B are adjustment parameters, η(T) is the viscosity and V_f is the free volume. This equation takes into consideration that a simple Arrhenius form cannot be used to describe the behavior of many glass systems at temperatures near the glass transition. Indeed, the apparent activation energy B, initially intended to be constant, seems to increase when the temperature decreases. This behavior, known as "super-Arrhenius", was described by the phenomenological equation of Vogel, Fulcher, Tammann and Hesse (VFTH) (section 1.1.1.3), whereas Williams et al. (1955) reported that Doolittle's equation was applicable to many glass-forming liquids in the vicinity of the glass transition.

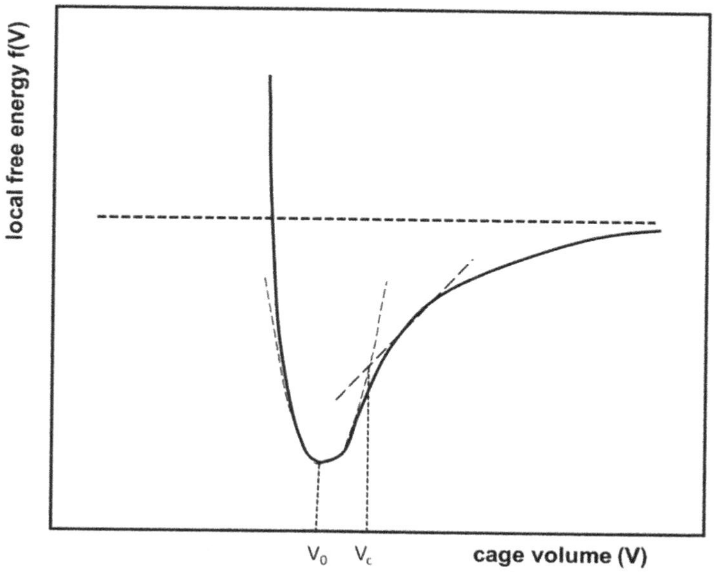

Figure 6.17. *Illustration of the potential energy as a function of volume (adapted from Shakespeare (2002))*

As a continuation of these works, Cohen and Turnbull et al. (1959) and Turnbull and Cohen (1961) proposed a model to link free volume and viscosity. In this model, it is assumed that the molecules of liquids are confined in cages constituted from neighboring atoms. The motion of atoms inside cages is governed by a function of potential energy. The profile of the potential energy is illustrated in Figure 6.17. At high temperatures, which are characteristic of the liquid region, the volumes of the cages are high. The decrease in temperature drives a reduction in volume, and the energy barrier becomes impossible to cross.

It can be assumed that two adjacent cages can shrink and expand to a certain extent if the total volume is conserved. No change in the total energy of the system during such an exchange of volume between cages is noted. This volume, which can be exchanged between cages, was defined as the free volume. When the size of the cage increases, the free volume available for the molecule also increases. Physically, this corresponds to the opening of the environment of a given molecule, which allows the molecule to leave its own cage, thus achieving diffusive movement. However, a minimum free volume is required for transportation to occur. Doolittle's equation, modified and rationalized by Cohen and Turnbull, yields:

$$\tau = A\, e^{\left(b/_{f} \right)} \quad [6.19]$$

where b is a constant and f is the fraction of the free volume defined on the basis of the equilibrium value of the specific volume at the corresponding temperature (V_∞), the "occupied" volume (V_0) and the free volume (V):

$$f = \frac{V - V_0}{V_\infty} = \frac{V_\infty - V_0}{V_\infty} + \frac{V - V_\infty}{V_\infty} \quad [6.20]$$

Macedo and Litovitz suggested that an energy term should be added to the free volume term to account for the thermal activation required for a particle to be able to migrate from one free volume pocket to another, which leads to:

$$\tau = A'\, e^{\left(b/_{f} + E/_{RT} \right)} \quad [6.21]$$

The free volume theory was applied to the glassy state (Kovacs et al. 1977) as follows:

$$f = \frac{V - V_0}{V_\infty} = f_T + \delta_V \quad [6.22]$$

$$f_T = f_g + \alpha_f \left(T - T_g \right) \quad [6.23]$$

$$\tau(T, \delta) = \tau_g \exp\left[\frac{b\, \delta/_{f_T}}{f_T + \delta} \right] \exp\left[\frac{\left(b/_{f_g} \right) \left(T - T_g \right)}{\left(f_g/_{\alpha_f} \right) + T - T_g} \right] \quad [6.24]$$

where f_T is the fraction of free volume at equilibrium, α_f is the coefficient of thermal expansion of the fraction of free volume, f_g is the fraction of free volume of the glassy state, b is a constant of the order of unity, and δ characterizes the distance to equilibrium, as defined in the context of the KAHR model.

The free volume theory can also be introduced into the Adam–Gibbs equation by postulating the existence of a term specific to the free volume per particle that can replace the term configurational entropy (Hodge 1994):

$$\tau_0\left(T, T_{f,V}\right) = A e^{\left(\frac{N_a \upsilon^* \Delta\mu}{k_B T\, V_f \left(T_{f,V} \right)} \right)} \quad [6.25]$$

where υ* is the minimal volume required for a rearrangement.

These works have contributed to many theories of free volume. Cohen and Grest (1979) used percolation theory to describe the passage of the glass transition. Chow (1984) linked the relaxation time to the statistical distribution of the energy associated with the formation of a vacancy in a given state. More recently, White and Lipson developed a model known as the locally correlated lattice (LCL) model to calculate the percentage of the theoretical free volume (White and Lipson 2009, 2016; Lipson and White 2014).

The LCL model is based on an image of a compressible fluid of molecules in the form of a chain, and the free volume is defined independently of the dynamic properties. According to investigations conducted on approximately 50 polymer systems, vitrification occurs when a critical free volume is reached, whose value varies linearly with temperature. The result is a predictive correlation between the free volume theoretically estimated in the liquid state and the experimentally measured glassy transition (White and Lipson 2015, 2016).

6.2.5. *Monitoring the mechanical properties*

6.2.5.1. *Linear viscoelastic region*

The first reference in the literature related to the effect of physical aging on mechanical properties was made by McLoughin and Tobolsky (1951). Notably, they presented the effects of the cooling rate on the relaxation of stress. Slow cooling leads to a slower relaxation of stress. They explain this through the variation in free volume: a system that cools more rapidly has a more significant free volume. Hence, it has the capacity to relax the stress faster (McLoughlin and Tobolsky 1951).

In these dilatometric works on Polyvinylacetate (PVAC) (Kovacs et al. 1963), Kovacs established a correlation between the volume relaxation behavior and the variation in complex moduli in the glassy state. The storage modulus G′ increases linearly when the volume approaches equilibrium by contracting. The opposite is observed upon volume expansion. These changes are attributed to the isothermal relaxation of the free volume and are interpreted in terms of a shift in the spectrum of mechanical relaxation along the logarithmic axis of time with changes in free volume, as illustrated in Figure 6.18. This analysis is shared by Turner (1964) in his study on PVC creep.

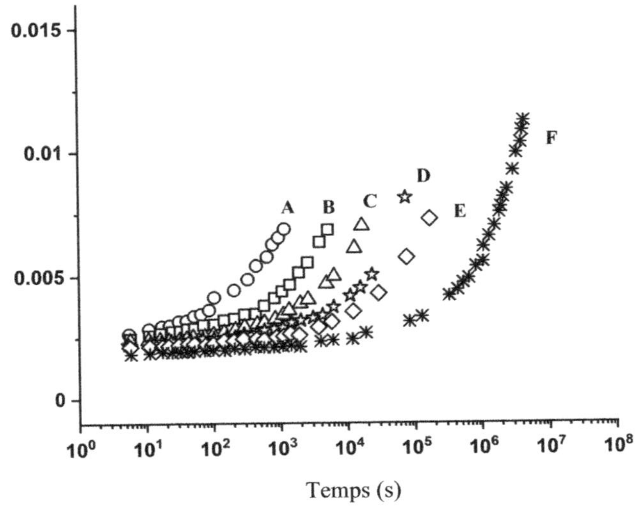

Figure 6.18. *Traction–creep curves for PVC at 60°C under 2 MPa stress. The storage durations at 60°C are as follows: (A) 1 h, (B) 4 h, (C) 17.5 h, (D) 72 h, (E) 168 h, (F) 2,016 h (adapted from Hutchinson (1995))*

Struik's (1978) works on low deformation creep confirm the shift of creep curves toward longer times as aging progresses.

Studies (Meyer et al. 1965; Petrie 1975) have been conducted by DMA using a torsion pendulum: an increase in aging time drives an increase in the storage modulus and a decrease in tan δ. A correspondence is found between the evolution of dynamic mechanical properties and enthalpy, which means that equilibrium is approached from a state with thermodynamic properties either in excess or in defect (as shown, for example, by the asymmetric approach experiment). Importantly, equilibrium is reached more rapidly for the mechanical properties than for the enthalpy. Struik (1977a, 1988) suggests that mechanical properties follow the behavior of volume. This suggestion relies on the idea that volume relaxation is faster than enthalpy relaxation.

6.2.5.2. *Nonlinear viscoelastic region*

In the linear viscoelastic region, the imposed stresses or strains are sufficiently weak to assume that the test does not significantly modify the structural state of the material. When higher stresses are applied, the material will be in the nonlinear viscoelastic region. Therefore, the interaction between the stress or strain imposed on the material and the glass structure will be highlighted.

In his review on the physical aging of epoxy systems, Odegard and Bandyopadhyay (2011) stated that there are divergent opinions on the influence of physical aging on mechanical properties.

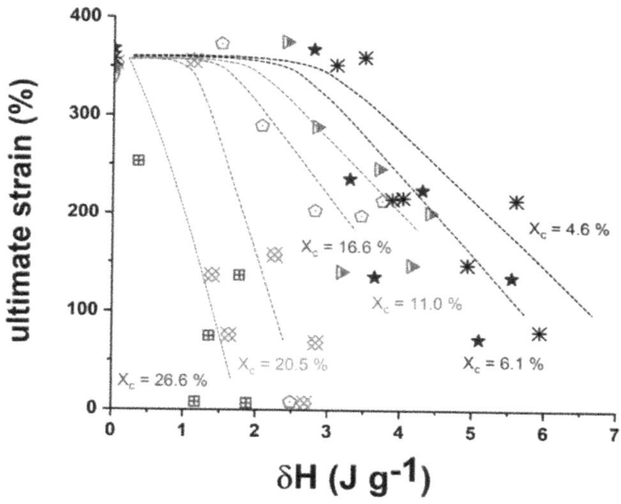

Figure 6.19. *Variation in the ultimate strain as a function of the recovery enthalpy (δH) for PET aged at 60°C, with crystallinity rate ranging between 4.6% and 26.6% (adapted from Zhou et al. (2007)). For a color version of this figure, see www.iste.co.uk/richaud/industrialpolymers1.zip*

In their study, Ophir et al. (1978) noted a decrease in the ultimate stress of epoxies with increasing aging time. Zhou et al. (2007) have shown, via traction tests, a decrease in the ultimate elongation with increasing aging, as illustrated in Figure 6.19.

6.3. Experimental methods for studying physical aging

6.3.1. *Differential scanning calorimetry*

Many studies have been conducted by DSC (Hodge 1994; Zhao et al. 2001; Borde et al. 2002).

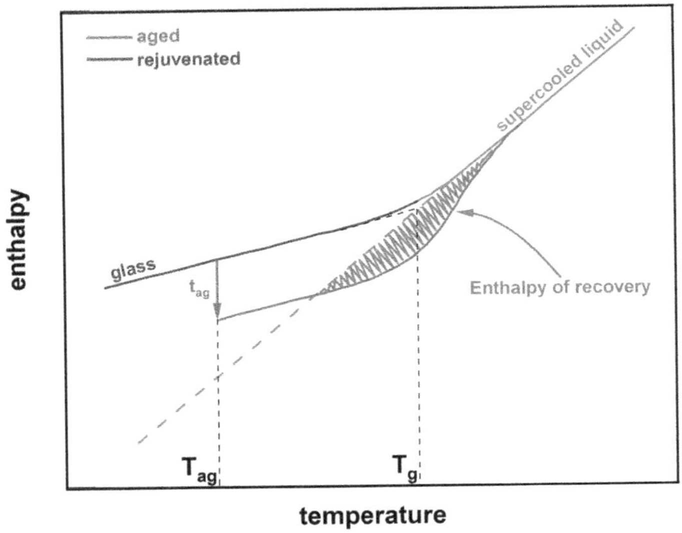

Figure 6.20. *Schematic illustration of the evolution of enthalpy during glass formation, physical aging and the recovery process (according to Quélennec (2022)). For a color version of this figure, see www.iste.co.uk/richaud/industrialpolymers1.zip*

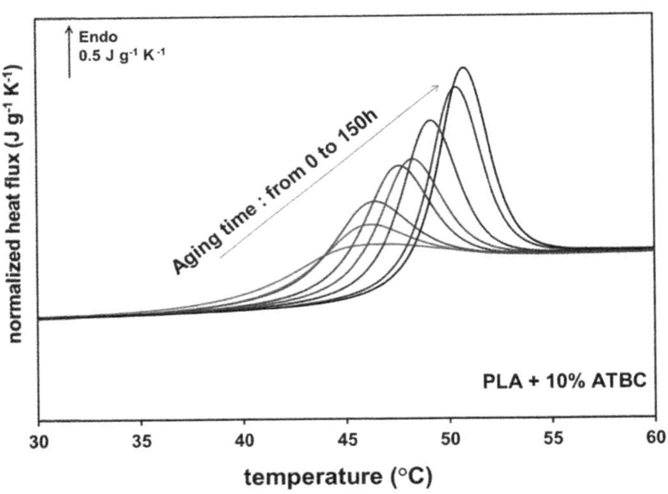

Figure 6.21. *Evolution of the DSC signature with aging time in the case of a plasticized PLA with 10% by mass of acetyltributylcitrate (adapted from Dobircau et al. (2015)). For a color version of this figure, see www.iste.co.uk/richaud/industrialpolymers1.zip*

DSC can be used to study the manifestation of aging by structural relaxation. The return to the liquid state of an aged system is delayed upon subsequent heating because of the lower energy state it reached (Figure 6.20).

Therefore, when an isothermal treatment is applied to a glass, a peak superposed to the glass transition appears upon subsequent heating. This endothermal peak is related to the enthalpy of recovery and indicates the excess energy released during aging. Figure 6.21 illustrates the evolution of the calorimetric signature with aging time.

The enthalpy of recovery of aged glass at a given temperature at time t is calculated by integrating the difference in heat capacity between aged glass and rejuvenated glass according to the following equation (Hodge 1994):

$$\Delta H(T_{ag}, t_{ag}) = \int_{T_1}^{T_2} [C_p^{ag}(T) - C_p^{r}(T)].dT \qquad [6.26]$$

where C_p^{ag} and C_p^r are the heat capacities of aged and rejuvenated glass, respectively, and T_1 and T_2 are arbitrary temperatures surrounding the glass transition temperature. At these temperatures, $C_p^{ag} = C_p^r$. The evolution of the fictive temperature is also a means to follow the kinetics of the enthalpy of recovery (Figure 6.22).

During physical aging, the fictive temperature evolves between two values, T_f' and T_{ag}. At zero aging time, the fictive temperature corresponds approximately to the T_g, except that it is determined at heating. This temperature is referred to as the limit fictive temperature T_f'.

The fictive temperature can be calculated with the equivalent areas method (equation [6.27]) developed by Moynihan et al. (1976b):

$$\int_{T_f}^{T_2} (Cp, l - Cp, g)dt = \int_{T_1}^{T_2} (Cp - Cp, g)dT \qquad [6.27]$$

where Cp, l is the heat capacity of the liquid, Cp, g is the heat capacity of the glass and Cp is the heat capacity of the sample. T_1 and T_2 are two arbitrary temperatures selected in the glassy state and in the liquid state, respectively, so that $Cp = Cp, g$ at T_1 and $Cp = Cp, l$ at T_2.

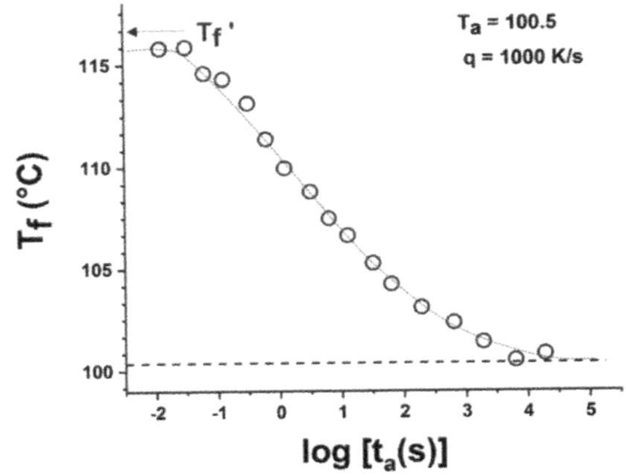

Figure 6.22. *Evolution of the fictive temperature with aging time in the case of a PS (adapted from Koh et al. (2016))*

The total enthalpy recovery ΔH_∞, expected after extrapolation from the line of the supercooled liquid, depends on the aging temperature T_{ag}, the glass transition temperature T_g, and the heat capacity of the glass. It is calculated based on the following equation:

$$\Delta H_\infty = \Delta C_p(T_g - T_{ag}) \qquad [6.28]$$

The isochrone representation of aging by structural relaxation is a function of the aging temperature.

Figure 6.23 illustrates, for three aging times, the typical signature of structural relaxation (Gunawan et al. 2006; Boucher et al. 2017; Perez-De-Eulate and Cangialosi 2018; Monnier et al. 2021), namely, an increase in the enthalpy of recovery by getting farther from glass transition, until reaching a maximum, then a more gradual decrease toward low temperatures (Figure 6.23(a)).

The opposite is observed by monitoring the fictive temperature (Figure 6.23(b)), namely, a decrease in the fictive temperature from the glass transition until it reaches a minimum, followed by an increase toward low temperatures.

The isothermal representation of aging by structural relaxation is illustrated in Figure 6.24. It is the time evolution of the enthalpy of recovery at a given aging temperature.

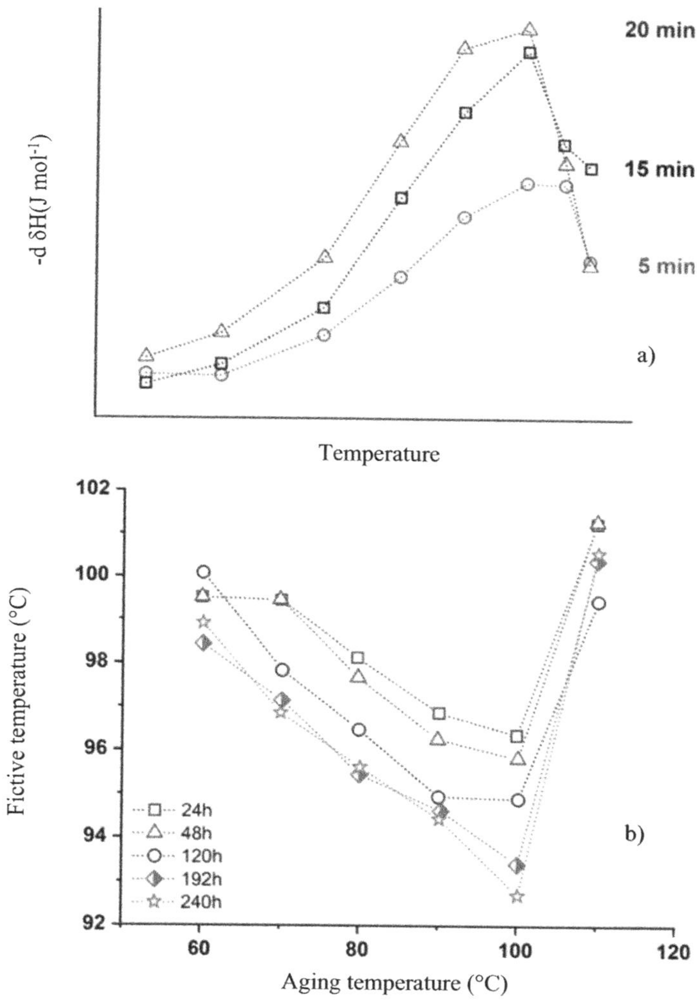

Figure 6.23. *(a) Profile of the enthalpy of recovery of aged acetaminophen for 5, 15 and 20 min as a function of temperature (adapted from Gunawan et al. (2006)); (b) variation in the fictive temperature of an aged epoxy resin for 24, 48, 120, 192 and 240 h at various temperatures (adapted from Fraga et al. (2003)). For a color version of this figure, see www.iste.co.uk/richaud/industrialpolymers1.zip*

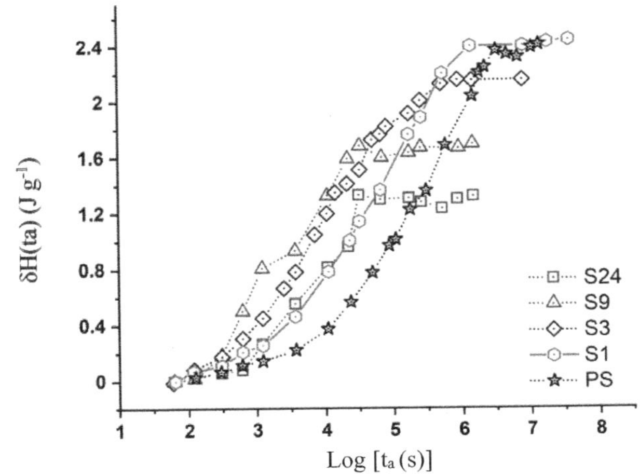

Figure 6.24. *Variation in the enthalpy of recovery of PS systems (adapted from Cangialosi et al. (2012)). For a color version of this figure, see www.iste.co.uk/richaud/industrialpolymers1.zip*

6.3.2. *Fast scanning calorimetry*

The study of aging by relaxation by differential calorimetry has rapidly shown its limits. Two temporal limits have appeared. The first limit is the quantification of aging for periods of time less than 1 min. The second limit is the quantification of aging far below the glass transition. Indeed, the scale of time required to reach equilibrium is not accessible.

The development of fast scanning calorimetry (FSC) is a major progress in overcoming this obstacle. Indeed, FSC can be used to apply scanning rates on the order of 1,000 K/s to nanometric samples (Schick and Mathot 2016). By means of this recent technology, polymer crystallization (van Drongelen et al. 2013), melting (Mathot et al. 2011) and the properties of amorphous polymers (Cebe et al. 2015) have been studied. Studies of the enthalpy of recovery by fast scanning calorimetry are consistent with those conducted by DSC after taking into account the static and dynamic thermal lag (Monnier et al. 2017a).

Fast scanning calorimetry can be used to study the starting point of aging, which is limited by the time scale of the experimental response. Most calculations of the enthalpy of recovery using standard DSC are not able to accurately determine the variations in enthalpy for aging times less than 1 min (Hutchinson and Kumar 2002;

Echeverría et al. 2003; Li and Simon 2006; Málek et al. 2009; Boucher et al. 2011a; Sato and Sprengel 2012; Cangialosi et al. 2013a; Koh and Simon 2013). Indeed, it is difficult to measure a small variation in enthalpy, not to mention the uncertainties related to the thermal history of the sample and to the thermal gradient in the sample. The relaxation times associated with higher temperatures are too short compared with the response time of standard DSC instruments (Amanuel et al. 2008; Boucher et al. 2011b, 2012a; Guo et al. 2011; Cangialosi et al. 2012; Ramakrishnan et al. 2016). To remedy this, Koh et al. (2015) used FSC to study polystyrene over a wider range of temperatures (up to T_g + 15°C) and shorter aging times (up to 0.01 s). High scanning rates were used to reach equilibrium and calculate the enthalpy of recovery at reasonable time scales for temperatures far from the glassy state. It was possible to study physical aging at a range of temperatures of 50°C below the glass transition temperature and over six decades in time (Koh et al. 2016).

6.3.3. *Broadband dielectric spectroscopy*

DRS uses the motion of charges under the effect of an electric field to study dielectric properties. The structural evolution of a system is quantified by the variation in the dielectric permittivity as a function of time or frequency.

In the literature, the study of aged systems revealed a modification of the dielectric spectrum: a shift in the α relaxation toward lower frequencies, as well as a decrease and shift in β relaxation.

6.3.3.1. *Modification of α relaxation*

During physical aging, the α process is slowed down, because of densification, which leads to a shift in the dielectric spectrum toward low frequencies. Historically, Nozaki was the first to monitor the evolution of α relaxation during the aging of PVAC (Nozaki and Mashimo 1987). Schlosser and Schönhals (1991) confirmed these observations.

Investigations have focused on other polymers (Boucher et al. 2012b) and small molecules (Leheny and Nagel 1998; Lukenheimer et al. 2005; Bhardwaj and Suryanarayanan 2012). When the aging time increases, the imaginary part of the permittivity ε″ decreases for a given frequency, as illustrated in Figure 6.25. This result, observed for aging temperatures near the glass transition, is associated with the slowing down of alpha relaxation (Wypych et al. 2005; Grigoriadi et al. 2019a).

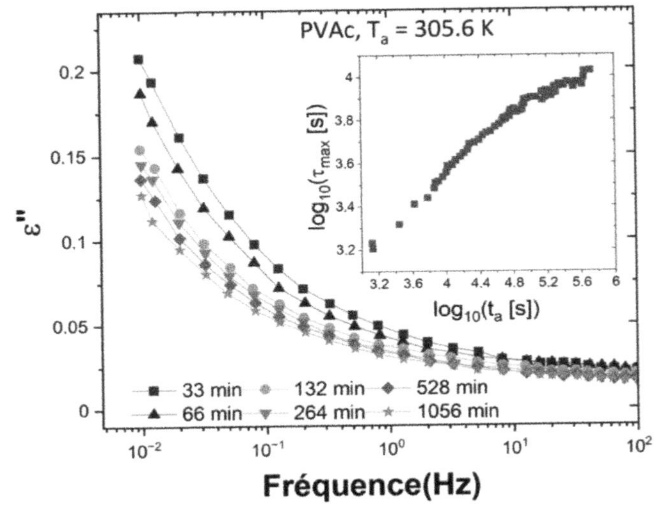

Figure 6.25. *Evolution of ϵ" as a function of the frequency for various aging times, t_{ag}, for PVAC at an aging temperature of T_{ag} = 305.6 K. The insert shows the evolution of the relaxation time as a function of the aging time. The latter is obtained by adjusting the α relaxation via the empirical equation of Havriliak-Negami (adapted from Boucher* et al. *(2012b)). For a color version of this figure, see www.iste.co.uk/richaud/industrialpolymers1.zip*

6.3.3.2. *Modification of β relaxation*

Glass densification upon physical aging impacts secondary relaxation, provided that the latter is correlated with changes in volume. Glass densification drives a decrease in the dielectric strength of β relaxation ($\Delta\varepsilon_\beta$), as shown in Figure 6.26. Haddad and Goldstein (1978) reported that an increase in aging time, and therefore in densification, led to a decrease in the dielectric strength (Haddad and Goldstein 1978; Casalini and Roland 2011).

Two interpretations are possible. This decrease in $\Delta\varepsilon_\beta$ could be explained by the heterogeneous localization of β relaxation in low-mobility islands when the glass densifies (Johari 1973).

Alternatively, according to nuclear magnetic resonance experiments, the localization of β relaxation would be homogeneous. The decrease in $\Delta\varepsilon_\beta$ would then be generated by the decrease in the amplitude of dipolar fluctuations. A comparison can be made with the behavior in DMA (Frank and Stuart 1968; Petrie 1976).

NOTE.– Importantly, dielectric relaxation spectroscopy can be used to probe mobilities at various scales because of its wide range of accessible frequencies.

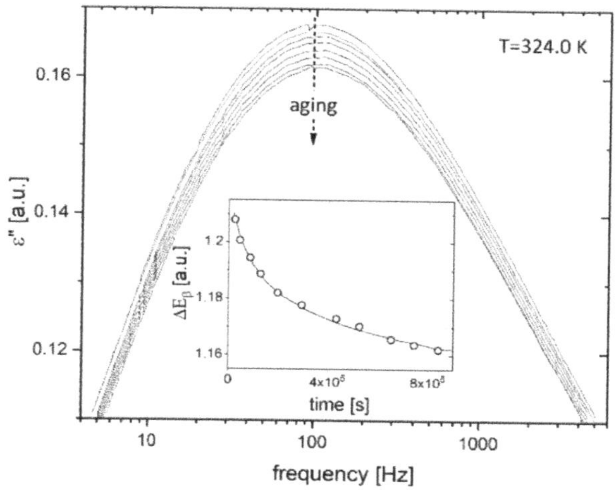

Figure 6.26. *Evolution of the imaginary permittivity of PMMA during aging at 324 K as a function of frequency. The aging time varies from 2.6 x 10^4 to 8.4 x 10^5 s. The insert shows the corresponding evolution of the dielectric strength of the peak during aging (adapted from Casalini and Roland (2011))*

6.3.4. *Thermally stimulated depolarization currents*

Thermally stimulated depolarization currents (TSDC) can be used to monitor α relaxation in the temporal domain. The principle of this method is presented in Figure 6.27. This method can be used to obtain the dielectric response of glass under stable thermodynamic conditions (Alegria et al. 1997).

Under isothermal conditions, glass polarization is performed with an electric field under a linear regime to avoid modifying the glass properties. The depolarization current resulting from field extinction is then observed. Depolarization is conducted at various aging times for short periods of time so that no significant modification of the thermodynamic state occurs.

The Struik method for the mechanical analysis of physical aging is transposed and applied. The Struik method consists of applying constant deformation to a previously aged sample at a predefined temperature and duration. A shift in the characteristic curve of the mechanical compliance as a function of creep time is observed, revealing that the relaxation time depends on the aging time imposed before applying the deformation (Struik 1977b, 1988). Applying this method to the thermostimulated current consists of monitoring depolarization after the application of an electric field.

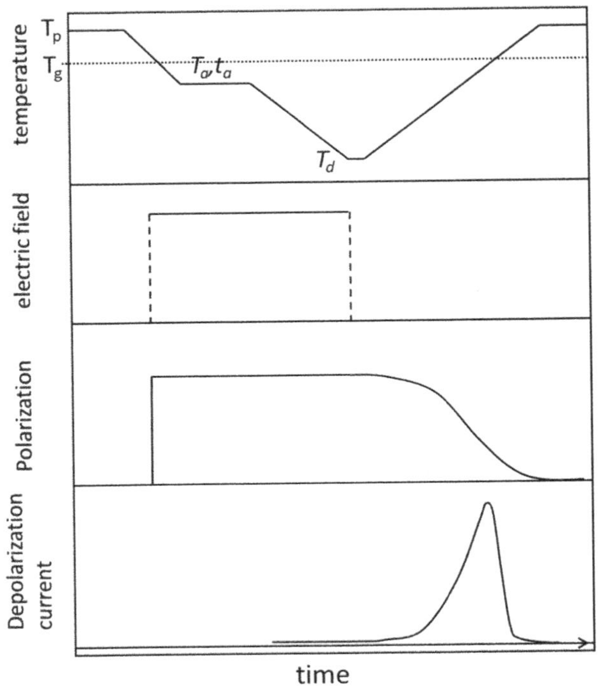

Figure 6.27. *Schematic representation of the temporal evolution of various quantities involved in TSDC measurements conducted to monitor physical aging. T_a corresponds to the aging temperature, ta corresponds to the aging time, T_p corresponds to the temperature of application of the electric field and T_d corresponds to the temperature at which the application of the electric field is stopped (adapted from Boucher et al. (2012b))*

Under non-isothermal conditions, polarization occurs above the glass transition temperature, and then the system is cooled to the temperature at which it undergoes isothermal aging. Depolarization is observed under heating (Schrader and Schonhals 1989; Saiter et al. 2003; Pinto et al. 2009; Arrese-Igor et al. 2014).

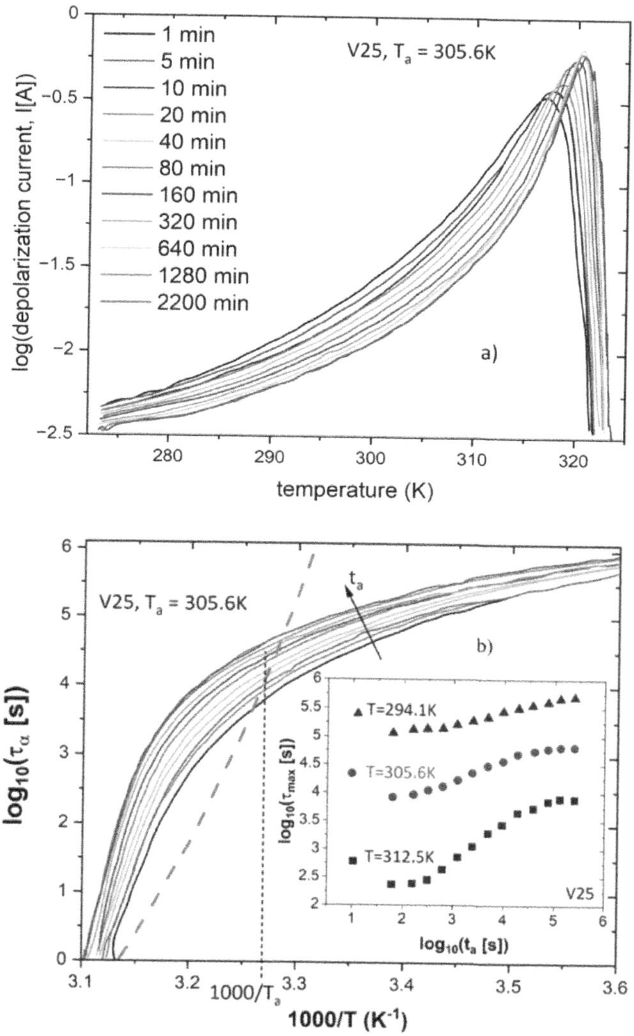

Figure 6.28. *Evolution of the depolarization current as a function of a) temperature and (b) relaxation time with aging time at T_{ag} = 305.6 K for a nanocomposite of PVAC with silica. The insert shows the evolution, with aging time, of the relaxation time at various temperatures after aging at T_{ag} = 305.6 K (adapted from Boucher et al. (2012b)). For a color version of this figure, see www.iste.co.uk/richaud/ industrialpolymers1.zip*

For aging near the glass transition, these two methods can be used to obtain information on the evolution of relaxation in the glassy state during physical aging (Alegria et al. 1996, 1997).

Figure 6.28 shows an example of the application of the non-isothermal method in the case of physical aging of a PVAC-silica nanocomposite.

The depolarization current undergoes a shift toward the highest temperatures during aging (Figure 6.28(a)). The relaxation time, which depends on the temperature and the aging time of the process, is calculated on the basis of the depolarization current (Figure 6.28(b)) (Alegria et al. 1996). Hence, the evolution of the relaxation time at an arbitrary temperature, including that of aging, can be obtained (Figure 6.28(b)).

6.3.5. *Dynamic mechanical analysis*

In DMA, the study of aged systems has shown a modification of the mechanical response: an increase in the storage modulus and a modification of the loss factor.

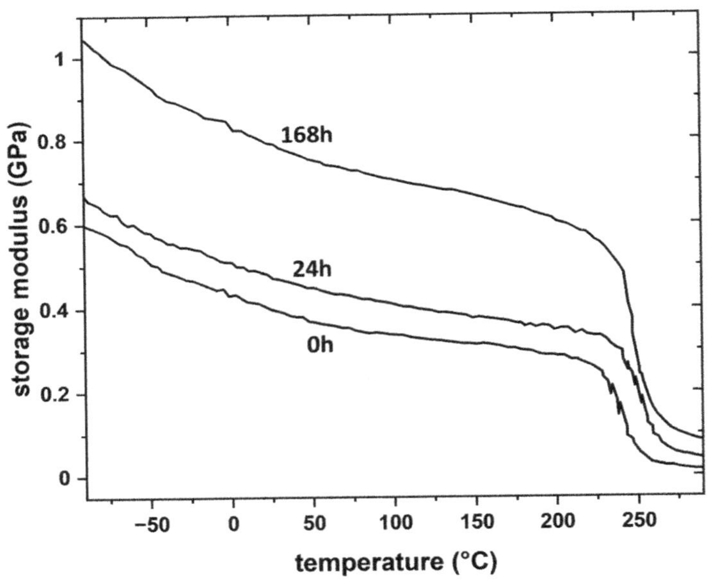

Figure 6.29. *Comparison of the storage modulus as a function of the temperature of TGDDM/EPN/DDS samples aged under isothermal conditions at 220°C between 0.24 and 168 h (adapted from Barral et al. (2000))*

The storage modulus increased with increasing aging time (Figure 6.29). This observation is related to the stiffening of the polymer chain during aging (Davis and Pethrick 1998; Barral et al. 1999, 2000). The decrease in the damping factor with aging time is due to a loss of mobility and of free volume when the system evolves toward equilibrium: the system has a reduced capacity to dissipate the energy. The damping factor also shrinks during aging (Kwon et al. 2010).

As can also be observed by dielectric spectroscopy, the loss modulus shifts toward high temperatures with increasing aging time (Davis and Pethrick 1998), as illustrated in Figure 6.30.

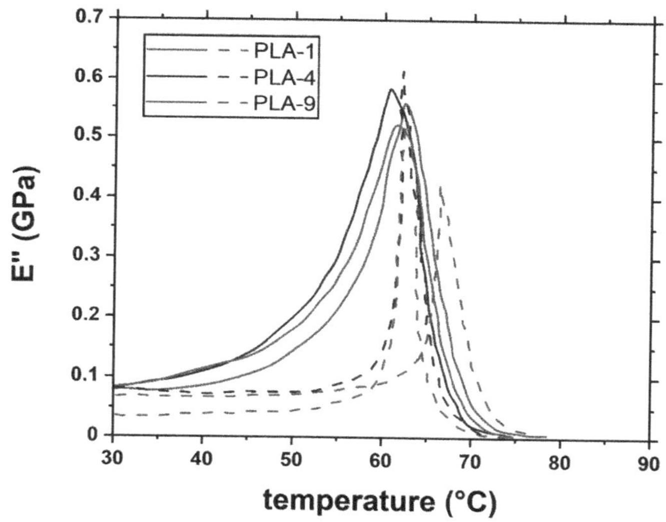

Figure 6.30. *Loss modulus as a function of temperature for aged and unaged PLA films. The solid lines correspond to unaged PLA films, and the dotted lines correspond to films aged at 45°C for 14 days (adapted from Kwon et al. (2010)). For a color version of this figure, see www.iste.co.uk/richaud/industrialpolymers1.zip*

6.3.6. *Nuclear magnetic resonance spectroscopy*

Nuclear magnetic resonance spectroscopy (NMRS) uses the resonance properties of atoms placed in a magnetic field. The magnetic field directs the nuclear spins of the atoms, and then a radio wave excites these spins at the resonance frequency. After excitation, the spins gradually return to the initial state. This relaxation depends on components such as spin-network (interactions of spins with other atoms) and spin-spin (interactions between spins) (Singh and Singh 2022).

The NMR can be used to study the dynamics of a chain by measuring the relaxation of the spin network. The magnetization decrease curve is adjusted by a KWW function, yielding an average relaxation time. An unaged system features a single exponential decrease in magnetization, whereas an aged system features a nonexponential decrease in magnetization (Pan et al. 2008). The simplest case of non-exponential recovery occurs if the spectral density is probed on dynamically non-equivalent sites or if motions or dynamic heterogeneities are present. The decrease in the magnetization values with aging time is attributed to microstructural changes that occur during physical aging. This implies that locally ordered domains are formed during physical aging. The formation of domains that are dynamically nonequivalent to other domains during aging was highlighted by X-ray scattering (Roe and Curro 1983).

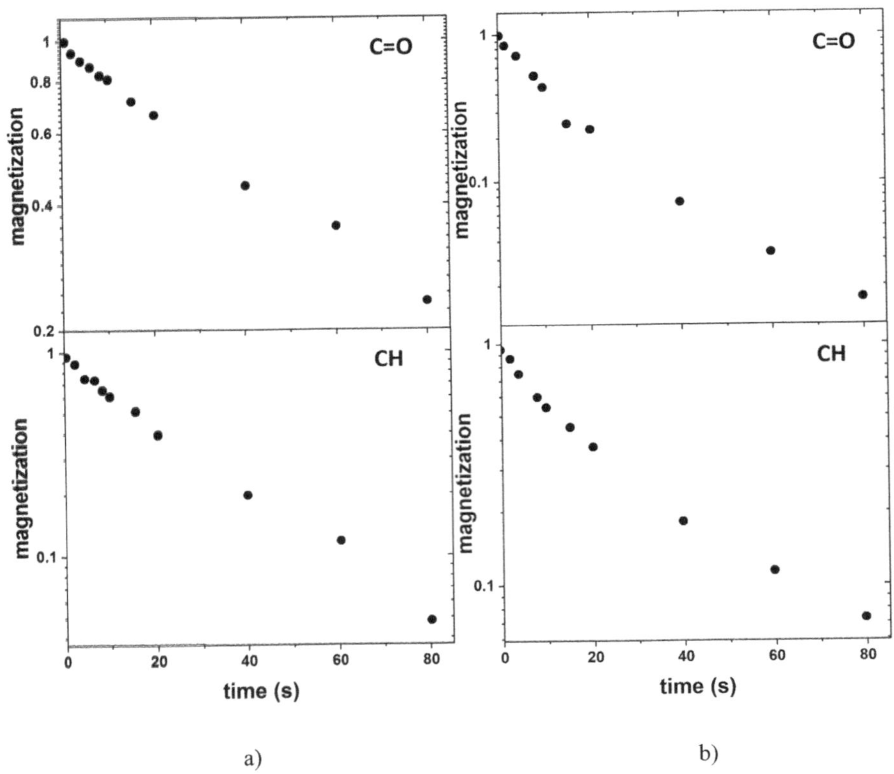

Figure 6.31. *Relaxation curves of the 13C spin network for PLLA carbonyl and methine (a) unaged and (b) aged at 40°C for 360 h. The solid lines are adjusted to the data using the KWW function (adapted from Pan et al. (2008))*

6.3.7. *Positron annihilation lifetime spectroscopy*

When an electron and a positron collide, they are annihilated, and two gamma photons are created. Prior to this annihilation, a quasistable state is formed, referred to as (ortho- or para-) "positronium", whose lifetime barely exceeds 100 ns. Positron annihilation time spectroscopy is based on this lifetime quantification.

This technique is adapted to quantify the free volume and its distribution (Wang et al. 1990; Viret 2003; Varol et al. 2020). The ortho-positroniums (o-Ps) are concentrated in the holes of the free volume. It is theoretically proven that the lifetime of positronium is proportional to the diameter of the hole. In his work on polycarbonate, Cangialosi et al. (2003) showed that the intensity of the probability of the presence of o-Ps decreases with aging time, reflecting a decrease in free volume (Figure 6.32).

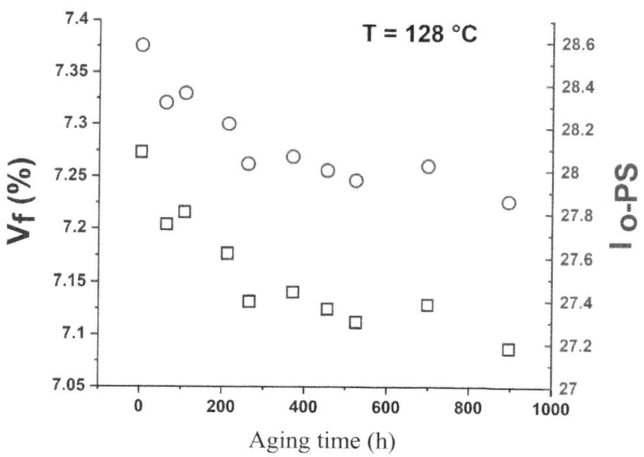

Figure 6.32. *Decrease in the intensity of annihilation of o-Ps (circles) and of free volume (squares) for samples of polycarbonate aged at 128°C (T_{ag} = T_g – 20°C) (adapted from Cangialosi et al. (2003)). For a color version of this figure, see www.iste.co.uk/richaud/industrialpolymers1.zip*

6.3.8. *Ellipsometry*

Ellipsometry is an optical technique for surface analysis based on the measurement of the change in the polarization state of light after reflection on a flat surface. Fresnel equations can be applied to determine the characteristic parameters of the polarization state of reflected light. Knowing the ratio of the reflection coefficients for polarizations parallel and perpendicular to the incidence plane, it is possible to deduce the ellipsometric angles characterizing the differences in amplitude and in phase with respect to the incident light. Models dividing matter into layers can then be used to determine optical constants such as the refractive index. The physical aging of thin films of polymers can be monitored by changes in the refractive index. The Lorentz–Lorenz equation (Huang and Paul 2007) is used to link the refractive index and the density of the polymer:

$$L = \frac{n^2 - 1}{n^2 + 2} = \frac{\rho \mathcal{N}_a \alpha}{3 M_0 \varepsilon_0} \quad [6.29]$$

where n is the refractive index, $\mathcal{N}_a$ is Avogadro's number, M_0 is the molar mass of the repeating unit of the polymer, ε_0 is the vacuum permittivity, ρ is the density of the polymer and α is the average polarizability of the repeating unit of the polymer.

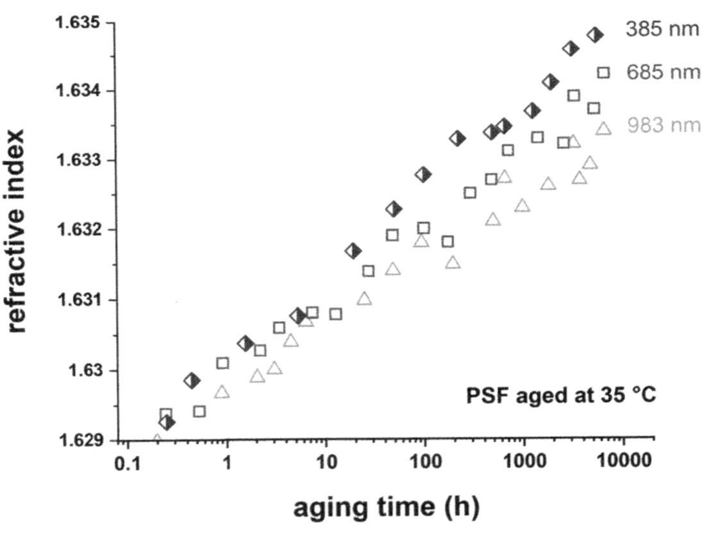

Figure 6.33. *Refractive index measured at 25°C for PSF thin films as a function of aging time at 35°C (adapted from Huang and Paul (2006)). For a color version of this figure, see www.iste.co.uk/richaud/industrialpolymers1.zip*

In their study, Huang and Paul (2006) have shown that the refractive index increases linearly with the logarithm of the aging time (Figure 6.33), confirming Robertson's observations from refractometry measurements (Robertson and

Wilkes 1998). The linear evolution is similar to that of density measurements (Tant and Wilkes 1981; Greiner and Schwarzl 1984). Consequently, the variation in the refractive index can be linked to that of the density.

6.3.9. *Raman spectroscopy*

The Raman effect is characterized by an inelastic light scattering effect. When light scatters through a sample, most scattered photons have the same energy as the incident photons. This is Rayleigh elastic scattering. During Raman inelastic scattering, energy transfer occurs, and the scattered photons have a lower energy (Stokes effect) or higher energy (anti-Stokes effect). This inelastic scattering phenomenon typically impacts vibrational levels, but Raman spectroscopy can also be used to characterize the phenomena impacting rotational levels.

Raman spectroscopy can be used to study the changes in the polymer structure at the nanometric scale via the Boson peak. The Boson peak mainly reflects the excess density of low-energy vibrational states compared with that predicted by Debye's theory. It appears as a low-frequency shouldering of the vibration band (Wypych et al. 2005).

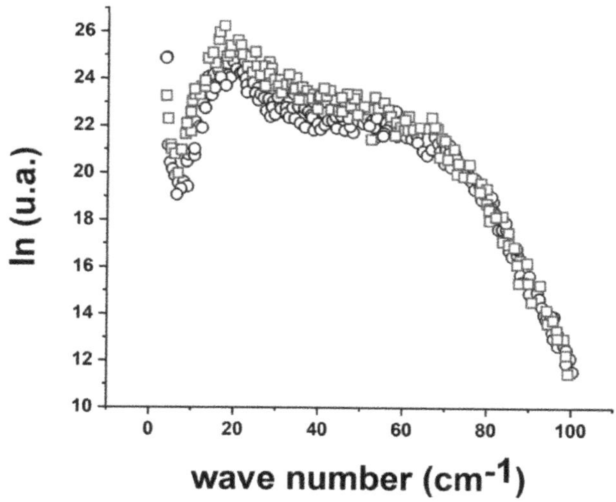

Figure 6.34. *Comparisons of the normalized LFRS (low-frequency Raman scattering) spectra of a reference sample and a sample annealed for three weeks at 95°C. The red squares and blue circles correspond to reference samples and aged samples, respectively (adapted from Wypych et al. (2005)). For a color version of this figure, see www.iste.co.uk/richaud/industrialpolymers1.zip*

The effect of aging is a decrease in the intensity of the Boson peak, which is attributed to a decrease in the degree of disorder of the polymer nanostructure. The effect of aging on the Boson peak is stronger when the aging temperature decreases (Duval et al. 2006).

6.4. Physical aging in glassy systems

6.4.1. *Thermosetting polymers*

Thermosetting polymers are mainly used for high-performance and/or long-term applications in the fields of construction (structural materials, coatings, etc.) and transportation (structural parts, structural elements in the aerospace industry, etc.). Aging by structural relaxation may cause mechanical degradation, which may have negative consequences for the lifetime of the product during its application.

Many studies have been conducted on the impact of cross-linking density on physical aging using calorimetric analyses. These works show that when the cross-linking density decreases, the endothermal peak temperature decreases (Hutchinson et al. 1996; Cook et al. 1999; Morancho and Salla 1999; Kawakami and Watanabe 2008), the fictive temperature decreases (Choy and Plazek 1986; Plazek and Frund 1990), the enthalpic recovery rate increases (Hutchinson et al. 1996), the total recovery enthalpy increases (Cook et al. 1999) and the time required to reach structural equilibrium decreases (Lee and McKenna 1988, 1990).

These observations can be explained by the greater number of molecular rearrangements available for low cross-linking densities. As there are even more available molecular motions, the capacity of the network to reduce its free volume and its conformational enthalpy by relaxation is enhanced. The network topography, which is driven by the cross-linking density, reduces the number of changes in molecular conformations produced during aging, which is consistent with the slowing down of the relaxation kinetics. Lee and McKenna (1988) applied a superposition principle conversion–temperature–aging time to describe the influence of the cross-linking density on the behavior of epoxy under physical aging.

6.4.2. *Amorphous thermoplastic polymers*

Many works have focused on the physical aging of amorphous thermoplastic polymers. In particular, there are studies on the impact of molar mass, tacticity, macromolecular architecture and plasticization.

Some works highlighted that an increase in molar mass reduces the enthalpy of recovery (Andreozzi et al. 2005; Soma et al. 2012). Indeed, the greater the molar mass is, the lower the mobility due to chain entanglement. Moreover, the dependence of physical aging on molar mass is linked to the molar mass between two physical nodes of entanglement. This mass between entanglements (M_e) is equal to half of the critical mass. The critical mass corresponds to the molar mass beyond which the viscosity varies with the mass according to a power law with an exponent of 3.4. Figure 6.35 illustrates the dependence of the enthalpy relaxation rate R_H on the molar mass. The enthalpy relaxation rate is defined as the inflection point of the curve ΔH (T_{ag}, t_{ag})/Δt_{ag}. A sudden drop in the enthalpy relaxation rate when approaching the mass between entanglements is observed (Málek 1998).

A study on the prediction of aging for polymers with high molar masses revealed that the prediction by models overestimated the state of relaxation reached, which suggests that entanglements reduce the mobility of chains during the relaxation of polymers (Andreozzi et al. 2005).

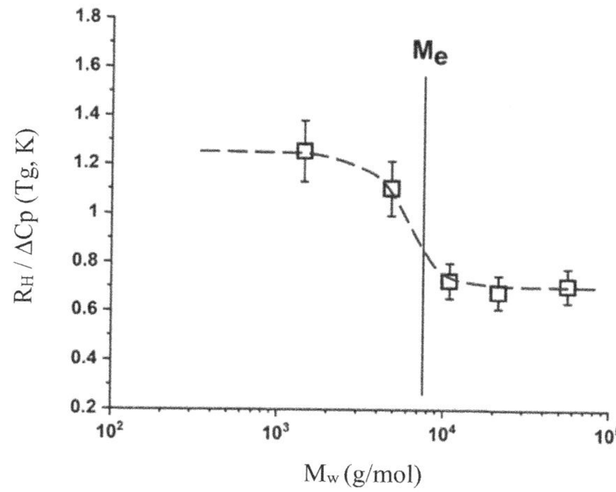

Figure 6.35. *Enthalpy relaxation rate (R_H) (divided by the jump in heat capacity at the glass transition) as a function of the molar mass for samples of PMMA aged at $T_{ag} = T_g - 4$ K. The dashed line is a guide for the eyes (adapted from Andreozzi et al. (2005))*

The effect of tacticity on physical aging has also been studied. A syndiotactic polymer features slower enthalpy recovery kinetics than the same atactic or isotactic polymer (Grigoriadi et al. 2019b). This result is presented in Figure 6.36. Chain flexibility differs depending on polymer tacticity. This can explain the observed difference in kinetics. This statement relies on a study on small-angle neutron scattering, concluding that the chains of syndiotactic polystyrene are stiffer than the chains of atactic and isotactic polystyrene (Takebe et al. 2009). Syndiotactic sequences stiffen the polymer chain (Woo and Chang 2011) and slow down the reorganization of chains. Therefore, a slower aging kinetics is observed.

The study of star-shaped polystyrene highlighted the influence of macromolecular architecture on the kinetics of physical aging (Frieberg et al. 2012a, 2012b, 2017; Green et al. 2015). Compared with analogous linear polymers, star-shaped polystyrene with functionalities ranging between 2 and 16 features slower aging kinetics (Frieberg et al. 2020). The relaxation rate decreases when the functionality increases for molar masses comparable between systems and identical arm lengths in the same system (Figure 6.37(a)). Consequently, the increase in functionality reduces chain mobility. Furthermore, relaxation is slower for molecular segments near the nucleus or for branching points of star-shaped polymers. This can be observed through the slowing of the aging kinetics for shorter arm lengths with identical functionality (Figure 6.37(b)). This effect widely outperforms the differences in kinetics induced by the difference in molar mass by itself, as proven by the behavior of a linear PS (Figure 6.38). It can be deduced that the density of monomers is greater in the region closer to the nucleus and that it increases with functionality (Uhrig and Mays 2005).

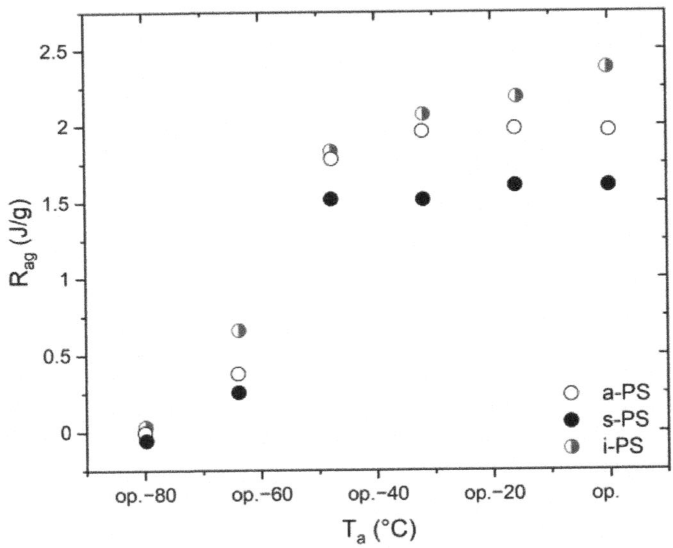

Figure 6.36. *Aging rate ($R_{ag} = d(\Delta H(T_{ag}, t_{ag}))/d(\ln(t_{ag}))$) as a function of the aging temperature for isotactic PS (half-filled circles), syndiotactic PS (filled circles) and atactic PS (empty circles) (adapted from Grigoriadi et al. (2019b))*

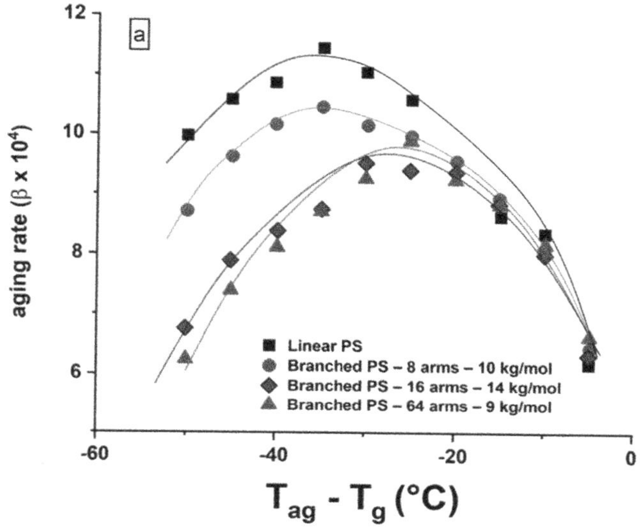

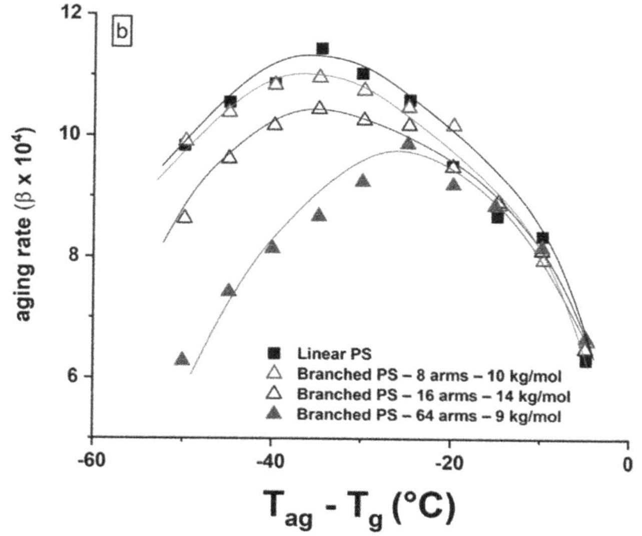

Figure 6.37. *Aging rates were calculated from volumetric experiments as a function of the aging temperature. (a) Influence of functionality; (b) influence of the molar mass of the arms (adapted from Frieberg et al. (2020)). For a color version of this figure, see www.iste.co.uk/richaud/industrialpolymers1.zip*

The length of the polymer side chains has an effect on physical aging. A study of 3-polymethyl(α-n-alkyl) acrylate revealed that an increase in the length of the side chain accelerates the kinetics of enthalpic recovery. Indeed, an increase in the length of side chains leads to an increase in the distance between the main chains and a decrease in the force of interchain interactions (Godard et al. 1998).

The addition of a plasticizing agent to the polymer weakens the attraction between the polymer chains. Therefore, the mobility of the polymer chains increases (Kucera et al. 2013). A decrease in the glass transition is generally observed. Adding plastic to a polymer accelerates the kinetics of physical aging. The increase in chain mobility explains this acceleration (Avolio et al. 2015; Dobircau et al. 2015).

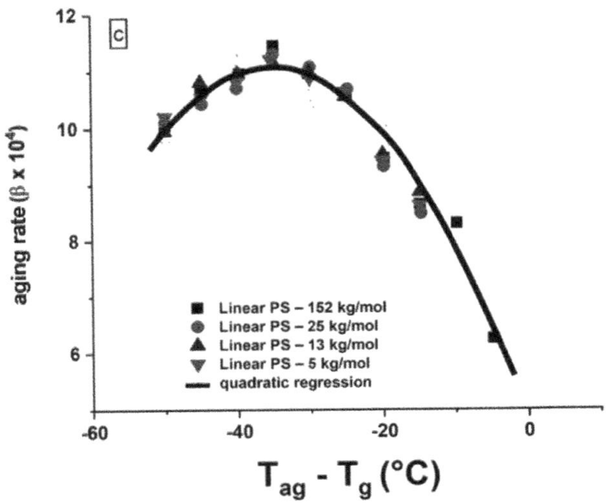

Figure 6.38. *Aging rates were calculated on the basis of volumetric experiments as a function of the aging temperature. Comparison of the influence of the molar mass of a linear polymer (adapted from Frieberg et al. (2020)) (continuation). For a color version of this figure, see www.iste.co.uk/richaud/industrialpolymers1.zip*

6.4.3. *Semicrystalline polymers*

In semicrystalline polymers, the amorphous and crystalline regions coexist. The presence of crystalline regions induces stress on the amorphous regions, which reduces the mobility of the latter. It is thus observed a slowing down in the physical aging with the increase in the crystallinity rate (Tant et al. 1981; Beckmann et al. 1997; Zhao et al. 2001; Krishnaswamy et al. 2003; Zhou et al. 2007; Righetti et al.

2017). The signature of physical aging was observed beyond the glass transition by Struik. The origin of this observation can be a strong slowing down of the mobility of the amorphous region (Struik 1989a, 1989b), as well as the consequence of the heterogeneity of the relaxational dynamics (Monnier et al. 2017b).

Figure 6.39 shows an example of the effect of physical aging of a semicrystalline polymer system impacted by the crystallization conditions. The lower the crystallization temperature, the higher the enthalpy of recovery. This might be explained by the higher crystallinity rates after high-temperature crystallization. The study interprets this result as an effect of the microstructure.

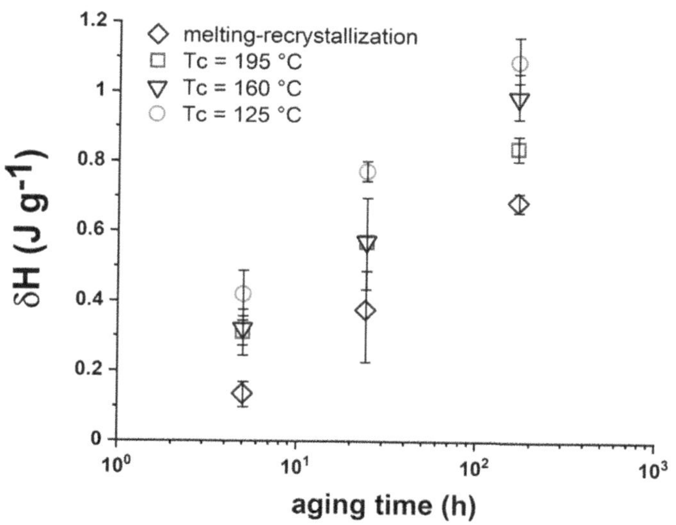

Figure 6.39. *Enthalpy of recovery (δH) plotted as a function of the aging time for four samples of semicrystalline polyphenylene sulfide. The aging was conducted at $T_{g\ onset}$ –15°C (adapted from Krishnaswamy et al. (2003)). For a color version of this figure, see www.iste.co.uk/richaud/industrialpolymers1.zip*

Moreover, the amorphous regions can be divided into two fractions: the rigid amorphous fraction (RAF) and the mobile amorphous fraction (MAF) (Suzuki et al. 1985). This subdivision depends on the degree of coupling with the crystalline region, which impacts the mobility of the polymer chain (Wunderlich 2003). RAF is present in most semicrystalline polymers (Di Lorenzo and Righetti 2018) but also exists in polymer systems, with many interfaces presenting alterations of the glass dynamics formation (Cangialosi et al. 2016; Lee et al. 2017). The contribution of the

RAF to physical aging was studied on semicrystalline PLLA: the RAF contributes to physical aging by local movements at a small scale (Righetti et al. 2017).

In their works on polyethylene terephthalate (PET), Montserrat and Cortés (1995) reported a double peak of recovery for samples of low crystallinity and a single peak of recovery for samples with high crystallinity. Depending on the crystallinity rate, the volume occupied by spherulites is not the same. When the crystallinity is low, the spherulites do not occupy the entire volume. Hence, two amorphous regions with different mobilities can exist. Two recovery peaks can then be observed, one corresponding to the contribution of the interspherulitic amorphous material and the other to the intrasherulitic amorphous material. In contrast, when the crystallinity is high, the spherulites occupy all of the available space. A single peak can then be noted, characteristic of the relaxation of the intrasspherical amorphous material. Similar observations have been made by Delpouve et al. (2014) on polylactic acid.

Crystallization under stress occurs when the macromolecules are ordered under deformation of some polymers and elastomers. The orientation of macromolecular chains combined with the presence of a crystalline phase impacts physical aging. Indeed, Lizundia et al. (2016) reported that the enthalpy of recovery of a PLLA decreases with increasing stretching ratio. Stretching also causes a shift in the calorimetric signature of the glass transition. This is associated with the stress of the chains induced by the orientation, as well as by the increase in RAF. Therefore, the crystallization induced by stretching would delay the structural relaxation.

6.4.4. *Copolymers*

Copolymers can be classified depending on the layout of the constituent units along the polymer chain (Jenkins et al. 1996). Linear copolymers are constituted of a single main chain and include alternate copolymers, statistical copolymers, periodic copolymers and random copolymers. Branched copolymers are constituted of a single main chain with one or more polymer side chains and include block copolymers or graft copolymers.

The physical aging of various distributions of sequences with random, alternate and block copolymers was studied on the basis of polyether sulfone and poly(1,4-phenylene ether-ether-sulfone). These works showed that the alternate copolymer has a faster relaxation process to reach equilibrium and a greater mobility than random and block copolymers. This is explained by the positioning of the ether bond, which allows, during aging by structural relaxation, a rearrangement toward a more ordered structure than that of random and block copolymers (Zhou and Jiang 2006).

Thus, no generalization of the behavior of copolymers during physical aging is possible. Indeed, the response of copolymers depends on the conditions in which the constitutional units are combined.

6.4.5. *Composites and nanocomposites*

Studies on fiber-reinforced composites showed that the addition of fibers and the ratio of fibers had no significant effect on physical aging. The responses of the composites from these studies are similar to those of the pure matrix (Issoupov et al. 2002; Leveque et al. 2005; Stocchi et al. 2006; Hu 2007). No influence on the resistance/strength of the fiber/epoxy interface has been noted (Mendelsy et al. 2002). Physical aging did not modify the characteristics of charge transfer of fiber-reinforced composites.

The effect of nanofillers on aging has been widely studied. However, the behavior of nanocomposites is difficult to predict, as there is a disparity between the results of various works.

Studies have shown a reduction and even a disappearance of physical aging with the addition of nanofillers, whereas other studies have shown an acceleration of the aging process with the addition of nanofillers. The materials can be classified according to the influence of the addition of nanofillers on the glass transition. The considered systems are those in which the glass transition temperature increases, those in which the glass transition temperature does not vary and those in which the glass transition temperature decreases.

6.4.5.1. *Nanocomposites featuring an increase in glass transition temperature*

An increase in glass transition, measured at the midpoint of the jump in heat capacity ($T_{g,\ \text{mid point}}$), was noted for the nanocomposites with the addition of fillers. A broadening of the glass transition was observed without any impact on the appearance of the glass transition ($T_{g,\ \text{onset}}$) (Lee and Lichtenhan 1998). The physical aging of these systems proved to be significantly slowed down (Lee and Lichtenhan 1998; Rittigstein and Torkelson 2006) or even suppressed (Priestley et al. 2007). This is due to the reduction in segment dynamics of the polymer by absorption of the polymer at the surface of the nanofillers. The proposed hypothesis is that there is a reduced mobility layer at the polymer/nanofiller interface (Lu and Nutt 2003a, 2003b).

6.4.5.2. *Nanocomposites featuring a decrease in glass transition temperature*

A decrease in the glass transition was noted for nanocomposites with a high content of charges (Boucher et al. 2011c, 2012b; Cangialosi et al. 2012). The study of Cangialosi et al. (2012) on the effect of the surface/volume ratio of the silica charge on physical aging showed an acceleration of the latter in terms of time to reach equilibrium. In these studied systems, nanofillers were coated with a silanized agent to reduce and/or prevent the formation of hydrogen bonds between the nanofiller and the matrix. This clarification explains the difference from the results of the Priestley study (Priestley et al. 2007), which revealed an increase in the T_g and also a slowing down in physical aging for a PMMA/silica system. Similar results have been obtained for PS/silica (silanized) nanocomposites (Boucher et al. 2012b) and PS/gold nanocomposites (Boucher et al. 2011c). In these two cases, a decrease in the glass transition temperature and also an acceleration in physical aging have been observed.

NOTE.– Nevertheless, there is currently no obvious correlation between aging kinetics and the evolution of the glass transition in nanocomposites. Indeed, certain studies have shown invariance in the glass transition temperature combined with an acceleration of physical aging, although the causes of these phenomena have not been clearly identified (D'amore et al. 1999; Amanuel et al. 2008; Boucher et al. 2010).

6.4.6. *Polymers and model systems: chalcogenide glasses*

Chalcogenide glasses are composed of at least one chalcogen element, namely, tellurium, sulfur or selenium. Owing to the simplicity of its structure, selenium glass is a model material for studying the complex behaviors related to structural relaxation (Cortes et al. 1998; Echeverría et al. 2003; Svoboda et al. 2007).

The physical aging in chalcogenide glasses depends on the glass composition. The coordination number per atom (n_c) impacts the response of glass to physical aging. In the case of glasses referred to as "under stress" ($n_c < 3$, dimension of the space), physical aging is visible, unlike glasses under stoichiometric conditions ($n_c = 3$), where no manifestation of physical aging has been detected after 20 years (Golovchak et al. 2009). The closer the coordination number is to 3, the slower the recovery kinetics (Golovchak et al. 2008, 2010).

The literature reports on the existence of a window of composition for which the recovery signature in the non-reversing heat flow signal, measured by MT-DSC, is negligible (Figure 6.40). This window is referred to as a "window of reversibility" by some authors (Chakravarty et al. 2005; Golovchak et al. 2008). The glasses formed in this window of composition do not undergo physical aging. This is explained by their greater degree of organization (Chakravarty et al. 2004; Vempati and Boolchand 2004; Boolchand et al. 2005; Bhosle et al. 2011). However, a recent study on chalcogen glasses by Raman spectroscopy measurements has not validated this hypothesis (King et al. 2021).

The stability of glasses aged for 30 and 40 years, which reached a state of equilibrium, was proven by DSC. With this technique, an enthalpic recovery in two stages was observed for aging far from the glass transition, whereas an enthalpic recovery in a single stage was observed near the glass transition. A phenomenon of crystallization for glasses at equilibrium was noted for aging temperatures near the glass transition (Morvan et al. 2021).

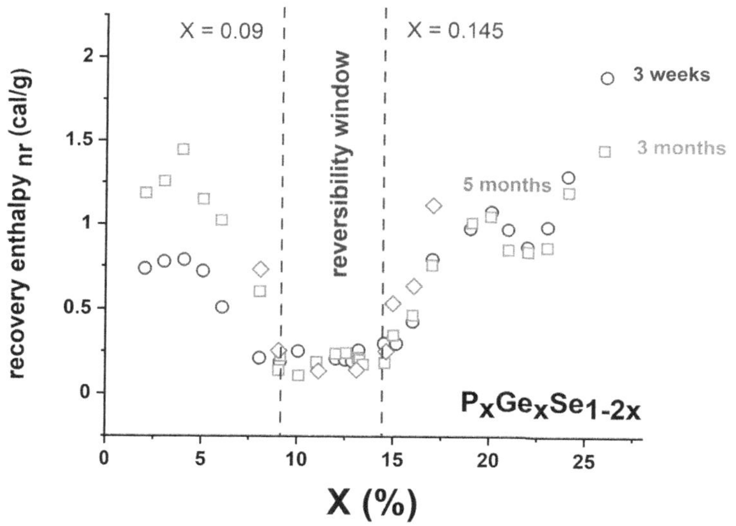

Figure 6.40. *Evolution of the recovery enthalpy calculated on the basis of the non-reversing heat flow signal (ΔH_{nr}) of glasses of composition $Ge_xP_xSe_{1-2x}$ showing the window of reversibility in the zone 0.09 < x < 0.145 (adapted from Chakravarty et al. (2004)). For a color version of this figure, see www.iste.co.uk/richaud/industrialpolymers1.zip*

The composition of chalcogenide glasses, and therefore the coordination number, impacts physical aging. Notably, an increase in the coordination number slows down the aging kinetics. Owing to the recent development of fast scanning calorimetry, it was possible to highlight kinetics in two stages for aging far from the glass transition. These observations, which were conducted on these model glasses, indicate research opportunities for development in the field of polymers: is it possible to establish a correlation between the coordination number and the cross-linking density or between the chalcogenide content and the chemical composition of polymers? Moreover, could two-step kinetics, as well as crystallization following aging by structural relaxation, be observed in polymer systems? These questions are currently being debated in the scientific community and will be addressed in the last part of this chapter.

6.5. For a better comprehension of physical aging

In recent years, the physical aging of glass systems has been the subject of in-depth studies. The results and hypotheses formulated to explain physical aging by structural relaxation are the subject of debate within the scientific community. This last part focuses on the questions being debated through recently conducted works.

6.5.1. *Nature of relaxation dynamics in the glassy state*

From a kinetic point of view, one of the controversies is the divergence or absence of VFT behavior in the glassy state near the glass transition state. Studies have shown departures of VFT behavior from relaxation times at equilibrium, suggesting a milder dependence on temperature at a larger scale (O'Connel and McKenna 1999; Simon et al. 2001; Cangialosi et al. 2004; Hecksher et al. 2008; Boucher et al. 2011a). In contrast, other studies show a divergence of the relaxation time following the VFT empirical law (Wagner and Richert 1997; Richert et al. 2013). Importantly, none of these works address the case of a system aged until equilibrium at more than 15°C below the glass transition.

In their study, Boehm et al. (1981) analyzed a glass with a composition of $Ag_7I_4AsO_4$. The glass with a glass transition temperature of 63°C was aged for one year at ambient temperature to reach equilibrium. To increase the temperature range to 10°C, where equilibrium is reached, several millennia are needed to reach equilibrium.

Hence, this discussion on divergence is not placed in the deep glass regime. Getting data on glasses at equilibrium at temperatures significantly below the glass transition is a major challenge for understanding the glassy state. Experimentally, glasses may require geological time scales to reach equilibrium far below the glass transition determined in the laboratory (Dudowicz et al. 2008; McKenna 2008;

Ediger 2017; Yoon et al. 2017). The objective is to be able to probe the aging of glasses at equilibrium or have the capacity to reach equilibrium within an acceptable time scale compared with the experimental time scale of the laboratory. Glasses of geological age and ultrastable glasses are good candidates to meet this challenge.

Amber is a macromolecular material formed from polymerized vegetal resins fossilized for millions of years (Sprincz and Beck 1981; Lydżba-Kopczyńka et al. 2012). Amber is assumed to be very stable because of its age and can be considered a model for exploring the physical properties of glasses (Pérez-Castañeda et al. 2013, 2014).

A Miocene amber was studied by calorimetry (Zhao et al. 2013). Amber, aged 20 million years, has a density 2.1% higher than that of the thermally rejuvenated material. This material was reported to have a fictive temperature of 43.6°C below the glass transition temperature of the thermally rejuvenated material. This wide range of temperatures can be used to evaluate the validity of behaviors during the glass transition. Mechanical measurements, by applying the Struik method, have shown an adjustment of experimental data to the VFT equation for temperatures above the glass transition, whereas the experimental data depart from the extrapolation of the VFT equation for temperatures below. The parabolic Chaudler model was used to describe the divergence of the dynamics with respect to the VFT behavior (Elmatad et al. 2010). In this model, the T_x parameter describes the temperature transition between the Arrhenius and super-Arrhenius behaviors. Under glass transition, a deviation of the VFT behavior toward another behavior was observed. This low-temperature behavior is likely of the Arrhenius type and has a very high activation energy (McKenna and Simon 2017).

NOTE.– The chemically unstable character of amber raises the question of the potential evolution of the structure over time, which would disturb the study of physical aging. For this reason, these results are still under debate in the scientific community.

The development of ultrastable glasses by physical vapor deposition is a research path for probing subglass dynamics. In his works, Ediger (Kearns et al. 2007; Swallen et al. 2007) created ultrastable glass with high kinetic stability, a low fictive temperature and a density similar to that of extremely aged glass. The difference between the fictive temperature and the glass transition temperature of an ordinary glass of the same composition, prepared by conventional means, is approximately 30°C. Yoon et al. (2017) created an ultrastable Teflon with a fictive temperature of 57°C below the glass transition temperature. This fictive temperature is close to the Kauzmann temperature. The creep experiments, with the creation of a master curve, showed a divergence of VFT behavior, as observed for amber. The experimental data followed the parabolic Chaudler model (Elmatad et al. 2010).

6.5.2. *An equilibration kinetics in several stages*

The scientific community does not agree on reaching the equilibrium line extrapolated upon physical aging. Indeed, studies agree that the equilibrium line is not reached during aging by structural relaxation despite a long aging time (Gómez Ribelles et al. 1995; Andreozzi et al. 2005), whereas other studies have shown that the equilibrium line is reached after a long aging time (Boucher et al. 2011c; Koh and Simon 2013).

However, there are diverging perspectives on how equilibrium is reached. Indeed, studies have shown that data have reached a plateau, which does not correspond to an equilibrium enthalpy but sometimes corresponds to only 20% of the equilibrium enthalpy (Boucher et al. 2010, 2011a). The recovery kinetics is then conducted in two stages (Brunacci et al. 1997; Mcgonigle et al. 2005; Cangialosi et al. 2013b). This observation of polymer systems echoes that of arsenic selenide glass (Golovchak et al. 2012, 2016), selenium glass (Morvan et al. 2021) and inorganic glass, where an equilibrium plateau is reached at 600 K during the glass transition (Welch et al. 2013). This observation is supported by other studies showing that enthalpic recovery is a single-stage process (Zhao and McKenna 2012; Gao et al. 2013; Koh et al. and Simon Gao and Simon 2015; McKenna and Simon 2017). Cangialosi et al. (2013b) mentioned a difference in the molecular weight of the polystyrenes used to explain the difference between studies.

Cangialosi interpreted the kinetics in two stages on the basis of the nature of movements through which physical aging occurs. According to his interpretation, local and cooperative movements coexist; cooperative movements predominate near the glass transition, whereas local movements predominate far from the glass transition. The results obtained for selenium (Morvan et al. 2021) support this interpretation, given the dependence on temperature of the number of stages in the recovery process.

Furthermore, the works of Napolitano (Song et al. 2022) highlighted the existence of a slow Arrhenius process (SAP), a relaxation mechanism that is assumed to be universally identifiable in nanoconfined polymers, detectable by dielectric spectroscopy in the same range of temperature as α relaxation, and characterized by a constant activation energy on the order of 100 kJ. This process is completely differentiated by a secondary relaxation, as its dielectric signature in the imaginary permittivity is a narrower peak than that of α relaxation and therefore much finer than those generally attributed to secondary relaxations. Furthermore, this process is slower than α relaxation when the relaxing system is in a liquid state.

On the other hand, it is faster in the glassy state, which is the reason why it might serve as an activation process of the cooperative mechanism when glass structural relaxation is conducted following several enthalpic decays. In any case, the existence of SAP is a strong argument to state that there are several intermediate scales of mobility preceding the interchain cooperative movement, as it is usually represented, for example, to justify the viscous slowing down.

Experiments with an asymmetric approach present the obtained plateau as a minimum in the energy landscape, suggesting the possibility of other "degradations" of the enthalpic relaxation for long aging times involving different mechanisms for reaching equilibrium (Boucher et al. 2011a). This hypothesis of the contribution of various mechanisms to structural relaxation is shared by Righetti in his study on amorphous and semicrystalline PLLA (Righetti and Mele 2019).

A comparison of the dynamics between the RAF and the MAF highlighted the contribution of the RAF to physical aging (Righetti et al. 2017). These works conclude that, near the glass transition, large-scale cooperative movements participate in structural relaxation. At intermediate aging temperatures, non-cooperative local movements, involving the entire structural unit, act simultaneously with structural relaxation.

Finally, in the deep glassy state, structural relaxation would be exclusively activated by movements at a small scale. These works propose a dependence of aging by structural relaxation on temperature. This idea is shared by Monnier et al. (2020) in a study on PLLA. A 10 min isochrone aging process was conducted on an amorphous PLLA system and systems with different conditions of crystallization. The variation in the recovery enthalpy as a function of the aging temperature is rescaled as a function of the quantity of the MAF. Hence, it appears that at aging temperatures near the glass transition, the enthalpy of recovery depends on the crystallization temperature and therefore on the quantity of the RAF, whereas at aging temperatures far from the glass transition, the enthalpy of recovery is independent of the crystallization temperature and therefore of RAF. In terms of mobility, this can be reflected by the fact that, after the glass transition, RAF and MAF mobilities differ, whereas RAF and MAF mobilities are equal at temperatures far from the glass transition. These works support the dependence of the recovery kinetics on the temperature.

6.5.3. *Sensitivity of aging by structural relaxation to scale effects and the behavior of a confined polymer*

The dimensions of the confined polymers are reduced until they reach the characteristic sizes for which behavioral modifications are observed. Many studies have investigated the effect of confinement on the glass transition. Kremer synthesized these studies. In his review on the confinement of polystyrene, there was no clear link between confinement and the glass transition (Kremer et al. 2015). Indeed, depending on the method used to determine the glass transition or on the nature of confinement (polymers deposited on a substrate or stress by a multilayer system), the latter varies up to 80°C, which indicates the complexity of the study of such systems.

Polymer confinement geometries are classified into three categories: one dimension (thin film), two dimensions (nanotube, nanopore) and three dimensions (nanoparticle, nanosphere and nanocomposite). The number of studies on the physical aging of confined glasses has increased in recent years (Priestley et al. 2009; Cangialosi et al. 2013b, 2016). These studies revealed a decrease in the glass transition temperature of confined glasses compared with that of massive glasses. This implies that confined glass is able to reach equilibrium at lower temperatures than massive materials. Calorimetric techniques show an acceleration to reach the equilibrium for thin films of polymers (Koh et al. 2008; Boucher et al. 2012a), nanocomposites (Amanuel et al. 2008; Boucher et al. 2011b, 2012b; Cangialosi et al. 2012; Ramakrishnan et al. 2016) and nanospheres (Guo et al. 2011; Mathlouthi et al. 2017). Enthalpic recovery acceleration in a confined glass was observed for thin films of toluene (Sepúlveda et al. 2011).

The effect of confinement on the enthalpy of recovery kinetics was studied at confinement scales above several nanometers on the basis of the thickness profiles of thin films. The results confirm that kinetic acceleration extends at distances from the interface above 100 nm (Priestley 2005; Rowe et al. 2009).

Studies on the impact of polymer confinement by a substrate have shown different results. A study on the physical aging of polystyrene confined by a mica substrate did not reveal the influence of a decrease in thickness on physical aging, in contrast to a similar measurement on a free-standing film of polystyrene (Gray et al. 2012).

A study related to the aging of polystyrene confined by a rubber substrate did not show a dependence of the aging rate by structural relaxation on thickness, despite a decrease of 25°C in the glass transition (Rausher et al. 2013). Finally, a study of polystyrene confined by polycarbonate revealed a decrease in the enthalpy of recovery with decreasing thickness of the polystyrene film (Langhe et al. 2012). This observation was explained by the existence of a mixture phase at the interphase of the two films (interdiffusion).

The studies of aging in confined multilayer systems did not show identical behaviors either. Indeed, the results differ depending on the studied systems. In a study of a multilayer polycarbonate and polymethyl methacrylate system, the authors reported that enthalpic recovery was accelerated in the confined polycarbonate, whereas no effect was observed in polymethyl methacrylate (Arabeche et al. 2022).

In a study on a multilayer polystyrene and polylactic acid system, a decrease in the enthalpy of recovery was observed for polylactic acid when it was constrained by polystyrene (Monnier et al. 2018).

6.5.4. *Possibility of crystallization following aging by structural relaxation*

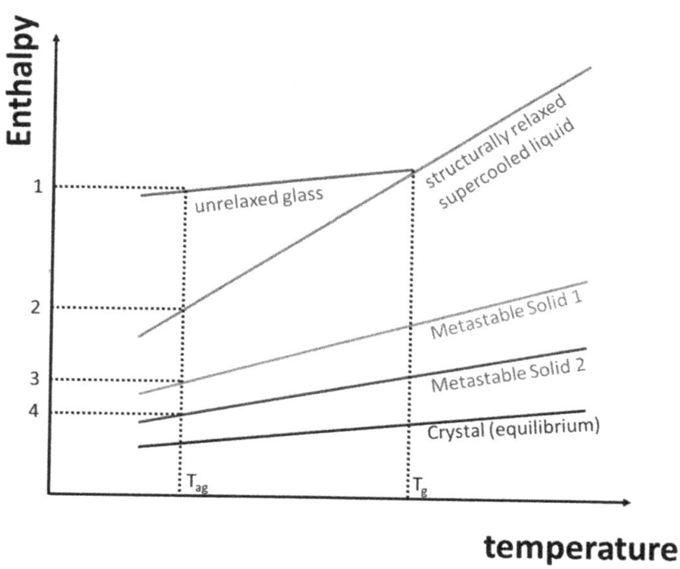

Figure 6.41. *Schematic representation of the enthalpic diagram, illustrating the evolution of various phases during aging at T_{ag} close to the T_g (adapted from Schawe et al. (2022)). For a color version of this figure, see www.iste.co.uk/richaud/industrialpolymers1.zip*

The characterization of crystallization by DSC can be limited by the possibility of crystal reorganization during measurement. FSC is one solution to reduce this inconvenience. Indeed, an increase in the heating rate reduces the possibility of reorganization (Minakov et al. 2004; Schawe 2015). In the glassy state, the crystallization of glasses and semicrystalline polymers is controlled by diffusion and nucleation. During structural relaxation, crystal precursors can be formed. The first reference in the literature for this observation is the study of PET by Yeh and Geil (1967). More recent studies have shown that aging by structural relaxation accelerated the subsequent cold crystallization. Indeed, during structural relaxation, crystal seeds are formed. In their study, Di Lorenzo et al. (2017) showed by calorimetric techniques that crystallization of polybutylene succinate was possible during structural relaxation. The behavior during structural relaxation of several semicrystalline polymers was studied by FSC. The result was that, during aging, the structure is formed in a sequential order: after achieving structural relaxation, homogeneous nucleation seems to be followed by crystal growth (Zhuravlev et al. 2011; Androsch et al. 2014, 2018; Papageorgiou et al. 2014; Androsch and Schick 2016; Quattrosoldi et al. 2020). In a recent study, the influence of the confinement of a nitrile-butadiene hydrogenated copolymer on the crystallization mechanism (Schawe and Wrana 2020) was observed by FSC. An analysis of the crystallization and relaxation shows that the kinetics of both is identical. Schawe proposes that crystallization is therefore not controlled by nucleation but rather by diffusion and ends with the formation of RAF. Indeed, he observed that crystallization in the copolymer causes segregation of non-cristallizable segments in the macromolecule. This leads to a reduction in mobility in the vicinity of the formed crystals and, consequently, an increased quantity of the RAF. The study of a metallic glass showed that, during structural relaxation, several processes occur sequentially (Figure 6.41): structural relaxation, a period of incubation for nucleation, the formation of seeds, the crystallization of a metastable crystalline phase and, finally, an enhancement of the crystalline structure by the growth of seeds reaching a more stable crystalline structure. The period of nucleation describes the stability of the supercooled liquid and its resistance to crystallization. The incubation time of nucleation depends on the temperature. An activation barrier is therefore defined. This finding indicates that at very low temperatures, nucleation may occur in unrelaxed glasses (Schawe and Löffler 2022).

6.5.5. *Predicting the influence of environmental factors on physical aging*

In long-term applications, polymer materials are used as structural and coating materials. To obtain the most realistic possible prediction in relation to experimental data, modifications of the existing prediction models have been proposed. These improvements are presented below.

6.5.5.1. *Modification of TNM models*

From an experimental point of view, structural relaxation is a well-known phenomenon, some of its characteristics being established, such as its reversibility, its non-exponential character and its nonlinearity (Struik 1977b; Hodge 1994; Schapery and Sun 2000). Several phenomenological models have been proposed and previously described (Drozdov 1999, 2001; Gutzow et al. 2000; Schapery and Sun 2000). The TNM model is the most widely used approach to describe nonlinearity. Studies on polymers showed differences between predictions and calorimetric measurements (Scherer 1986; Tribone et al. 1986; McKenna 1989; Cowie and Ferguson 1993; Hodge 1994; Grassia and Simon 2012), particularly overestimations by models of enthalpies during aging (Cowie and Ferguson 1993; Cowie et al. 1996). Following these observations, it is possible to reconsider hypotheses of the TNM model. The manner in which nonlinearity is introduced in the model by the fictive temperature is one of the hypotheses to explain the gaps between measurements and predictions (Moynihan et al. 1991). Several studies have facilitated the development of configurational entropy approaches to extend the TNM model. An improvement was noted when comparing the experimental results and the predictions of the modified model (Gómez Ribelles et al. 1995, 1997; Meseguer Duenas et al. 1997; Andreozzi et al. 2004). This modification is intended to separate the dependence on the equilibrium temperature of the relaxation time from its behavior out of equilibrium. These modifications have limits, as these improvements are only valid for polymers of low molecular weight. The notion of entanglement in polymers with high molecular weight with different segmental and terminal relaxations is excluded (Andreozzi et al. 2003).

More recently, Koh et al. (2015) proposed a modification of the TNM model, assuming that the relaxation time is a different function of temperature and structure. This assumption is rooted in the literature (McKenna et al. 1999; Kolla and Simon 2005; Simon and Bernazzani 2006) and presents this difference between the relaxation time for a given structure and the relaxation time at equilibrium as an uneven symmetric function at the start of equilibrium. This TNM model, thus modified, contains two adjustable parameters, with the others being fixed to correspond to the dependence on the temperature of the shifting factors in the conditions of equilibrium above the glass transition. There is no separate "adjustable" parameter to define the dependence of the relaxation time on the glass structure (as the x parameter in the original TNM model). This modification of the model is even more different from that of the TNM model for the equation of the relaxation time. Indeed, the equation of the relaxation time follows a WLF dependence and does not change into an Arrhenius dependence on temperature under the glass transition, as is the case for various experimental observations (Badrinarayanan and Simon 2007; Hecksher et al. 2008; Zhao and McKenna 2012). This difference, induced by the modification of the model, is problematic for temperatures far from the glass transition (Grassia and Simon 2012) and cannot be used to make reliable predictions in this range of temperature.

These modifications of the TNM model do not provide an accurate and reliable prediction of aging for all glass systems. The development of fast scanning calorimetry is essential for understanding and modeling structural relaxation. On the basis of the enthalpy of recovery data in the deep glassy state, a model of return to equilibrium kinetics can be considered.

6.5.5.2. *Influence of environmental factors*

Polymer materials used as structural coating materials are intended for long-term application. These materials then undergo both chemical (oxidative, UV, and water) and physical (structural relaxation) aging.

Thermo-oxidation and UV-based aging were studied for many polymers. However, to the best of our knowledge, there is no study related to the impact of chemical aging on structural relaxation aging. As previously presented in this chapter, aging by structural relaxation has negative effects on the mechanical properties at the macroscopic scale. It therefore appears essential to understand whether chemical aging impacts structural relaxation and how. Currently, there is no reliable method for time prediction of this type of aging. This might be a limiting factor for applications that are demanding in terms of durability.

6.6. Conclusion

All of the polymers in the glassy state undergo physical aging by structural relaxation. From a thermodynamic point of view, this unfolds identically to what is observed for all glass families, that is, a decrease in excess quantities, such as the enthalpy or specific volume. Nevertheless, due to their macromolecular nature, polymer systems have several structural characteristics that impact the kinetic aspect of relaxation and, consequently, the evolution of macroscopic properties. Hence, in thermoplastic polymers, the hindrance will slow down the aging phenomenon, as chain mobility will be reduced. This may result from the increase in molar mass, from a difference in tacticity, or from the presence of a crystalline phase. In contrast, any structural modification susceptible to induce an increase in flexibility of the polymer chains, such as an increase in the length of the side chain or the addition of a plasticizing agent, accelerates physical aging. The influence of macromolecular mobility is also found in thermoset polymers, where, for example, the resistance and the barrier will evolve all the more slowly as the cross-linking density increases. Notably, when the structure and formulation become more complex, it is far more difficult to predict, even qualitatively, the relaxation kinetics because of the existence of antagonistic effects. This explains why the behavior of copolymers and nanocomposites remains so uncertain.

As already presented in this chapter, many experimental methods can be used to monitor aging by structural relaxation. This study focuses on the disorder degree (Raman spectroscopy) as well as on the free volume (PALS) or the storage modulus (DMA). Calorimetric techniques are still the most commonly used techniques today, as they facilitate a clear examination of the evolution of the system enthalpy while being relatively easy to use. Nevertheless, these calorimetric approaches, like all other empirical approaches, are confronted with the problem of time of analysis. Physical aging at ambient temperature may unfold over geological time scales and is incompatible with laboratory measurements. Therefore, the use of models proves to be quasi-indispensable for predictive goals. There are many models, but the most widespread remains TNM model. As for the concept of fictive temperature introduced by Tool, it is also a major player in studies dealing with aging by structural relaxation.

The combination of experimental methods and theoretical models relies essentially on the hypothesis that aging by structural relaxation unfolds following the same mechanisms at any distance of the glass transition. This hypothesis is highly questionable to the extent that it seems extremely difficult today to consider a description of the phenomenon when it unfolds far from the glassy state. Fast scanning calorimetry offers interesting perspectives in terms of aging acceleration. This contributed to the emergence of elements that respond to two great questions intrinsic to aging by structural relaxation, namely, the existence of various relaxational dynamics in the glassy state and glass crystallization at thermodynamic equilibrium. In the next few years, it can be expected that some of the questions presented in the fifth section of this chapter find more assertive and validated responses owing to the technological progress in high-speed calorimetry.

Therefore, even if our understanding of physical aging is only partial, there has been significant progress in recent years, which is worth mentioning. This newly acquired knowledge will be of primary importance for predicting the behavior of polymers under multiple aging conditions, at a time when the question of the future of plastic materials is increasingly under scrutiny.

6.7. References

Adam, G. and Gibbs, J.H. (1965). On the temperature dependence of cooperative relaxation properties in glass-forming liquids. *The Journal of Chemical Physics*, 43(1), 139–146.

Alder, B.J. and Wainwright, T.E. (1962). Phase transition in elastic disks. *Physical Review*, 127, 359.

Alegria, A., Goitiandia, L., Colmenero, J. (1996). On the interpretation of the TSDC results in the study of the α-relaxation of amorphous polymers. *Polymer*, 37, 2915–2923.

Alegria, A., Goitiandia, L., Telleria, I., Colmenero, J. (1997). Alpha-relaxation in the glass-transition range of amorphous polymers. 2. Influence of physical aging on the dielectric relaxation. *Macromolecules*, 30, 3881–3887.

Amanuel, S., Gaudette, A.N., Sternstein, S.S. (2008). Enthalpic relaxation of silica-polyvinyl acetate nanocomposites. *Journal of Polymer Science Part B: Polymer Physics*, 46, 2733–2740.

Andrade, E.N. da C. (1934). XLI. A theory of the viscosity of liquids – Part I. *The London, Edinburgh, and Dublin Philosophical Magazine and Journal of Science*, 17(112), 497–511.

Andreozzi, L., Faetti, M., Giordano, M., Palazzuoli, D., Zulli, F. (2003). Enthalpy relaxation in polymers: A comparison among different multiparameter approaches extending the TNM/AGV model. *Macromolecules*, 36, 7379–7387.

Andreozzi, L., Faetti, M., Zulli, F., Giordano, M. (2004). Connecting shear stress relaxation and enthalpy recovery in polymers through a modified TNM approach. *Macromolecules*, 37, 8010–8016.

Andreozzi, L., Faetti, M., Giordano, M., Zulli, F. (2005). Molecular-weight dependence of enthalpy relaxation of PMMA. *Macromolecules*, 38, 6056–6067.

Andreozzi, L., Faetti, M., Salmerón Sanchez, M., Gómez Ribelles, J.L. (2008). Phenomenological theory of structural relaxation based on a thermorheologically complex relaxation time distribution. *The European Physical Journal E*, 27(1), 87–97.

Androsch, R. and Schick, C. (2016). Interplay between the relaxation of the glass of random L/D-lactide copolymers and homogeneous crystal nucleation: Evidence for segregation of chain defects. *The Journal of Physical Chemistry B*, 120(19), 4522–4528.

Androsch, R., Schick, C., Schmelzer, J.WP. (2014). Sequence of enthalpy relaxation, homogeneous crystal nucleation and crystal growth in glassy polyamide 6. *European Polymer Journal*, 53, 100–108.

Androsch, R., Zhuravlev, E., Schmelzer, J.W.P., Schick, C. (2018). Relaxation and crystal nucleation in polymer glasses. *European Polymer Journal*, 102, 195–208.

Angell, C.A. (1985). Spectroscopy simulation and scattering, and the medium range order problem in glass. *J. Non-Cryst. Solids*, 73(1/3), 1–17.

Angell, C.A. (1991). Thermodynamic aspects of the glass transition in liquids and plastic crystals. *Pure Appl. Chem.*, 63(10), 1387–1392.

Angell, C.A. (1997). Why C1 = 16–17 in the WLF equation is physical – and the fragility of polymers. *Polymer*, 38(26), 6261–6266.

Aouadi, A., Dreyfus, C., Massot, M., Pick, R.M. (2000). Light scattering study of the liquid–glass transition of meta-toluidine. *J. Chem. Phys.*, 112(22), 9860.

Arabeche, K., Delbreilh, L., Baer, E. (2022). Correction to: Physical aging of multilayer polymer films–influence of layer thickness on enthalpy relaxation process, effect of confinement. *Journal of Polymer Research*, 29(1), 1.

Araujo, S., Batteux, F., Li, W., Butterfield, L., Delpouve, N., Esposito, A., Tan, L., Saiter, J.-M., Negahban, M. (2018). A structural interpretation of the two components governing the kinetic fragility from the example of interpenetrated polymer networks. *Journal of Polymer Science Part B: Polymer Physics*, 56, 1393–1403.

Araujo, S., Delpouve, N., Delbreilh, L., Papkov, D., Dzenis, Y., Dargent, E. (2019). Dielectric and calorimetric signatures of chain orientation in strong and tough ultrafine electrospun polyacrylonitrile. *Polymer*, 178, 121638.

Arrese-Igor, S., Alegria, A., Colmenero, J. (2014). Chain dynamics on crossing the glass transition: Nonequilibrium effects and recovery of the temperature dependence of the structural relaxation. *ACS Macro Letters*, 3, 1215–1219.

Avolio, R., Castaldo, R., Gentile, G., Ambrogi, V., Fiori, S., Avella, M., Cocca, M., Errico, M.E. (2015). Plasticization of poly (lactic acid) through blending with oligomers of lactic acid: Effect of the physical aging on properties. *European Polymer Journal*, 66, 533–542.

Badrinarayanan, P. and Simon, S.L. (2007). Origin of the divergence of the timescales for volume and enthalpy recovery. *Polymer*, 48, 1464–1470.

Barral, L., Cano, J., Lopez, J., Lopez-Bueno, I., Nogueira, P., Abad, M.J., Ramirez, C. (1999). Physical aging of an epoxy/cycloaliphatic amine resin. *European Polymer Journal*, 35(3), 403–411.

Barral, L., Cano, J., Lopez, J., Lopez-Bueno, I., Nogueira, P., Abad, M.J., Ramirez, C. (2000). Physical aging of a tetrafunctional/phenol novolac epoxy mixture cured with diamine. DSC and DMA measurements. *Journal of Thermal Analysis and Calorimetry*, 60, 391–399.

Beckmann, J., Gregory, B., McKenna, B.G., Landes, D., Bank, H., Robert, A.B. (1997). Physical aging kinetics of syndiotactic polystyrene as determined from creep behavior. *Polymer Engineering & Science*, 37, 1459–1468.

Berthier, L. and Biroli, G. (2011). Theoretical perspective on the glass transition and amorphous materials. *Reviews of Modern Physics*, 83, 587.

Berthier, L. and Kob, W. (2011). Static point-to-set correlations in glass-forming liquids. *Phys. Rev. E*, 85, 011102.

Bhardwaj, S.P. and Suryanarayanan, R. (2012). Use of dielectric spectroscopy to monitor molecular mobility in glassy and supercooled trehalose. *Journal Physical Chemistry B*, 116, 11728–11736.

Bhosle, S., Gunasekera, K., Chen, P., Boolchand, P., Micoulaut, M., Massobrio, C. (2011). Meeting experimental challenges to physics of network glasses: Assessing the role of sample homogeneity. *Solid State Communications*, 151(24), 1851–1855.

Biroli, G., Bouchaud, J.P., Cavagna, A., Grigera, T.S., Verrocchio, P. (2008). Thermo-dynamic signature of growing amorphous order in glass-forming liquids. *Nat. Phys*, 4, 771–775.

Boehm, L., Ingram, M.D., Angell, C.A. (1981). Test of a year-annealed glass for the Cohen-Grest percolation transition. *Journal of Non-Crystalline Solids*, 44, 305–313.

Boolchand, P., Lucovsky, G., Phillips, J.C., Thorpe, M.F. (2005). Self-organization and the physics of glassy networks. *Philosophical Magazine*, 85(32), 3823–3838.

Borde, B., Bizot, H., Vigier, G., Buléon, A. (2002). Calorimetric analysis of the structural relaxation in partially hydrated amorphous polysaccharides. II. Phenomenological study of physical ageing. *Carbohydrate Polymers*, 48, 111–123.

Boucher, V.M., Cangialosi, D., Alegría, A., Colmenero, J. (2010). Enthalpy recovery of PMMA/silica nanocomposites. *Macromolecules*, 43(18), 7594–7603.

Boucher, V.M., Cangialosi, D., Alegría, A., Colmenero, J. (2011a). Enthalpy recovery of glassy polymers: Dramatic deviations from the extrapolated liquidlike behavior. *Macromolecules*, 44, 8333–8342.

Boucher, V.M., Cangialosi, D., Alegría, A., Colmenero, J., González-Irun, J., Liz-Marzan, L.M. (2011b). Physical aging in PMMA/silica nanocomposites: Enthalpy and dielectric relaxation. *Journal of Non-Crystalline Solids*, 357, 605–609.

Boucher, V.M., Cangialosi, D., Alegría, A., Colmenero, J, Pastoriza-Santos, I., Liz-Marzan, L.M. (2011c). Physical aging of polystyrene/gold nanocomposites and its relation to the calorimetric T_g depression. *Soft Matter*, 7, 3607–3620.

Boucher, V.M., Cangialosi, D., Alegría, A., Colmenero, J. (2012a). Enthalpy recovery in nanometer to micrometer thick polystyrene films. *Macromolecules*, 45, 5296–5306.

Boucher, V.M, Cangialosi, D., Alegría, A., Colmenero, J. (2012b). Time dependence of the segmental relaxation time of poly (vinyl acetate)-silica nanocomposites. *Physical Review E, Nonlinear, Soft Matter Physics*, 86, 041501.

Boucher, V.M., Cangialosi, D., Alegría, A., Colmenero, J. (2017). Complex nonequilibrium dynamics of stacked polystyrene films deep in the glassy state. *The Journal of Chemical Physics*, 146(20), 203312.

Brun, C. (2011). Des liquides surfondus aux verres : étude des corrélations à et hors équilibre. PhD Thesis, Université Paris-Sud, Paris.

Brunacci, A., Cowiet, J.M.G., Ferguson, R., McEwen, I.J. (1997). Enthalpy relaxation in glassy polystyrenes: 2. *Polymer*, 38, 3263–3268.

Cangialosi, D., Schut, H., van Veen, A., Picken, S.J. (2003). Positron annihilation lifetime spectroscopy for measuring free volume during physical aging of polycarbonate. *Macromolecules*, 36, 142–147.

Cangialosi, D., Wübbenhorst, M., Schut, H., van Veen, A., Picken, S.J. (2004). Dynamics of polycarbonate far below the glass transition temperature: A positron annihilation lifetime study. *Physical Review B*, 69, 134206.

Cangialosi, D., Boucher, V., Alegría, A., Colmenero, J. (2012). Enhanced physical aging of polymer nanocomposites: The key role of the area to volume ratio. *Polymer*, 53, 1362–1372.

Cangialosi, D., Boucher, V.M., Alegría, A., Colmenero, J. (2013a). Direct evidence of two equilibration mechanisms in glassy polymers. *Physical Review Letters*, 111, 095701.

Cangialosi, D., Boucher, V., Alegría, A., Colmenero, J. (2013b). Physical aging in polymers and polymer nanocomposites: Recent results and open questions. *Soft Matter*, 9, 8619–8630.

Cangialosi, D., Alegría, A., Colmenero, J. (2016). Effect of nanostructure on the thermal glass transition and physical aging in polymer materials. *Progress in Polymer Science*, 54(55), 128–147.

Casalini, R. and Roland, C.M. (2011). Aging of a low molecular weight poly (methyl methacrylate). *Journal of Non-Crystalline Solids*, 357(2), 282–285.

Cebe, P., Partlow, B.P., Kaplan, D.L., Wurm, A., Zhuravlev, E., Schick, C. (2015). Using flash DSC for determining the liquid state heat capacity of silk fibroin. *Thermochimica Acta*, 615, 8–14.

Chakravarty, S., Georgiev, D.G., Boolchand, P., Micoulaut, M. (2004). Ageing, fragility and the reversibility window in bulk alloy glasses. *Journal of Physics: Condensed Matter*, 17(1), L1.

Chakravarty, S., Georgiev, D.G., Boolchand, P., Micoulaut, M. (2005). Ageing, fragility and the reversibility window in bulk alloy glasses. *Journal of Physics: Condensed Matter*, 17(1), L1–7. doi: 10.1088/0953-8984/17/1/L01.

Chow, T.S. (1984). Kinetics of free volume and physical aging in polymer glasses. *Macromolecules*, 17(11), 2336–2340.

Choy, I. and Plazek, D.J. (1986). The physical properties of bisphenol-A-based epoxy resins during and after curing. *Journal of Polymer Science Part B: Polymer Physics*, 24, 1303–1320.

Cohen, M.H. and Grest, G.S. (1979). Liquid-glass transition, a free-volume approach. *Physical Review B*, 20(3), 1077–1098.

Cohen, M.H. and Turnbull, D. (1959). Molecular transport in liquids and glasses. *The Journal of Chemical Physics*, 31, 1164–1169.

Cook, W.D., Mehrabi, M., Edward, G.H. (1999). Ageing and yielding in model epoxy thermosets. *Polymer*, 40, 1209–1218.

Cortes, P., Montserrat, S., Ledru, J., Saiter, J.M. (1998). Enthalpy relaxation in Gex Se1− x chalcogenide glasses. *Journal of Non-crystalline Solids*, 235, 522–526.

Cowie, J.M.G. and Ferguson, R. (1993). Physical ageing of poly(methyl methacrylate) from enthalpy relaxation measurements. *Polymer*, 34, 2135–2141.

Cowie, J.M.G., Harris, S., Mcewen, I.J. (1996). Physical ageing in poly(vinyl acetate) 1. Enthalpy relaxation. *Journal of Polymer Science*, 35, 1107–1116.

D'Amore, A., Caprino, G., Nicolais, L., Marino, G. (1999). Long-term behaviour of PEI and PEI-based composites subjected to physical aging. *Composites Science and Technology*, 59(13), 1993–2003.

Dalle-Ferrier, C. (2009). Transition vitreuse et hétérogénéités dynamiques dans les liquides moléculaires et les polymères. PhD Thesis, Université Paris-Sud, Paris.

Datin, P. (2019). Manipulation optique de molécules pour l'étude de la transition vitreuse. PhD Thesis, Université Paris-Sud, Paris.

Davis, W.J. and Pethrick, R.A. (1998). Investigation of physical ageing in polymethylmethacrylate using positron annihilation, dielectric relaxation and dynamic mechanical thermal analysis. *Polymer*, 39(2), 255–266.

Delpouve, N., Arnoult, M., Saiter, A., Dargent, E., Saiter, J.M. (2014). Evidence of two mobile amorphous phases in semicrystalline polylactide observed from calorimetric investigations. *Polymer Engineering & Science*, 54(5), 1144–1150.

Di Lorenzo, M.L. and Righetti, M.C. (2018). Crystallization-induced formation of rigid amorphous fraction. *Polymer Crystallization*, 1, e10023.

Di Lorenzo, M.L., Androsch, R., Righetti, M.C. (2017). Low-temperature crystallization of poly (butylene succinate). *European Polymer Journal*, 94, 384–391.

Dobircau, L., Delpouve, N., Herbinet, R., Domenek, S., Le Pluart, L., Delbreilh, L., Ducruet, V., Dargent, E. (2015). Molecular mobility and physical ageing of plasticized poly(lactide). *Polymer Engineering & Science*, 55(4), 858–865.

Donth, E. (1996). Characteristic length of the glass transition. *Journal of Polymer Science: Part B: Polymer Physics*, 34, 2881–2892.

van Drongelen, M., Meijer-Vissers, T., Cavallo, D., Portale, G., Vanden Poel, G., Androsch, R. (2013). Microfocus wide-angle X-ray scattering of polymers crystallized in a fast scanning chip calorimeter. *Thermochimica Acta*, 563, 33–37.

Drozdov, A.D. (1999). A model of traps for structural relaxation in glassy polymers. *Physics Letters A*, 258, 158–170.

Drozdov, A.D. (2001). A model for structural relaxation in amorphous glassy polymers based on the aggregation-fragmentation theory. *Journal of Polymer Science Part B: Polymer Physics*, 39, 1312–1325.

Dudowicz, J., Freed, K.F., Douglas, J.F. (2008). Generalized entropy theory of polymer glass formation. *Advances in Chemical Physics*, 137, 125.

Dudowicz, J., Douglas, J.F., Freed, K.F. (2015). The meaning of the "universal" WLF parameters of glass-forming polymer liquids. *J. Chem. Phys.*, 142(1), 014905.

Duval, E., Etienne, S., Simeoni, G., Mermet, A. (2006). Physical aging effect on the boson peak and heterogeneous nanostructure of a silicate glass. *Journal of Non-Crystalline Solids*, 352(42/49), 4525–4530.

Dyre, J.C. (2006). The glass transition and elastic models of glass-forming liquids. *Reviews of Modern Physics*, 78, 953.

Echeverría, I., Kolek, P.L., Plazek, D.J., Simon, S.L. (2003). Enthalpy recovery, creep and creep–recovery measurements during physical aging of amorphous selenium. *Journal of Non-Crystalline Solids*, 324, 242–255.

Ediger, M.D. (2000). Spatially heterogeneous dynamics in supercooled liquids. *Annu. Rev. Phys. Chem.*, 51(1), 99–128.

Ediger, M.D. (2017). Perspective: Highly stable vapor-deposited glasses. *The Journal of Chemical Physics*, 147(21), 210901.

van Ekeren, P.J. (1998). Thermodynamic background to thermal analysis and calorimetry. In *Handbook of Thermal Analysis and Calorimetry: Vol. 1. Principles and Practice*, Brown, M.E. (ed.). Elsevier, Amsterdam.

Elmatad, Y.S., Chandler, D., Garrahan, J.P. (2010). Corresponding states of structural glass formers. II. *The Journal of Physical Chemistry B*, 114(51), 17113–17119.

Fraga, F., Castro-Diaz, C., Rodríguez-Núñez, E., Martínez-Ageitos, J.M. (2003). Physical aging for an epoxy network diglycidyl ether of bisphenol A/m-xylylenediamine. *Polymer*, 44(19), 5779–5784.

Frank, W. and Stuart, H. (1968). Umwandlungserscheinungen in amorphen Hochpolymeren beim tempern unterhalb der Glastemperatur. *Kolloid-Zeitschrift*, 225, 1–6.

Franz, S. and Semerjian, G. (2011). Analytical approaches to time and length scales in models of glasses. In *Dynamical Heterogeneities in Glasses, Colloids, and Granular Media*, Berthier, L., Biroli, G., Bouchaud, J.P., Cipelletti, L., van Saarloos, W. (eds). Oxford University Press, New York.

Frieberg, B., Glynos, E., Green, P.F. (2012a). Structural relaxations of thin polymer films. *Physical Review Letters*, 108, 268304.

Frieberg, B., Glynos, E., Sakellariou, G., Green, P.F. (2012b). Physical aging of star-shaped macromolecules. *ACS Macro Letters*, 1, 636–640.

Frieberg, B.R., Glynos, E., Stathouraki, M., Sakellariou, G., Green, P.F. (2017). Glassy dynamics of polymers with star-shaped topologies: Roles of molecular functionality, arm length, and film thickness. *Macromolecules*, 50, 3719–3725.

Frieberg, B.R., Glynos, E., Sakellariou, G., Tyagi, M., Green, P.F. (2020). Effect of molecular stiffness on the physical aging of polymers. *Macromolecules*, 53, 7684–7690.

Fulcher, G.S. (1925). Analysis of recent measurements of the viscosity of glasses. *J. Am. Ceram. Soc.*, 8(6), 339–355.

Gao, S. and Simon, S.L. (2015). Measurement of the limiting fictive temperature over five decades of cooling and heating rates. *Thermochim. Acta*, 603(9), 123–127.

Gao, S., Koh, Y.P., Simon, S.L. (2013). Calorimetric glass transition of single polystyrene ultrathin films. *Macromolecules*, 46, 562–570.

Garden, J.L., Guillou, H., Richard, J., Wondraczek, L. (2012). Non-equilibrium configurational Prigogine-Defay ratio. *J. Non-Equilib. Thermodyn.*, 37(2), 143–177.

Godard, M.E., Saiter, J.M., Cortés, P., Montserrat, S., Hutchinson, J.M., Burel, F., Bunel, C. (1998). Enthalpy relaxation in polymethyl (α-n-alkyl) acrylates: Effect of length of alkyl chain. *Journal of Polymer Science Part B: Polymer Physics*, 36(4), 583–593.

Goldstein, M. (1976). Viscous liquids and the glass transition. V. Sources of the excess specific heat of the liquid. *The Journal of Chemical Physics*, 64(11), 4767–4774.

Golovchak, R., Jain, H., Shpotyuk, O., Kozdras, A., Saiter, A., Saiter, J.M. (2008). Experimental verification of the reversibility window concept in binary As-Se glasses subjected to a long-term physical aging. *Physical Review B*, 78(1), 014202.

Golovchak, R., Kozdras, A., Shpotyuk, O., Kozyukhin, S.A., Saiter, J.M. (2009). Long-term ageing behaviour in Ge–Se glasses. *Journal of Materials Science*, 44(15), 3962–3967.

Golovchak, R., Kozyukhin, S.A., Kozdras, A., Shpotyuk, O.I., Novotortsev, V.M. (2010). Physical aging of chalcogenide glasses. *Inorganic Materials*, 46(8), 911–913.

Golovchak, R., Kozdras, A., Balitska, V., Shpotyuk, O. (2012). Step-wise kinetics of natural physical ageing in arsenic selenide glasses. *Journal of Physics: Condensed Matter*, 24, 505106.

Golovchak, R., Kozdras, A., Shpotyuk, O., Balitska, V. (2016). Crossover between cooperative and fractal relaxation in complex glass-formers. *Journal of Physics: Condensed Matter*, 28, 355101.

Gomez Ribelles, J.L. and Monleon Pradas, M. (1995). Structural relaxation of glass-forming polymers based on an equation for configurational entropy. 1. DSC Experiments on polycarbonate. *Macromolecules*, 28(17), 5867–5877.

Gomez Ribelles, J.L., Monleon Pradas, M., Vidaurre Garayo, A., Romero Colomer, F., Más Estellés, J., Meseguer Duenas, J.M. (1995). Structural relaxation of glass-for-ming polymers based on an equation for configurational entropy. 2. Structural relaxation in polymethacrylates. *Macromolecules*, 28(17), 5878–5885.

Gomez Ribelles, J.L., Monleon Pradas, M., Vidaurre Garayo, A., Romero Colomer, F., Más Estellés, J., Meseguer Dueñas, J.M. (1997). Structural relaxation of glass-for-ming polymers based on an equation for configurational entropy. 3. On the states attained at infinite time in the structural relaxation process. Results on poly(ether imide). *Polymer*, 38, 963–969.

Grassia, L. and Simon, S.L. (2012). Modeling volume relaxation of amorphous polymers: Modification of the equation for the relaxation time in the KAHR model. *Polymer*, 53, 3613–3620.

Grassia, L., Koh, Y.P., Rosa, M., Simon, S.L. (2018). Complete set of enthalpy recovery data using flash DSC: Experiment and modeling. *Macromolecules*, 51(4), 1549–1558.

Gray, L.A.G., Yoon, S.W., Pahner, W.A., Davidheiser, J.E., Roth, C.B. (2012). Importance of quench conditions on the subsequent physical aging rate of glassy polymer films. *Macromolecules*, 45(3), 1701–1709.

Green, P.F., Glynos, E., Frieberg, B. (2015). Polymer films of nanoscale thickness: Linear chain and star-shaped macromolecular architectures. *MRS Communications* 5, 423–434.

Greiner, R. and Schwarzl, F.R. (1984). Thermal contraction and volume relaxation of amorphous polymers. *Rheologica Acta*, 23(4), 378–395.

Grigoriadi, K., Putzeys, T., Wubbenhorst, M., van Breemen, L.C., Anderson, P.D., Hutter, M. (2019a). Effect of low-temperature physical aging on the dynamic transitions of atactic polystyrene in the glassy state. *Journal of Polymer Science Part B: Polymer Physics*, 57, 1394–1401.

Grigoriadi, K., Westrik, J.B.H.M., Vogiatzis, G.G., van Breemen, L.C.A., Anderson, P.D., Hütter, M. (2019b). Physical ageing of polystyrene: Does tacticity play a role? *Macromolecules*, 52, 5948–5954.

Gujrati, P.D. (2010). Nonequilibrium thermodynamics: Structural relaxation, fictive temperature, and Tool-Narayanaswamy phenomenology in glasses. *Phys. Rev. E*, 81, 051130.

Gunawan, L., Johari, G.P., Shanker, R.M. (2006). Structural relaxation of acetaminophen glass. *Pharmaceutical Research*, 23(5), 967–979.

Guo, Y., Zhang, C., Lai, C., Priestley, R.D., D'Acunzi, M., Fytas, G. (2011). Structural relaxation of polymer nanospheres under soft and hard confinement: Isobaric versus isochoric conditions. *ACS Nano*, 5, 5365–5373.

Gutzow, I., Ilieva, D., Babalievski, F., Yamakov, V. (2000). Thermodynamics and kinetics of the glass transition: A generic geometric approach. *The Journal of Chemical Physics*, 112, 10941–10948.

Haddad, J. and Goldstein, M. (1978). Viscous liquids and the glass transition. VIII. Effect of fictive temperature on dielectric relaxation in the glassy state. *Journal of Non-Crystalline Solids*, 30, 1–22.

Hakme, C., Stevenson, I., David, L., Boiteux, G., Seytre, G., Schönhals, A. (2005). Uniaxially stretched poly(ethylene naphthalene 2,6-dicarboxylate) films studied by broadband dielectric spectroscopy. *J. Non-Cryst. Solids*, 351(33/36), 2742–2752.

Hecksher, T., Nielsen, A.I., Olsen, N.B., Dyre, J.C. (2008). Little evidence for dynamic divergences in ultraviscous molecular liquids. *Nature Physics*, 4, 737–741.

Hodge, I.M. (1994). Enthalpy relaxation and recovery in amorphous materials. *J. Non-Cryst. Solids*, 169(3), 211–266.

Hodge, I.M. (1995). Physical aging in polymer glasses. *Science*, 267(5206), 1945–1947.

Hu, H.W. (2007). Physical aging in long term creep of polymeric composite laminates. *Journal of Mechanics*, 23, 245–252.

Huang, Y. and Paul, D.R. (2006). Physical aging of thin glassy polymer films monitored by optical properties. *Macromolecules*, 39, 1554–1559.

Huang, Y. and Paul, D.R. (2007). Effect of molecular weight and temperature on physical aging of thin glassy poly(2,6-dimethyl-1,4-phenylene oxide) films. *Journal of Polymer Science Part B: Polymer Physics*, 45, 1390–1398.

Hutchinson, J.M. (1995). Physical aging of polymers. *Prog. Polym. Sci.*, 20(4), 703–760.

Hutchinson, J.M. (2009). Determination of the glass transition temperature. Methods correlation and structural heterogeneity. *J. Therm. Anal. Calorim.*, 98, 579–589.

Hutchinson, J.M. and Kumar, P. (2002). Enthalpy relaxation in polyvinyl acetate. *Thermochimica Acta*, 391, 197–217.

Hutchinson, J.M. and Ruddy, M. (1988). Thermal cycling of glasses. II. Experimental evaluation of the structure (or nonlinearity) parameter x. *Journal of Polymer Science Part B: Polymer Physics*, 26(11), 2341–2366.

Hutchinson, J.M., McCarthy, D., Montserrat, S., Cortes, P. (1996). Enthalpy relaxation in a partially cured epoxy resin. *Journal of Polymer Science: Part B Polymer Physics*, 34, 229–239.

Hutchinson, J.M., Montserrat, S., Calventus, Y., Cortes, P. (2000). Application of the Adam–Gibbs equation to the non-equilibrium glassy state. *Macromolecules*, 33(14), 5252–5262.

Issoupov, V., Demont, C., Lacabanne, O.V., Startsev, V., Dinguirard, M. (2002). Thermally stimulated creep (TSCr) study of viscoelastic behavior and physical aging of a polymeric matrix composite for spacecraft structures. *Journal of Applied Polymer Science*, 85, 342–350.

Jaeger, G. (1998). The Ehrenfest classification of phase transitions: Introduction and evolution. *Arch. Hist. Exact. Sci.*, 53, 51–81.

Jenkins, A.D., Kratochvíl, P., Stepto, R.F.T., Suter, U.W. (1996). Glossary of basic terms in polymer science (IUPAC Recommendations 1996). *Pure and Applied Chemistry*, 68, 2287–2311.

Johari, G.P. (1973). Intrinsic mobility of molecular glasses. *Journal of Chemical Physics*, 58, 1766–1770.

Kaliappan, S.K. (2007). Characterization of physical properties of polymers using AFM force-distance curves. PhD Thesis, Université de Siegen, Siegen.

Kausch, H.H., Heymans, N., Plummer, C.J., Decroly, P. (2001). *Matériaux polymères : propriétés mécaniques et physiques. Principe de mise en œuvre*. Presses polytechniques et universitaires romandes, Lausanne.

Kauzmann, W. (1948). The nature of the glassy state and the behavior of liquids at low temperatures. *Chem. Rev.*, 43(2), 219–256.

Kawakami, H. and Watanabe, M. (2008). Changes in the thermomechanical behavior of epoxy glasses under the state of strain aging. *Journal of Applied Polymer Science*, 107, 2095–2100.

Kearns, K.L., Swallen, S.F., Ediger, M.D., Wu, T., Yu, L. (2007). Influence of substrate temperature on the stability of glasses prepared by vapor deposition. *The Journal of Chemical Physics*, 127, 154702.

King, E.A., Sen, S., Takeda, W., Boussard-Pledel, C., Bureau, B., Guin, J.P., Lucas, P. (2021). Extended aging of Ge–Se glasses below the glass transition temperature. *The Journal of Chemical Physics*, 154(16), 164502.

Kob, W. and Andersen, H.C. (1994). Scaling behavior in the β-relaxation regime of a super-cooled Lennard-Jones mixture. *Phys. Rev. Lett.*, 73, 1376.

Koh, Y.P. and Simon, S.L. (2008). Structural relaxation of stacked ultrathin polystyrene films. *Journal of Polymer Science Part B: Polymer Physics*, 46, 2741–2753.

Koh, Y.P. and Simon, S.L. (2013). Enthalpy recovery of polystyrene: Does a long-term aging plateau exist? *Macromolecules*, 46, 5815–5821.

Koh, Y.P., Grassia, L., Simon, S.L. (2015). Structural recovery of a single polystyrene thin film using nanocalorimetry to extend the aging time and temperature range. *Thermochimica Acta*, 603, 135–141.

Koh, Y.P., Gao, S., Simon, S.L. (2016). Structural recovery of a single polystyrene thin film using flash DSC at low aging temperatures. *Polymer*, 96, 182–187.

Kohlrausch, R. (1854). Theorie des elektrischen rückstandes in der leidenerflasche. *Ann. Phys.*, 167(1), 179–214.

Kolla, S. and Simon, S.L. (2005). The τ-effective paradox: New measurements towards a resolution. *Polymer*, 46, 733–739.

Kovacs, A.J. (1964). Transition vitreuse dans les polymères amorphes. Étude phénoménologique. In *Fortschritte der hochpolymeren-forschung (Advances in Polymer Science)*, Ferry, J.D., Kern, W., Natta, G. et al. (eds). Springer, Berlin.

Kovacs, A.J. (1981). A multiparameter approach for structural recovery of glasses and its implication for their physical properties. *Ann. N. Y. Acad. Sci.*, 371(1), 38–66.

Kovacs, A.J., Stratton, R.A., Ferry, J.D. (1963). Dynamic mechanical properties of polyvinyl acetate in shear in the glass transition temperature range. *The Journal of Physical Chemistry*, 67(1), 152–161.

Kovacs, A.J., Hutchinson, J.M., Aklonis, J.J. (1977). Isobaric volume and enthalpy recovery of glasses (I) A critical survey of recent phenomenological approaches. In *The Structure of Non-Crystalline Materials*, Gaskell, P.H. (ed). Taylor and Francis, London.

Kovacs, A.J., Aklonis, J.J., Hutchinson, J.M., Ramos, A.R. (1979). Isobaric volume and enthalpy recovery of glasses. II. A transparent multiparameter theory. *J. Polym. Sci. Pol. Phys.*, 17(7), 1097–1162.

Kremer, F., Tress, M., Mapesa, E.U. (2015). Glassy dynamics and glass transition in nanometric layers and films: A silver lining on the horizon. *Journal of Non-Crystalline Solids*, 407, 277–283.

Krishnaswamy, R.K., Geibel, J.F., Lewis, B.J. (2003). Influence of semicrystalline morphology on the physical aging characteristics of poly(phenylene sulfide). *Macromolecules*, 36, 2907–2914.

Kucera, S.A., Felton, L.A., McGinity, J.W. (2013). Physical aging in pharmaceutical polymers and the effect on solid oral dosage form stability. *International Journal of Pharmaceutics*, 457(2), 428–436.

Kunal, K., Robertson, C.G., Pawlus, S., Hahn, S.F., Sokolov, A.P. (2008). Role of chemical structure in fragility of polymers: A qualitative picture. *Macromolecules*, 41(19), 7232–7238.

Kwon, M., Lee, S.C., Jeong, Y.G. (2010). Influences of physical aging on enthalpy relaxation behavior, gas permeability, and dynamic mechanical property of polylactide films with various D-isomer contents. *Macromolecular Research*, 18(4), 346–351.

Lacey, A.A., Price, D.M., Reading, M. (2006). Theory and practice of modulated temperature differential scanning calorimetry. In *Modulated Temperature Differential Scanning Calorimetry*, Reading, M. and Hourston, D.J. (eds). Springer, Berlin.

Langhe, D.S., Murphy, T.M., Shaver, A., LaPorte, C., Freeman, B.D., Paul, D.R., Baer, E. (2012). Structural relaxation of polystyrene in nanolayer confinement. *Polymer*, 53(9), 1925–1931.

Lee, A. and Lichtenhan, J.D. (1998). Viscoelastic responses of polyhedral oligosilsesquioxane reinforced epoxy systems. *Macromolecules*, 31(15), 4970–4974.

Lee, A. and McKenna, G. (1988). Effect of crosslink density on physical ageing of epoxy networks. *Polymer*, 29, 1812–1817.

Lee, A. and McKenna, G. (1990). Viscoelastic response of epoxy glasses subjected to different thermal treatments. *Polymer Engineering and Science*, 30(7), 431–435.

Lee, J., Mangalara, J.H., Simmons, D.S. (2017). Correspondence between the rigid amorphous fraction and nanoconfinement effects on glass formation. *Journal of Polymer Science Part B: Polymer Physics*, 55, 907–918.

Leheny, R.L. and Nagel, S.R. (1998). Frequency-domain study of physical aging in a simple liquid. *Physical Review B*, 57, 5154.

Leveque, D., Schieffer, A., Mavel, A., Maire, J. (2005). Analysis of how thermal aging affects the long-term mechanical behavior and strength of polymer matrix composites. *Composites Science and Technology*, 65, 395–401.

Li, Q. and Simon, S.L. (2006). Enthalpy recovery of polymeric glasses: Is the theoretical limiting liquid line reached? *Polymer*, 47, 4781–4788.

Lipson, J.E.G. and White, R.P. (2014). Connecting theory and experiment to understand miscibility in polymer and small molecule mixtures. *Journal of Chemical & Engineering Data*, 59(10), 3289–3300.

Lixon, C., Delpouve, N., Saiter, A., Dargent, E., Grohens, Y. (2008). Evidence of cooperative rearranging region size anisotropy for drawn PET. *Eur. Polym. J.*, 44(11), 3377–3384.

Lizundia, E., Larrañaga, A., Vilas, J.L., León, L.M. (2016). Three-dimensional orientation of poly (L-lactide) crystals under uniaxial drawing. *RSC Advances*, 6(15), 11943–11951.

Lorenz, B., Persson, B.N.J., Fortunato, G., Giustiniano, M., Baldoni, F. (2013). Rubber friction for tire tread compound on road surfaces. *Journal of Physics: Condensed Matter*, 25(9), 095007.

Lu, H. and Nutt, S. (2003a). Enthalpy relaxation of layered silicate-epoxy nanocomposites. *Macromolecular Chemistry and Physics*, 204(15), 1832–1841.

Lu, H. and Nutt, S. (2003b). Restricted relaxation in polymer nanocomposites near the glass transition. *Macromolecules*, 36(11), 4010–4016.

Lunkenheimer, P., Wehn, R., Schneider, U., Loidl, A. (2005). Glassy aging dynamics. *Physical Review Letters*, 95, 055702.

Lunkenheimer, P., Kastner, S., Köhler, M., Loidl, A. (2010). Temperature development of glassy α-relaxation dynamics determined by broadband dielectric spectroscopy. *Phys. Rev. E*, 81, 051504.

Lydżba-Kopczyńska, B.I., Bogusław, G., Jan, C., Sachanbiński, M. (2012). Provenance investigations of amber jewelry excavated in Lower Silesia (Poland) and dated back to Early Iron Age. *Journal of Raman Spectroscopy*, 43, 1839–1844.

Málek, J. (1998). Volume and enthalpy relaxation rate in glassy materials. *Macromolecules*, 31, 8312–8322.

Málek, J., Svoboda, R., Pustková, P., Čičmanec, P. (2009). Volume and enthalpy relaxation of A-Se in the glass transition region. *Journal of Non-Crystalline Solids*, 355, 264–272.

Mathlouthi, C., Hugenell, F., Delpech, F., Rharbi, Y. (2017). Heat capacity of confined polystyrene in close-packed particles. *Macromolecules*, 50, 472–481.

Mathot, V., Pyda, M., Pijpers, T., Vanden Poel, G., van de Kerkhof, E., van Herwaarden, S., Herwaarden, F., Leenaers, A. (2011). The flash DSC 1, a power compensation twin-type, chip-based fast scanning calorimeter (FSC): First findings on polymers. *Thermochim. Acta*, 522(1/2), 36–45.

Mcgonigle, E.A., Cowie, J.M.G., Arrighi, V., Pethrick, R.A. (2005). Enthalpy relaxation and free volume changes in aged styrene copolymers containing a hydrogen bonding co-monomer. *Journal of Materials Science*, 40, 1869–1881.

McKenna, G.B. (1989). Glass formation and glassy behavior. In *Comprehensive Polymer Science Polymer Properties*, Booth, C. and Prime, C. (eds). Pergamon, Oxford.

McKenna, G.B. (2008). Diverging views on glass transition. *Nature Physics*, 4(9), 673–673.

McKenna, G.B. and Simon, S.L. (2017). 50th anniversary perspective: Challenges in the dynamics and kinetics of glass-forming polymers. *Macromolecules*, 50, 6333–6361.

McKenna, G.B., Vangel, M.G., Rukhin, A.L., Leigh, S.D., Lotz, B., Straupe, C. (1999). The t-effective paradox revisited: An extended analysis of Kovacs' volume recovery data on poly(vinyl acetate). *Polymer*, 40, 5183–5205.

McLoughlin, J.R. and Tobolsky, A.V. (1951). Effect of rate of cooling on stress relaxation of polymethyl methacrylate. *Journal of Polymer Science*, 7(6), 658–658.

Menczel, J.D., Judovits, L., Bruce Prime, R., Bair, H.E., Reading, M., Swier, S. (2009). Differential scanning calorimetry (DSC). In *Thermal Analysis of Polymers: Fundamentals and Applications*, Menczel, J.D. and Bruce Prime, R. (eds). John Wiley & Sons, New York.

Mendelsy, D.A., Leterrier, Y., Manson, J.A.E., Nairn, J.A. (2002). The influence of internal stresses on the microbond test II: Physical aging and adhesion. *Journal of Composite Materials*, 36(14), 1655–1676.

Meseguer Duenas, J.M., Vidaurre Garayo, A., Romero Colomer, F., Mas Estelles, J., Gomez Ribelles, J.L., Monleon Pradas, M. (1997). Structural relaxation of glass-forming polymers based on an equation for configurational entropy. Structural relaxation in styrene-acrylonitrile copolymer. *Journal of Polymer Science Part B: Polymer Physics*, 35, 2201–2217.

Meyer, H.H., Mangin, P.M.F., Ferry, J.D. (1965). Dynamic mechanical properties of poly(n-butyl methacrylate) near its glass transition temperature. *Journal of Poly-mer Science Part A: General Papers*, 3(11), 4000–4000.

Minakov, A.A., Mordvintsev, D.A., Schick, C. (2004). Melting and reorganization of poly (ethylene terephthalate) on fast heating (1000 K/s). *Polymer*, 45(11), 3755–3763.

Minakov, A.A., Morikawa, J., Zhuravlev, E., Ryu, M., van Herwaarden, A.W., Schick, C. (2019). High-speed dynamics of temperature distribution in ultrafast (up to 10^8 K/s) chip-nanocalorimeters, measured by infrared thermography of high resolution. *J. Appl. Phys.*, 125, 054501.

Moghaddam, S.M., Quelennec, B., Delpouve, N., Atawa, B., Delbreilh, L., Saiter-Fourcin, A., Passaglia, E., Fiori, S. (2021). Fragility of short-chain poly (lactic acid)s derivatives by combining dielectric spectroscopy and fast scanning calorimetry. *Journal of Polymer Science*, 59, 1571–1577.

Monnier, X., Saiter, A., Dargent, E. (2017a). Physical aging in PLA through standard DSC and fast scanning calorimetry investigations. *Thermochimica Acta*, 648, 13–22.

Monnier, X., Saiter, A., Dargent, E. (2017b). Vitrification of PLA by fast scanning calorimetry: Towards unique glass above critical cooling rate? *Thermochimica Acta*, 658, 47–54.

Monnier, X., Nassar, F.S., Domenek, S., Guinault, A., Sollogoub, C., Dargent, E., Delpouve, N. (2018). Reduced physical aging rates of polylactide in polystyrene/polylactide multilayer films from fast scanning calorimetry. *Polymer*, 150, 1–9.

Monnier, X., Delpouve, N., Saiter-Fourcin, A. (2020). Distinct dynamics of structural relaxation in the amorphous phase of poly (L-lactic acid) revealed by quiescent crystallization. *Soft Matter*, 16(13), 3224–3233.

Monnier, X., Marina, S., Lopez de Pariza, X., Sardón, H., Martin, J., Cangialosi, D. (2021). Physical aging behavior of a glassy polyether. *Polymers*, 13(6), 954.

Montserrat, S. and Cortés, P. (1995). Physical ageing studies in semicrystalline poly(ethylene terephthalate). *Journal of Materials Science*, 30, 1790–1793.

Morancho, J.M. and Salla, J.M. (1999). Relaxation in partially cured samples of an epoxy resin and of the same resin modified with a carboxyl-terminated rubber. *Polymer*, 40(10), 2821–2828.

Morineau, D. and Alba-Simionesco, C. (1998). Hydrogen-bond-induced clustering in the fragile glass-forming liquid m-toluidine: Experiments and simulations. *J. Chem. Phys.*, 109(19), 8494.

Morvan, A., Delpouve, N., Vella, A., Saiter-Fourcin, A. (2021). Physical aging of selenium glass: Assessing the double mechanism of equilibration and the crystallization process. *Journal of Non-Crystalline Solids*, 570, 121013.

Moynihan, C.T., Macedo, P.B., Montrose, C.J., Gupta, P.K., DeBolt, M.A., Dill, J.F., Dom, B.E., Drake, P.W., Easteal, A.J., Elterman, P.B. et al. (1976a). Structural relaxation in vitreous materials. *Ann. N.Y. Acad. Sci.*, 279(1), 15–35.

Moynihan, C.T., Easteal, A.J., DeBolt, M.A., Tucker, J. (1976b). Dependence of the fictive temperature of glass on cooling rate. *J. Am. Ceram. Soc.*, 59(1/2), 12–16.

Moynihan, C.T., Crichton, S.N., Opalka, S.M. (1991). Linear and non-linear structural relaxation. *Journal of Non-Crystalline Solids*, 131(133), 420–434.

Narayanaswamy, O.S. (1971). A model of structural relaxation in glass. *J. Am. Ceram. Soc.*, 54(10), 491–498.

Nozaki, R. and Mashimo, S. (1987). Dielectric relaxation measurements of poly (vinyl acetate) in glassy state in the frequency range 10– 6–106 Hz. *The Journal of Chemical Physics*, 87(4), 2271–2277.

Núñez, L., Gómez-Barreiro, S., Gracia-Fernández, C.A., Núñez, M.R. (2004). Use of the dielectric analysis to complement previous thermoanalytical studies on the system diglycidyl ether of bisphenol A/1,2 diamine cyclohexane. *Polymer*, 45(4), 1167–1175.

O'Connell, P.A. and McKenna, G.B. (1999). Arrhenius-type temperature dependence of the segmental relaxation below T_g. *The Journal of Chemical Physics*, 110, 11054–11060.

Odegard, G.M. and Bandyopadhyay, A. (2011). Physical aging of epoxy polymers and their composites. *Journal of Polymer Science Part B: Polymer Physics*, 49, 1695–1716.

Ophir, Z.H., Emerson, J.A., Wilkes, G.L. (1978). Sub-T_g annealing studies of rubber-modified and unmodified epoxy systems. *Journal of Applied Physics*, 49, 5032–5038.

Pan, P., Bo, Z., Tungalag, D., Koji, Y., Tadashi, S., Masataka, T., Yoshio, I. (2008). Conformational and microstructural characteristics of poly(L-lactide) during glass transition and physical aging. *The Journal of Chemical Physics*, 18, 129–140.

Papageorgiou, D.G., Zhuravlev, E., Papageorgiou, G.Z., Bikiaris, D., Chrissafis, K., Schick, C. (2014). Kinetics of nucleation and crystallization in poly (butylene succi-nate) nanocomposites. *Polymer*, 55(26), 6725–6734.

Paroli, R.M. and Penn, J. (1994). Measuring the glass transition temperature of EPDM roofing materials: Comparison of DMA, TMA, and DSC techniques. In *Assignment of the Glass Transition*, Seyler, R.J. (ed.). American Society for Testing and Materials, Baltimore.

Pérez-Castañeda, T., Rafael, J., Jiménez, R., Miguel, A.R. (2013). Low-temperature thermal properties of a hyperaged geological glass. *Journal of Physics: Condensed Matter*, 25, 295402.

Pérez-Castañeda, T., Rafael, J., Jiménez, R., Miguel, A.R. (2014). Two-level systems and boson peak remain stable in 110-million-year-old amber glass. *Physical Review Letters*, 112, 165901.

Perez-De-Eulate, N.G. and Cangialosi, D. (2018). Double mechanism for structural recovery of polystyrene nanospheres. *Macromolecules*, 51(9), 3299–3307.

Petrie, S. (1975). *Polymeric Materials: Relationships Between Structure and Mechanical Behavior*. American Society for Metals, Detroit.

Petrie, S. (1976). The effect of excess thermodynamic properties versus structure formation on the physical properties of glassy polymers. *Journal of Macromolecular Science, Part B*, 12, 225–247.

Phalen, R.N., Cousins, K., Usher, T., Young, M., Zhang, R. (2016). Revising the formulization of the Narayanaswamy equation using the Adam-Gibbs theory. *J. Non-Cryst. Solids*, 453, 59–65.

Pijpers, T.F.J., Mathot, V.B.F., Goderis, B., Scherrenberg, R.L., van der Vegte, E.W. (2002). High-speed calorimetry for the study of the kinetics of (de)vitrification, crystallization, and melting of macromolecules. *Macromolecules*, 35(9), 3601–3613.

Pinto, S.S., Ramos, J.J.M., Diogo, H.P. (2009). The slow molecular mobility in poly (vinyl acetate) revisited: New contributions from thermally stimulated currents. *European Polymer Journal*, 45, 2644–2652.

Plazek, D.J. and Frund, Z.N. (1990). Epoxy resins (DGEBA): The curing and physical aging process. *Journal of Polymer Science Part B: Polymer Physics*, 28, 431–448.

Priestley, R.D. (2005). Structural relaxation of polymer glasses at surfaces, interfaces, and in between. *Science*, 309, 456–459.

Priestley, R.D. (2009). Physical aging of confined glasses. *Soft Matter*, 5, 919–926.

Priestley, R.D., Rittigstein, P., Broadbelt, L.J., Fukao, K., Torkelson, J.M. (2007). Evidence for the molecular-scale origin of the suppression of physical ageing in confined polymer: Fluorescence and dielectric spectroscopy studies of polymer–silica nanocomposites. *Journal of Physics: Condensed Matter*, 19(20), 205120.

Prigogine, I. and Defay, R. (1954). *Chemical Thermodynamics*. Longmans, London.

Qin, Q. and McKenna, G.B. (2006). Correlation between dynamic fragility and glass transition temperature for different classes of glass forming liquids. *J. Non-Cryst. Solids*, 352(28/29), 2977–2985.

Quattrosoldi, S., Androsch, R., Janke, A., Soccio, M., Lotti, N. (2020). Enthalpy relaxation, crystal nucleation and crystal growth of biobased poly (butylene isophthalate). *Polymers*, 12(1), 235.

Quélennec, B. (2022). Investigation des dynamiques relaxationnelles dans des systèmes époxy pour apporter une meilleure compréhension des effets de vieillissement physiques et chimiques dans le cadre d'applications structurales. PhD Thesis, Université de Rouen Normandie, Rouen.

Ramakrishnan, V., Harsiny, S., Goossens, J.G.P., Hoeks, T.L., Peters, G.W.M. (2016). Physical aging in polycarbonate nanocomposites containing grafted nanosilica particles: A comparison between enthalpy and yield stress evolution. *Journal of Polymer Science Part B: Polymer Physics*, 54, 2069–2081.

Ramos, A.R., Hutchinson, J.M., Kovacs, A.J. (1984). Isobaric thermal behavior of glasses during uniform cooling and heating. III. Predictions from the multiparameter KAHR model. *J. Polym. Sci. Pol. Phys.*, 22(9), 1655–1696.

Rauscher, P.M., Pye, J.E., Baglay, R.R., Roth, C.B. (2013). Effect of adjacent rubbery layers on the physical aging of glassy polymers. *Macromolecules*, 46(24), 9806–9817.

Richert, R., Lunkenheimer, P., Kastner, S., Loidl, A. (2013). On the derivation of equilibrium relaxation times from aging experiments. *The Journal of Physical Chemistry B*, 117, 12689–12694.

Righetti, M.C. and Mele, E. (2019). Structural relaxation in PLLA: Contribution of different scale motions. *Thermochimica Acta*, 67, 157–161.

Righetti, M.C., Gazzano, M., Delpouve, N., Saiter, A. (2017). Contribution of the rigid amorphous fraction to physical ageing of semi-crystalline PLLA. *Polymer*, 125, 241–253.

Ritland, H.N. (1956). Limitations of the fictive temperature concept. *J. Am. Ceram. Soc.*, 39(12), 403–406.

Rittigstein, P. and Torkelson, J.M. (2006). Polymer–nanoparticle interfacial interactions in polymer nanocomposites: Confinement effects on glass transition temperature and suppression of physical aging. *Journal of Polymer Science Part B: Polymer Physics*, 44(20), 2935–2943.

Robertson, C.G. and Wilkes, G.L. (1998). Refractive index: A probe for monitoring volume relaxation during physical aging of glassy polymers. *Polymer*, 39, 2129–2133.

Roe, R.J. and Curro, J.J. (1983). Small-angle x-ray scattering study of density fluctuation in polystyrene annealed below the glass transition temperature. *Macromolecules*, 16(3), 428–434.

Roland, C.M. (ed.) (2011). Viscoelastic behavior of rubbery materials. In *Cooperative Local Dynamics – The Glass Transition Zone*. Oxford University Press, New York.

Roseveare, W.E., Powell, R.E., Eyring, H. (1941). The structure and dynamics of liquids. *Journal of Applied Physics*, 12(9), 669–679.

Rowe, B.W., Pas, S.J., Hill, A.J., Suzuki, R., Freeman, B.D., Paul, D.R. (2009). A variable energy positron annihilation lifetime spectroscopy study of physical aging in thin glassy polymer films. *Polymer*, 50, 6149–6156.

Saiter, A. (2002). Contribution à l'étude de la relaxation principale α et des cinétiques de relaxation dans des polymers vitreux : fragilité et approche entropiques. PhD Thesis, Université de Rouen, Rouen.

Saiter, J.M., Dargent, E., Kattan, M., Cabot, C., Grenet, J. (2003). Fragility index of drawn or annealed poly (ethylene terephthalate) films studied by thermally stimulated depolarisation currents. *Polymer*, 44, 3995–4001.

Sato, K. and Sprengel, W. (2012). Element-specific study of local segmental dynamics of polyethylene terephthalate upon physical aging. *The Journal of Chemical Physics*, 137, 104906.

Schapery, R. and Sun, C. (2000). *Time Dependent and Nonlinear Effects in Polymers and Composites*. ASTM International, West Conshohocken.

Schawe, J.E.K. (2015). Analysis of non-isothermal crystallization during cooling and reorganization during heating of isotactic polypropylene by fast scanning DSC. *Thermochimica Acta*, 603, 85–93.

Schawe, J.E.K. and Löffler, J.F. (2022). Kinetics of structure formation in the vicinity of the glass transition. *Acta Materialia*, 226, 117630.

Schawe, J.E.K. and Wrana, C. (2020). Competition between structural relaxation and crystallization in the glass transition range of random copolymers. *Polymers*, 12(8), 1778.

Scherer, G.W. (1986). Volume relaxation far from equilibrium. *Journal of the American Ceramic Society*, 69, 374–381.

Schick, C. and Mathot, V. (2016). *Fast Scanning Calorimetry*. Springer, Berlin.

Schlosser, E. and Schönhals, A. (1991). Dielectric relaxation during physical ageing. *Polymer*, 32(12), 2135–2140.

Schrader, S. and Schonhals, A. (1989). *Relaxation in Polymers.* Springer, Berlin.

Sepúlveda, A., Leon-Gutierrez, E., Gonzalez-Silveira, M., Rodríguez-Tinoco, C, Clavaguera-Mora, M.T., Rodríguez-Viejo, J. (2011). Accelerated aging in ultrathin films of a molecular glass former. *Physical Review Letters*, 107, 025901.

Shakespeare, W. (2002). The glass transition phenomenon. In *Structural Chemistry of Glasses*, Rao, K.J. (ed.). Elsevier, Amsterdam, 77–135.

Simon, S.L. and Bernazzani, P. (2006). Structural relaxation in the glass: Evidence for a path dependence of the relaxation time. *Journal of Non-Crystalline Solids*, 352, 4763–4768.

Simon, S.L., Sobieski, J.W., Plazek, D.J. (2001). Volume and enthalpy recovery of polystyrene. *Polymer*, 42, 2555–2567.

Singh, M.K. and Singh, A. (2022). Nuclear magnetic resonance spectroscopy. In *Characterization of Polymers and Fibres, the Textile Institute Book Series*, Singh, M.K. and Singh, A. (eds). Woodhead Publishing, Sawston.

Soma, H., Nishitsuji, S., Inoue, T. (2012). Molecular weight dependence in a relaxation phenomenon at glassy state: Physical aging of polycarbonate. *Polymer*, 53, 895–896.

Song, Z., Rodriguez-Tinoco, C., Mathew, A., Napolitano, S. (2022). Fast equilibration mechanisms in disordered materials mediated by slow liquid dynamics. *Science Advances*, 8(15), eabm7154.

Sprincz, E. and Beck, C.W. (1981). Classification of the amber beads of the Hungarian Bronze Age. *Journal of Field Archaeology*, 8, 469–485.

Stocchi, A., Pellicano, A., Rossi, J.P., Bernal, C., Montemartini, P. (2006). Physical and water aging of glass fiber-reinforced plastic pipes. *Composite Interfaces*, 13(8/9), 685–697.

Struik, L.C.E. (ed.) (1977a). Aging after complicated thermal histories. In *Physical Ageing in Amorphous Polymers and Other Materials*. Elsevier, Amsterdam.

Struik, L.C.E. (1977b). Physical aging in plastics and other glassy materials. *Polymer Engineering and Science*, 17(3), 165–173.

Struik, L.C.E. (1978). *Physical Aging in Amorphous Polymers and Other Materials*. Elsevier, New York.

Struik, L.C.E. (1988). Dependence of relaxation times of glassy polymers on their specific volume. *Polymer*, 29(8), 1347–1353.

Struik, L.C.E. (1989a). Mechanical behaviour and physical ageing of semi-crystalline polymers: 3. Prediction of long term creep from short time tests. *Polymer*, 30, 799–814.

Struik, L.C.E. (1989b). Mechanical behaviour and physical ageing of semi-crystalline polymers: 4. *Polymer*, 30, 815–830.

Suzuki, H., Grebowicz, J., Wunderlich, B. (1985). Glass transition of poly(oxymethylene). *British Polymer Journal*, 17, 1–3.

Svoboda, R., Pustková, P., Málek, J. (2007). Relaxation behavior of glassy selenium. *Journal of Physics and Chemistry of Solids*, 68(5/6), 850–854.

Swallen, S.F., Kearns, K.L., Mapes, M.K., Kim, Y.S., McMahon, R.J., Ediger, M.D., Satija, S. (2007). Organic glasses with exceptional thermodynamic and kinetic stability. *Science*, 315, 353–356.

Takebe, T., Yamasaki, K., Funaki, K., Malanga, M. (2009). Properties of syndiotactic polystyrene. In *Syndiotactic Polystyrene: Synthesis, Characterization, Processing, and Applications*, Schellenberg, J. (ed.). John Wiley and Sons, New York.

Tamann, G. and Hesse, W. (1926). Die abhängigkeit der viskosität von der temperatur bei unterkühlten flüssigkeiten. *Z. Anorg. Allg. Chem.*, 156(1), 245–257.

Tant, M.R. and Wilkes, G.L. (1981). Physical aging studies of semicrystalline poly(ethylene terephthalate). *Journal of Applied Polymer Science*, 26(9), 2813–2825.

Tarjus, G., Kivelson, S.A., Nussinov, Z., Viot, P. (2005). The frustration-based approach of supercooled liquids and the glass transition: A review and critical assessment. *Journal of Physics: Condensed Matter*, 17(50), R1143.

Tool, A.Q. (1946). Relation between inelastic deformability and thermal expansion of glass in its annealing range. *J. Am. Ceram. Soc.*, 29(9), 240–253.

Tool, A.Q. and Eichlin, C.G. (1925). Variations in glass caused by heat treatment. *J. Am. Ceram. Soc.*, 8(1), 1–17.

Tool, A.Q. and Eichlin, C.G. (1931). Variations caused in the heating curves of glass by heat treatment. *J. Am. Ceram. Soc.*, 14(4), 276–308.

Tool, A.Q., Lloyd, D.B., Merritt, G.E. (1930). Dimensional changes caused in glass by heating cycles. *J. Am. Ceram. Soc.*, 13(9), 632–654.

Tribone, J.J., O'Reilly, J.M., Greener, J. (1986). Analysis of enthalpy relaxation in poly(methyl methacrylate): Effects of tacticity, deuteration, and thermal history. *Macromolecules*, 19, 1732–1739.

Turnbull, D. and Cohen, M.H. (1961). Free-volume model of the amorphous phase: Glass transition. *The Journal of Chemical Physics*, 34, 120. doi: 10.1063/1.1731549.

Turner, S. (1964). Creep in thermoplastics-unplasticized PVC. *British Plastics*, 37, 682–685.

Uhrig, D. and Mays, J.W. (2005). Experimental techniques in high-vacuum anionic polymerization. *Journal of Polymer Science Part A: Polymer Chemistry*, 43, 6179–6222.

Varol, N., Delpouve, N., Araujo, S., Domenek, S., Guinault, A., Golovchak, R., Ingram, A., Delbreilh, L., Dargent, E. (2020). Amorphous rigidification and cooperativity drop in semi-crystalline plasticized polylactide. *Polymer*, 194, 122373.

Vempati, U. and Boolchand, P. (2004). The thermally reversing window in ternary Gex PxS1 − 2x glasses. *Journal of Physics: Condensed Matter*, 16(44), S5121.

Viret, J. (2003). Caractérisation de matériaux avancés à faible permittivité par Spectroscopie Temporelle d'Annihilation de Positons (PALS). Report, PALS.

Vogel, H. (1921). The temperature dependence law of the viscosity of fluids. *Phys. Z.*, 22, 645–646.

Wagner, H. and Richert, R. (1997). Thermally stimulated modulus relaxation in polymers: Method and interpretation. *Polymer*, 38(23), 5801–5806.

Wang, Y.Y., Nakanishi, H., Jean, Y.C., Sandreczki, T.C. (1990). Positron annihilation in amine-cured epoxy polymers–pressure dependence. *Journal of Polymer Science Part B: Polymer Physics*, 28(9), 1431–1441.

Welch, R.C., Smith, J.R., Potuzak, M., Guo, X., Bowden, B.F., Kiczenski, T.J., Allan, D.C., King, E.A., Ellison, A.J., Mauro, J.C. (2013). Dynamics of glass relaxation at room temperature. *Physical Review Letters*, 110, 265901.

White, R.P. and Lipson, J.E.G. (2009). Chain fluids: Contrasts of theoretical and simulation approaches, and comparison with experimental alkane properties. *The Journal of Chemical Physics*, 131(7), 074109.

White, R.P. and Lipson, J.E.G. (2015). Free volume in the melt and how it correlates with experimental glass transition temperatures: Results for a large set of polymers. *ACS Macro Lett.*, 4(5), 588–592.

White, R.P. and Lipson, J.E.G. (2016). Polymer free volume and its connection to the glass transition. *Macromolecules*, 49(11), 3987–4007.

Williams, G. and Watts, D.C. (1970). Non-symmetrical dielectric relaxation behaviour arising from a simple empirical decay function. *Trans. Farad. Soc.*, 66, 80–85.

Williams, M.L., Landel, R.F., Ferry, J.D. (1955). The temperature dependence of relaxation mechanisms in amorphous polymers and other glass-forming liquids. *Journal of the American Chemical Society*, 77(14), 3701–3707.

Woo, E.M. and Chang, L. (2011). Tacticity in vinyl polymers. In *Encyclopedia of Polymer Science and Technology*, Herman, F.M. (ed.). John Wiley and Sons, New York.

Wunderlich, B. (2003). Reversible crystallization and the rigid-amorphous phase in semicrystalline macromolecules. *Progress in Polymer Science*, 28, 383–450.

Wypych, A., Duval, E., Boiteux, G., Ulanski, J., David, L., Mermet, A. (2005). Effect of physical aging on nano- and macroscopic properties of poly(methyl methacrylate) glass. *Polymer*, 46, 12523–12531.

Yeh, G.S.Y. and Geil, P.H. (1967). Crystallization of polyethylene terephthalate from the glassy amorphous state. *Journal of Macromolecular Science, Part B: Physics*, 1(2), 235–249.

Yoon, H., Koh, Y.P., Simon, S.L., McKenna, G.B. (2017). An ultrastable polymeric glass: Amorphous fluoropolymer with extreme fictive temperature reduction by vacuum pyrolysis. *Macromolecules*, 50(11), 4562–4574.

Zhao, J. and McKenna, G.B. (2012). Temperature divergence of the dynamics of a poly(vinyl acetate) glass: Dielectric vs. mechanical behaviors. *The Journal of Chemical Physics*, 136, 154901.

Zhao, J., Song, R., Zhang, Z., Linghu, X., Zheng, Z., Fan, Q. (2001). A study of the physical aging in semicrystalline poly(ethylene terephthalate) via differential scanning calorimetry. *Macromolecules*, 34, 343–345.

Zhao, J., Simon, S.L., McKenna, G.B. (2013). Using 20-million-year-old amber to test the super-Arrhenius behaviour of glass-forming systems. *Nat. Commun.*, 4, 1783.

Zheng, Y., Simon, S.L., McKenna, G.B. (2002). Modeling structural recovery in glasses: An analysis of the peak-shift method. *Journal of Polymer Science Part B: Polymer Physics*, 40(18), 2027–2036.

Zhou, X. and Jiang, Z. (2006). Effects of sequence distribution and physical aging on physical properties of PES/PEES random, block, and alternative copolymers. *Chemical Research in Chinese Universities*, 22(6), 787–791.

Zhou, H., Lofgren, E.A., Jabarin, S.A. (2007). Effects of microcrystallinity and morphology on physical aging of poly(ethylene terephthalate): Physical aging of PET. *Journal of Applied Polymer Science*, 106(5), 3435–3443.

Zhuravlev, E., Schmelzer, J.W.P., Wunderlich, B., Schick, C. (2011). Kinetics of nucleation and crystallization in poly (ε-caprolactone)(PCL). *Polymer*, 52(9), 1983–1997.

7

Several Aspects of Elastomer Fatigue

Stéphane MÉO[1], Alexis DELATTRE[2], Jean-Louis POISSON[3], Florian LACROIX[1] and Stéphane LEJEUNES[4]

[1]*LaMé, INSA, Université de Tours, France*

[2]*Airbus, Toulouse, France*

[3]*Baker Hugues, Houston, United States*

[4]*LMA, CNRS, AMU, Marseille, France*

7.1. Introduction

Elastomers are increasingly used in the industrial world. They exhibit high deformability while being (quasi-)incompressible and show significant damping when subjected to loading, a viscosity-related characteristic that is often sought after. In many cases, the mechanical stresses the material experiences are cyclic and long term. As a result, elastomers can be subjected to millions of cycles, which typically lead to the degradation of their performance properties. To predict, understand, and model such phenomena, the study of elastomers focuses on material fatigue, which can be viewed as aging under mechanical loading.

Although interesting, the aforementioned remarkable properties are often coupled with other not fully understood dissipative mechanical phenomena, and they generate many difficulties in the prediction of their behavior (large strains, incompressibility, heating, strong sensitivity to temperature, stress rate, Payne effect, etc.). Furthermore, being used under increasingly extreme conditions (wide ranges of temperature, amplitude and frequency), it is difficult to propose a model, that is, a

Aging of Industrial Polymers 1,
coordinated by Emmanuel RICHAUD.

law of behavior and a unique set of parameters, which can cover the entire field of use and/or predict the fatigue durability of real industrial parts.

The first point, that is, the model aspects and modeling, are beyond the scope of this chapter. There are various articles to which readers can refer:

– for hyperelasticity, that is, nonlinear elasticity: Mooney (1940); Hart-Smith (1966) for strain-invariant approaches; Ogden (1972) for models based on principal stretches; and Arruda and Boyce (1993) or Miehe (2000) for more physically motivated models;

– to account for the Mullins effect (mechanical softening phenomenon observed for cyclic loading and illustrated, for example, in Mullins (1969) and broadly described in Diani et al. (2009)): Govindjee and Simo (1991); Marckmann et al. (2002); Göktepe and Miehe (2005);

– for viscoelastic aspects (rate dependence coupled with (hyper)elasticity), several references that may be helpful to the reader can be mentioned:

- integral approaches: Coleman and Noll (1974); Christensen (1982); Morman (1988),

- microphysically justified approaches: Reese (2003); Göktepe and Miehe (2005),

- approaches referred to as "differential" (relying on the seminal works of Sidoroff (1973, 1975, 1976)), which include Tallec and Rahler (1994), Miehe and Keck (2000), Meo et al. (2002) and Martinez et al. (2011);

– with respect to thermomechanical coupling (elastomers having both high sensitivity to their thermal environment and dissipative nature generating heat under mechanical loading), several authors have considered and modeled it. Hence, Holzapfel and Reiter (1995) propose a thermomechanical framework with linear viscoelasticity. In terms of thermoelasticity, it is also possible to refer to Holzapfel and Simo (1996a, 1996b) and Lion (1997a, 1997b), to Holzapfel and Simo (1996a) and Boukamel et al. (2001) in case of thermo-visco-elasticity and to Lion (2000) in case of thermo-visco-plasticity.

Although not previously presented, all these *thermomechanical* phenomena (sometimes coupled to others, such as chemistry (e.g. Lejeunes et al. 2018)) must be considered when approaching the main subject of this chapter, namely, the study of elastomer fatigue behavior. If fatigue is considered the consequence of mechanical damage under cyclic loading, in relation to the literature, three approaches are typically used:

– crack propagation;

– crack initiation;

– those based on continuous damage mechanics.

This chapter focuses on initiation approaches (without pre-existing defects in the material), which are, from a historical *perspective*, the first ones to have been studied (Wöhler 1867).

7.2. Problem formulation

7.2.1. *Fatigue in general terms*

Subjecting a material, component or industrial structure to fatigue involves applying cyclic mechanical loading with varying amplitude and frequency. Even mechanical loading of relatively low amplitude (e.g. significantly below the limit of elasticity for a metallic material) can eventually lead to failure.

It is therefore clear that this problem is among the most important ones to comprehend for many industrial applications because, finally, anticipating fatigue behavior makes it possible to estimate the service life of a part under cyclic loading.

7.2.2. *Elastomer fatigue*

Generally speaking, predicting the fatigue behavior of a material is never simple, and this is all the more so for elastomer materials. Mars and Fatemi (2004) list the various factors that may influence the fatigue behavior of these materials (naturally, some factors are transposable to other families of materials, but some are clearly specific to elastomers):

– *History and type of loading*: fatigue uniaxial loading (tension, for example) can be described by various parameters (maximal, minimal, average or alternate stress or strain, load ratio) (section 7.2.2.3.1) that may influence fatigue behavior. For example, the load ratio (equations [7.4] and [7.5]) will be an essential factor for a crystallizable elastomer: for a tension–tension load, a strictly positive value of this quantity would tend to increase the lifetime; on a non-crystallizable material, an increase in the strain (or stress) amplitude may accelerate the end of life. The load path or its multiaxial character also has a very strong influence on the fatigue response (Poisson et al. 2018).

– *Experimental conditions and environment*: it seems obvious that temperature plays a role in such characterizations. For example, as in the case of static loading, the positioning of the test with respect to the glass transition temperature will be essential. The reinforcing side, attributed to crystallization and previously cited, could be reduced or enhanced at high temperature or a quite low temperature, respectively (Hosseini et al. 2016). Other effects of the environment, such as immersion in water, salted water and nitrogen, will also impact the fatigue response, particularly through property gradient effects (Bathias 2013; Narynbek Ulu et al. 2016).

– *Formulation and process*: these involve numerous compounds and technical operations that will impact the static and fatigue behavior of the material. For example, the specific surface, nature and quality of filler dispersion will be key points (Majesté 2017), as well as the process used to obtain the part (injection, compression, extrusion, etc.).

– *Intrinsic behavior of the material*: as previously noted, the ability to crystallize and the more or less dissipative nature (which, via heating, will lead to the second point of this list) should also be considered.

Though non-exhaustive, this list of factors that may influence the elastomer fatigue response highlights the complexity of accurately characterizing it. As stated in the introduction, three experimental paths can be identified: cracking approaches and damage determination or monitoring approaches, which will be briefly detailed, and initiation approaches, which will be explained in greater detail.

7.2.2.1. *Crack propagation approaches*

Crack propagation approaches are based on the hypothesis that an initial defect or crack exists. They rely on the seminal works of Griffith (1921) on metallic materials. These works were initially expanded to elastomers by Rivlin and Thomas (1953) and Thomas (1955, 1958), introducing the tearing energy released during propagation. Other works focused on the influence of experimental conditions (Greensmith and Thomas 1956), the role of loads (Greensmith 1957), the propagation rate (Thomas 1961; Lake and Lindley 1964, 1965) and the regimes it can follow (Thomas 1958; Lake and Lindley 1965; Mars 2001, 2009). Finally, the cracking process and its typology are also widely dependent on the studied elastomer material, particularly on its capacity to crystallize under stress (this point will be resumed further in this chapter) (Saintier 2001; Le Cam et al. 2004; Mars 2009).

7.2.2.2. *Continuous damage mechanics approaches*

Damage mechanics is a way to approach elastomer fatigue by interpreting this phenomenon as the appearance of microcracks or *microdamage*, which, by coalescence, will generate a macrocrack leading to rupture. To keep things simple, the real stress of the damaged material (σ) is "corrected" based on the healthy state stress ($\tilde{\sigma}$) via a damage variable D (Lemaitre et al. 2009)):

$$\sigma = (1 - D)\tilde{\sigma} \qquad [7.1]$$

where D equals 0 if the material is healthy and 1 if it is completely damaged, that is, ruptured. The laws of evolution of this variable can be obtained by approaches relying on the thermodynamics of irreversible processes and the context of generalized standard materials (Halphen and Son Nguyen 1975; Chrysochoos 2018).

In the article by Wang et al. (2002), a hyperelastic behavior law is coupled with a continuous damage variable. Nevertheless, within the framework of non-dissipative behavior, Ayoub et al. (2010) extended the field of application to multiaxial loading. Finally, cases of complex mechanical behaviors are addressed by Grandcoin Laneyrie (2008) and Delattre (2014).

7.2.2.3. *Initiation approaches*

7.2.2.3.1. Characterization tests

The experimental characterization to be implemented for such approaches relies on cyclic loads that are often sinusoidal (they may sometimes be triangular). The load is defined by:

– the type of stress control (controlled force) or strain control (controlled displacement);

– stress frequency;

– its amplitude;

– its maximum or minimum value.

Knowing that during a fatigue cycle, we have:

– an average stress/strain:

$$\sigma_{moy} = \frac{\sigma_{max} + \sigma_{min}}{2}\ ;\ \varepsilon_{moy} = \frac{\varepsilon_{max} + \varepsilon_{min}}{2} \qquad [7.2]$$

– an alternate stress/strain:

$$\sigma_{alt} = \frac{\sigma_{max} - \sigma_{min}}{2};\ \varepsilon_{alt} = \frac{\varepsilon_{max} - \varepsilon_{min}}{2} \qquad [7.3]$$

where σ_{max}, σ_{min}, ε_{max} and ε_{min} represent the maximal stress, minimal stress, maximal strain and minimal strain applied to the sample, respectively.

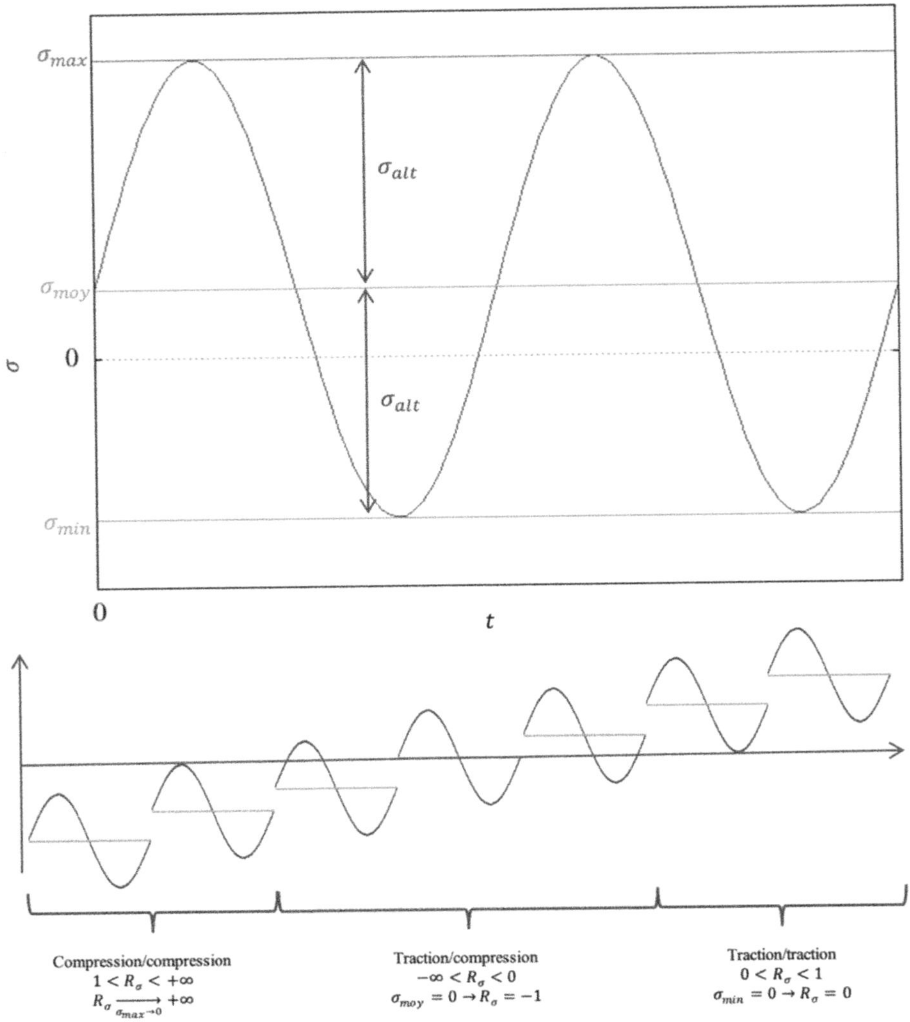

Figure 7.1. *Definition of fatigue load (stress control). For a color version of this figure, see www.iste.co.uk/richaud/industrialpolymers1.zip*

The load ratio can also be defined in terms of stress or strain as:

$$R_\sigma = \frac{\sigma_{\min}}{\sigma_{\max}} \qquad [7.4]$$

$$R_\varepsilon = \frac{\varepsilon_{\min}}{\varepsilon_{\max}} \qquad [7.5]$$

By stating the load ratio (and its type) and the maximal stress or strain (the minimal stress or strain, respectively), the minimal stress or strain (the maximal stress or strain, respectively) can be directly determined.

The load ratio facilitates a proper understanding of the type of applied fatigue cycle.

– Tension-tension:

$$0 < R_\sigma < 1 \qquad [7.6]$$

– Compression–compression:

$$1 < R_\sigma < +\infty \qquad [7.7]$$

– Tension-compression:

$$-\infty < R_\sigma < 0 \qquad [7.8]$$

Figure 7.1 summarizes the situations encountered as a function of the load ratio.

Fatigue behavior is typically represented using S–N (stress versus number of cycles) curves. Although the vertical axis originally represented stress, the concept of the S–N curve has been extended to include any predictive mechanical quantity – such as maximum stress, alternating stress or strain (this point will be discussed further below) – plotted as a function of the number of cycles required to cause specimen failure or, alternatively, to achieve significant degradation of a mechanical property, for example, a percentage loss of stiffness. Each point on the graph corresponds to a single test (excluding repeatability), defined by cyclic loading characterized, for instance, by maximum and minimum stress values applied at a given frequency, for which the number of cycles to failure (or another criterion, as previously described) is recorded.

The low-cycle fatigue (LCF) regime corresponds to conditions involving highly damaging mechanical stresses, which may result from significant plastic deformation (in metals) or successive pullouts (in elastomers) during each cycle. High-cycle fatigue (HCF), on the other hand, refers to a regime where mechanical stresses cause relatively minor damage. In metals, this regime is often characterized by the absence of macroscopic plastic deformation. Consequently, failure under fatigue loading in this regime is generally attributed to phenomena at smaller scales, such as thermomechanical effects, intrinsic material defects or imperfections in the specimen. Variability in fatigue life can be quite significant within this regime.

The shape of the S–N curve varies for each material, meaning that the distinctions between these two regimes may differ. This transition is often characterized by a "significant" change in behavior. Identifying this limit is often key in a fatigue study, and if the objective is only its determination, a complete S–N study may prove very (too) costly and finally useless. Faster methods may therefore be adopted (Le Saux et al. 2010; Cruanes 2015; Cruanes et al. 2019).

As a general rule, by adopting a *semilogarithmic* representation (linear on the abscissa and logarithmic on the ordinate), the *evolution* in these two distinct zones will often be linear.

In elastomer studies, the transition between these two zones is not necessarily easy to determine, and it is often only the HCF behavior, that is, the second linear zone, that is studied. The objective is therefore to identify an end-of-life criterion (section 7.2.2.3.3) that will allow a linear representation in a semilogarithmic plane of the S–N curve. Ultimately, the ability of this criterion to align the results along a straight line will be assessed alongside its robustness and relevance. This will often be measured by a correlation coefficient R^2 (the closer to 1 the latter is, the better the extrapolation).

7.2.2.3.2. Initiation criterion or end-of-life criterion

As previously mentioned, a Wöhler or S–N approach consists of plotting a series of points on a graph. The ordinate of each point represents the value of a criterion generated by the loading (which can be directly the applied stress or strain), whereas the abscissas correspond to the number of cycles until the end of life under that loading. Failure of a sample is not necessarily the most appropriate method adopted; some authors prefer to consider the end of life as the instant when the studied part loses its functionality. Thus, three main criteria are encountered in the literature:

– *Failure of the part*: the study presented in Cadwell et al. (1940) is one of the first studies to implement this approach in elastomers. It is one of the easiest criteria to implement but may be questionable for industrial applications, as components often lose their functionality well before failure. Moreover, it can lead to rather long testing times.

– *Optical detection of defects*, for which the end-of-life of the part is considered to occur when a crack reaches a predefined critical size. Naturally, this choice is complex and may be criticized as, in particular, only cracks that reach the surface are detectable, thereby limiting both the specimen geometries and their loading conditions (Robisson 2000; Saintier 2001).

– *Evolution of a macroscopic mechanical parameter*: this may be:

- a significant decrease in the maximal force during displacement-controlled loading. Mars (2001) adopted a 15 % drop in the maximum force (under displacement control) relative to the value of the 128th cycle (to ignore the smoothing of the first cycles related to the Mullins effect and to relaxation phenomena) (to exclude the initial softening caused by the Mullins effect and relaxation phenomena);

- a significant rate of change in the stiffness of the specimen (Ostoja-Kuczynski et al. 2003).

Each of these three approaches has advantages and drawbacks. The choice between them should be guided by the type of material, the application in which it is used and its operating environment.

7.2.2.3.3. Fatigue criteria

We have previously seen that the choice of the number of cycles corresponding to initiation, that is, the abscissa in the S–N representation, is not necessarily a straightforward decision. It is now natural to question the ordinate of such a representation: these are the predicitive criteria for fatigue. Numerous options exist and may either be obtained directly by "reading" the test conditions or may require calculations based on more or less complex concepts from continuum mechanics. Verron and Andriyana (2008) provide a good overview of what an initiation criterion should be:

– It should preferably be expressed in terms of quantities derived from continuum mechanics rather than from fracture mechanics to enable the prediction of the spatial distribution of fatigue via modeling (finite elements, for example) of an industrial part and its mechanical environment.

– It should be initiated based on observations of physical phenomena.

– It should be formulated from a well-established theoretical framework.

– It should be suitable for implementation in a finite element code without leading to excessive computational cost (some sort of postprocessing).

Before providing the most commonly used criteria to quantify and qualify the fatigue of elastomers, it is worth recalling a few tools from continuum mechanics that will be used in the definitions of these quantities. Classically, in the case of a large strain approach, if a medium undergoes a transformation from an initial configuration C_0 to a deformed configuration C_t, the transformation can be characterized by the gradient of transformation (with J denoting its determinant):

$$\bar{\bar{F}};\ J = det\bar{\bar{F}} \quad [7.9]$$

This can then be used for the definitions in the Lagrangian configuration, that is, in C_0, of:

$$\bar{\bar{C}} = \bar{\bar{F}}^T.\bar{\bar{F}} \quad [7.10]$$

$$\bar{\bar{E}} = \frac{1}{2}(\bar{\bar{C}} - \bar{\bar{1}}) \quad [7.11]$$

where $\bar{\bar{C}}$ and $\bar{\bar{E}}$ are the right Cauchy–Green tensor and the Lagrangian strain tensor, respectively.

In the Eulerian configuration, this yields:

$$\bar{\bar{B}} = \bar{\bar{F}}.\bar{\bar{F}}^T \quad [7.12]$$

$$\bar{\bar{A}} = \frac{1}{2}(\bar{\bar{1}} - \bar{\bar{B}}^{-1}) \quad [7.13]$$

where $\bar{\bar{B}}$ and $\bar{\bar{A}}$ are the left Cauchy–Green tensor and the Euler–Almansi strain tensor, respectively.

Typically, it will be possible to define stresses via $\bar{\bar{\pi}}$, $\bar{\bar{S}}$ or $\bar{\bar{\sigma}}$. $\bar{\bar{\pi}}$ is the first Piola–Kirchoff tensor (PK_1) representing the real cohesive force density in the material per unit of undeformed surface. $\bar{\bar{\sigma}}$ is the Cauchy stress tensor quantifying the real cohesive force density in the material per unit of deformed surface, and $\bar{\bar{S}}$ is the second Piola–Kirchoff tensor (PK_2), representing the cohesive force density transported in the initial configuration per unit of undeformed surface:

$$\bar{\bar{S}} = J\bar{\bar{F}}^{-1}.\bar{\bar{\sigma}}.\bar{\bar{F}}^{-1} = \bar{\bar{F}}^{-1}.\bar{\bar{\pi}} \quad [7.14]$$

Hence, in the case of simple tension by a force $\vec{F}$ carried by the axis $\vec{x}$ of a specimen of initial section S_0 and deformed section S_t, we have:

$$\bar{\bar{\pi}} = \begin{bmatrix} \frac{F}{S_0} & 0 & 0 \\ 0 & 0 & 0 \\ 0 & 0 & 0 \end{bmatrix}; \bar{\bar{\sigma}} = \begin{bmatrix} \frac{F}{S_t} & 0 & 0 \\ 0 & 0 & 0 \\ 0 & 0 & 0 \end{bmatrix} \quad [7.15]$$

It is now possible to quantify the main initiation criteria found in the literature:

– *The maximum principal strain over a cycle* (E_1 or, rarer cases, A_1 obtained from the diagonalization of $\bar{\bar{E}}$ or $\bar{\bar{A}}$, respectively) can be used as an initiation criterion. This choice is guided by the fact that, generally, a crack tends to initiate on a plane normal to the direction of the highest strain under tension. Another possible strain-based approach is to consider the *alternate strain* (equation [7.3]):

$$E_{1_{\text{alt}}} = \frac{E_{1_{\max}} - E_{1_{\min}}}{2}; A_{1_{\text{alt}}} = \frac{A_{1_{\max}} - A_{1_{\min}}}{2} \quad [7.16]$$

– *The maximal main stress over a cycle* (π_1 or σ_1) leads to observations similar to those mentioned for the first principal strain. As with strain-based approaches, it is also possible to use the *alternate stress*:

$$\pi_{1_{\text{alt}}} = \frac{\pi_{1_{\max}} - \pi_{1_{\min}}}{2}; \sigma_{1_{\text{alt}}} = \frac{\sigma_{1_{\max}} - \sigma_{1_{\min}}}{2} \quad [7.17]$$

Saintier (2001) proposes to consider only its positive part, assuming that crack propagation does not occur under compression:

$$\sigma_{\text{eff}} = \sigma_{\max}^{P} - \sup(0, \sigma_{\min}^{P}) \quad [7.18]$$

To better account for the loading path, the *strain energy density (SED)* can be used as a criterion. Ultimately, this approach considers that crack propagation is related to the mechanical energy applied to the material (Beatty 1964). It can be expressed, for example, as follows:

$$\max_{\text{cycle}} \int_t \bar{\bar{\pi}} : d\bar{\bar{F}} \quad [7.19]$$

Some limitations are related to the previous criterion, particularly its inability to predict crack orientation or its lack of accuracy in the case of multiaxial loading (Mars 2001). On the basis of this observation, the works presented in Mars (2001), Mars and Fatemi (2002), and Zine et al. (2006, 2011) propose characterizing fatigue through the *crack energy density (CED)*, which can be expressed in incremental form as follows:

$$dW_C = \frac{\rho}{\rho_0} \frac{\vec{R}^T.\bar{\bar{C}}.\bar{\bar{S}}.d\bar{\bar{E}}.\vec{R}}{\vec{R}^T.\bar{\bar{C}}.\vec{R}} \quad [7.20]$$

where $\frac{\rho}{\rho_0}$ is the ratio of the mass density under the deformed configuration (C_t) to that in the initial configuration (C_0) and where $\vec{R}$ is a unit vector normal to the crack plane.

Relying on configurational mechanics (Eshelby 1951, 1975) is another approach. Within this formalism, the *Eshelby stress tensor* is defined as follows:

$$\bar{\bar{\Sigma}} = W\bar{\bar{1}} - \bar{\bar{C}}.\bar{\bar{S}} \quad [7.21]$$

where W is the strain energy and $\bar{\bar{1}}$ is the second-order identity tensor. Verron et al. (2006), Andriyana and Verron (2007) and Verron and Andriyana (2008) then proposed the following criterion:

$$\Sigma^* = |\min[(\Sigma_i)_{i=1\ldots3}, 0]| \quad [7.22]$$

where Σ_i are the eigenvalues of $\bar{\bar{\Sigma}}$. For multiaxial complex loading, the authors also propose an incremental reformulation of their criterion.

The dissipated energy density (DED) over a loading cycle can also serve as an initiation criterion; on the basis of works of Payne and Whittaker (1971), Lacroix et al. (2005), Méo (2012) and Poisson (2012), the following expression is proposed for this criterion:

$$W_d = \int_{cycle} \bar{\bar{\pi}}: d\bar{\bar{F}} \quad [7.23]$$

7.2.2.3.4. Use

Once the S–N curve is obtained by plotting the predictive criterion (stress, strain, energy density, etc.) as a function of the end-of-life criterion (number of cycles to rupture, corresponding to a decrease in stiffness, etc.), two uses are possible:

– The fatigue behavior (often obtained from laboratory specimens subjected to uniaxial loading) can be considered in a “contemplative” way, interpreting this result as an intrinsic property of the studied elastomer and, ultimately, using it to differentiate it from another material.

– A more quantitative approach consists of transposing these results to a complex structure (industrial, for example) made from the studied material and subjected to complex loading (thus clearly going beyond the 1D framework). A prerequisite is the precise knowledge of the material's constitutive law which, when implemented in a finite element calculation, for instance, can yield spatial and temporal distributions of various mechanical fields and therefore of the previously mentioned criteria. The geometric and temporal location of the maximum value of the chosen criterion will be considered the probable crack initiation zone, and the value of the criterion will provide, by simple transfer onto the S–N curve, the number of cycles the structure can undergo before this initiation. The approach is illustrated in Figure 7.2.

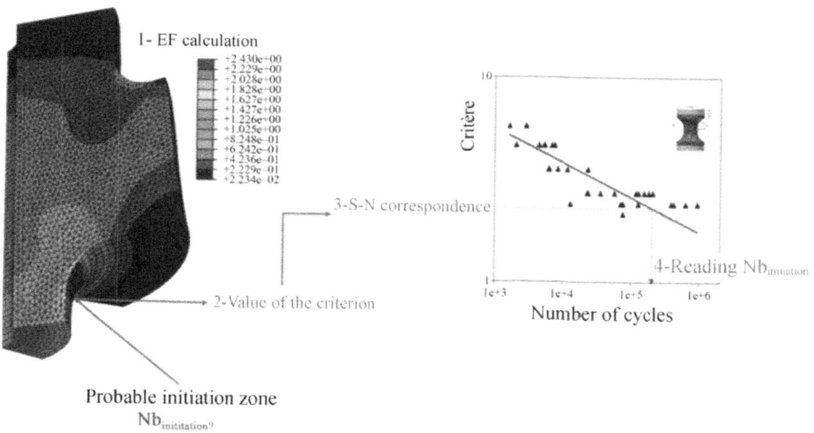

Figure 7.2. *Example of calculating the number of cycles to initiation on a structure under complex loading. (1) Computation of the spatial distribution of the chosen criterion; (2) identification of the maximum value of this quantity; and (3) transfer of this maximum onto the S–N curve and determination of the theoretical number of cycles corresponding to the fatigue life. For a color version of this figure, see www.iste.co.uk/richaud/industrialpolymers1.zip*

7.2.3. An example of a crack initiation study

7.2.3.1. Context

The study presented below was carried out on polychloroprene rubber (CR). The choice of this rubber is common in fatigue applications, given its good performance when faced with this type of phenomena, particularly in harsh environments (such as marine). A small portion of the results is presented below and can be exhaustively completed by referring to the works of Poisson (Méo 2012; Poisson 2012; Hosseini 2016).

As previously mentioned, the objective is to obtain a linear representation of the fatigue behavior (in a semilogarithmic plane). Hence, in this semilogarithmic plot, the more a criterion yields a linear trendline with a correlation coefficient (R^2) close to 1, the more reliable it is considered.

7.2.3.2. *Study under uniaxial loading (tension)*

7.2.3.2.1. Experimental conditions

In this study, stress-controlled loading at a frequency of 5 Hz is adopted. Table 7.1 summarizes the different force levels investigated. Tests are carried out at a temperature of 20°C on an INSTRON 8802 hydraulic fatigue testing machine.

F_{max}	R_σ	F_{max}	R_σ	F_{max}	R_σ	F_{max}	R_σ	F_{max}	R_σ
300 N	0.5	200 N	0.5	175 N	0.5	150 N	0.4	160 N, 125 N, 100 N	0.3
	0.4		0.4		0.4		0.3		0.2
250 N	0.5		0		0.3		0.2		0.1
	0.4		–0.1		0.2		0.1		0
			–0.3		0.1		0		–0.1
					0		–0.1		–0.3
					–0.1		–0.3		
					–0.3				

Table 7.1. *Experimental tension conditions*

7.2.3.2.2. Specimens

In a characterization program, the choice of specimens is not insignificant and may lead to variations in the results (Mars 2001). These variations may be caused by the presence of a geometric singularity, which results in both stress concentration and a loss of the uniaxial nature of the loading, or by an excessive large specimen volume, which, in the case of dissipative materials (as is our case), can lead to internal heating, which is likely to alter the material behavior and the nature of the phenomena involved (such as thermal degradation) (Lejeunes et al. 2018). We opted for an axisymmetric specimen geometry described in Figure 7.3, which is representative of an industrial application.

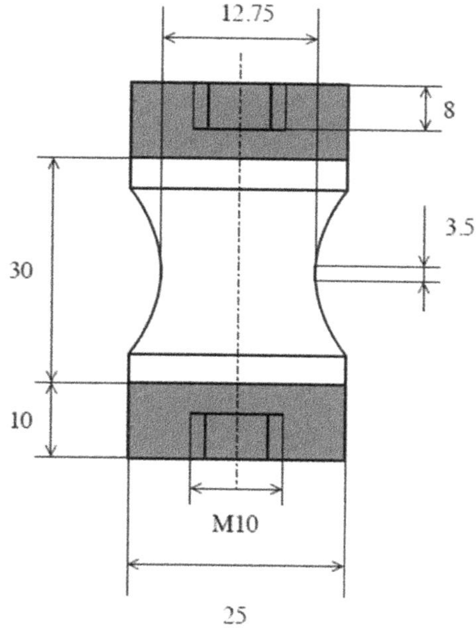

Figure 7.3. *Description of a "diabolo" specimen (dimension in mm)*

7.2.3.3. *Results*

7.2.3.3.1. S–N curves

Four criteria have been explored: the maximum main stress (π_{max}), the alternate main stress (π_{alt}), the SED and the DED. The results are presented in Figures 7.4–7.7.

The representations of alternate stress (Figure 7.5) and energy dissipation density (Figure 7.7) yield better results, particularly for high loading ratios. This analysis is supported mainly by the comparison of correlation coefficients R^2, which reflect the ability of each criterion to consolidate the fatigue data into a linear trend in the semilogarithmic plane.

One hypothesis for this enhanced performance of these criteria is their partial or full consideration of the loading history, which is partial in the case of alternate stress and full in the case of DED.

7.2.3.3.2. Haigh diagram

The analysis of the S–N curves based on the maximum stress (Figure 7.4) and SED (Figure 7.6) seems to highlight that the experimental results obtained for high loading ratios ($R_\sigma > 0.3$) do not appear to "align" well. This might suggest a change in the fatigue regime. To better understand this phenomenon, we adopt a Haigh diagram representation, which consists of plotting fatigue results as contour lines in a ($\pi_{\text{alt}}, \pi_{\text{max}}$) plane. This approach was adopted for natural rubber (NR), revealing enhanced fatigue resistance for load ratios $R_\sigma > 0$. This phenomenon was attributed to strain-induced crystallization of the material (Saintier 2001; Le Cam 2005). Figure 7.8 highlights a similar behavior, that is, a reinforcement, but for $R_\sigma > 0.2$. As with NR, this behavior is attributed to the strain-induced crystallization of our CR.

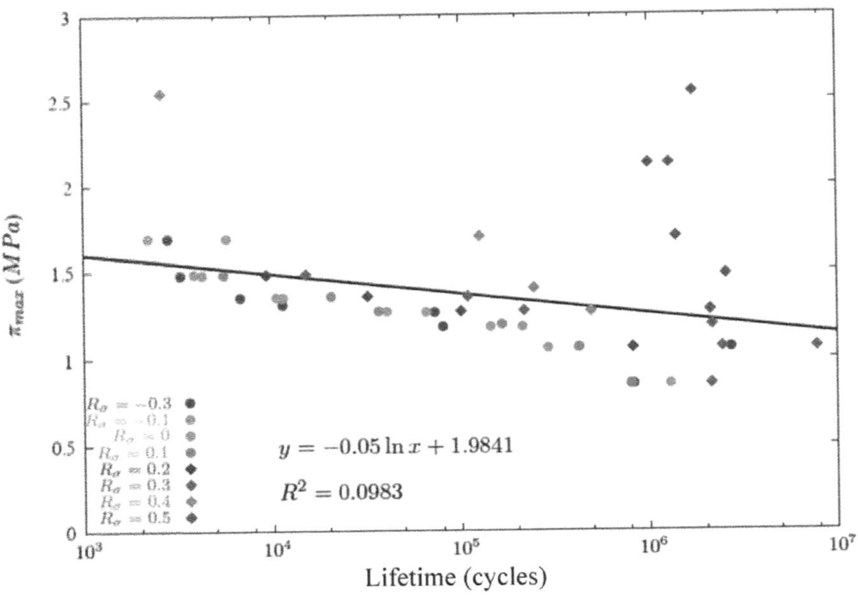

Figure 7.4. *S–N curve plotted against the maximum stress (π_{max}). Points: experimental results; black line: linear trendline. For a color version of this figure, see www.iste.co.uk/richaud/industrialpolymers1.zip*

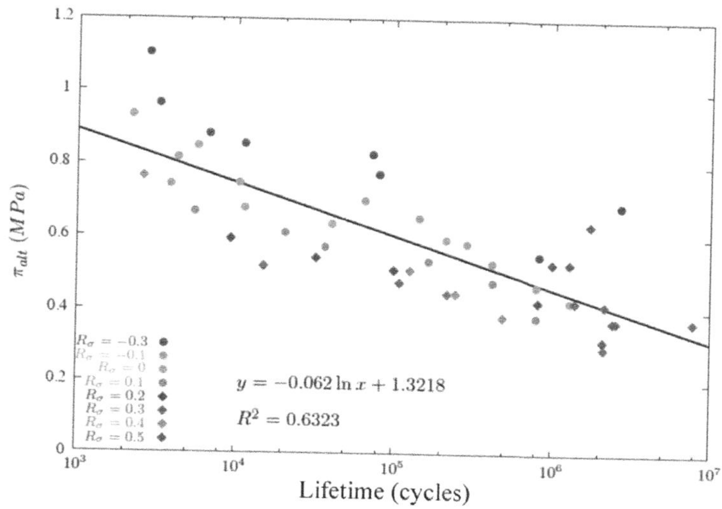

Figure 7.5. *S–N curve plotted against alternate stress (π_{alt}). Points: experimental results; black line: linear trendline. For a color version of this figure, see www.iste.co.uk/richaud/industrialpolymers1.zip*

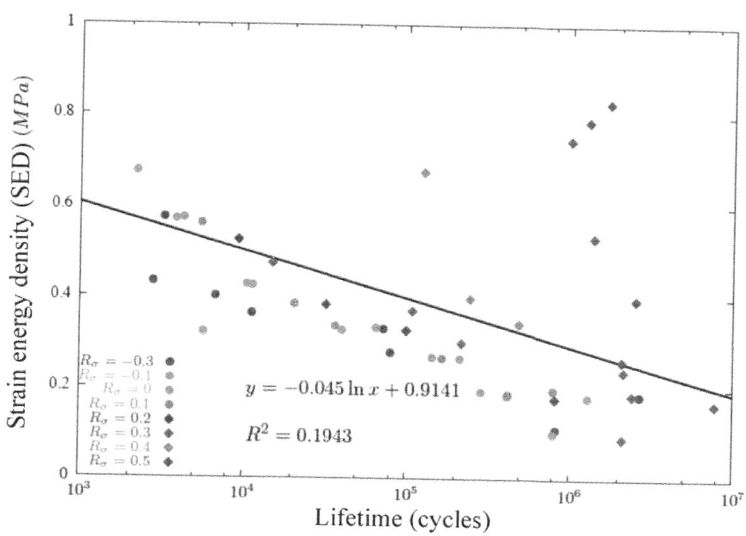

Figure 7.6. *S–N curve plotted against the strain energy density (SED). Points: experimental results; black line: linear trendline. For a color version of this figure, see www.iste.co.uk/richaud/industrialpolymers1.zip*

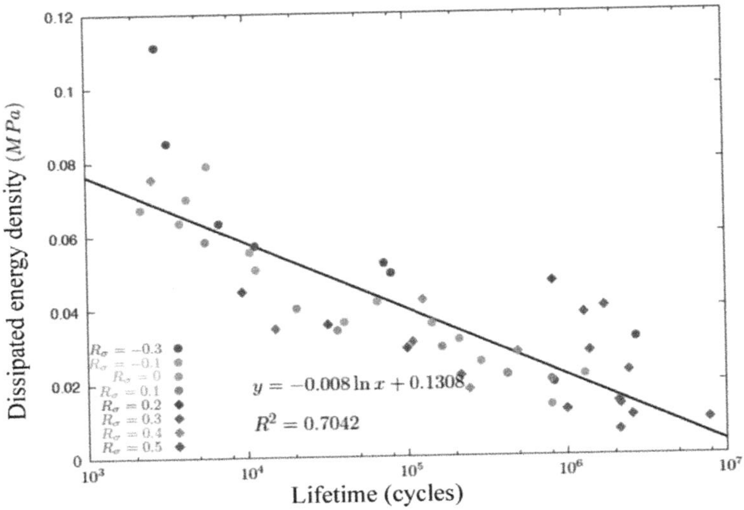

Figure 7.7. *S–N curve plotted against dissipated energy density (DED). Points: experimental results; black line: linear trendline. For a color version of this figure, see www.iste.co.uk/richaud/industrialpolymers1.zip*

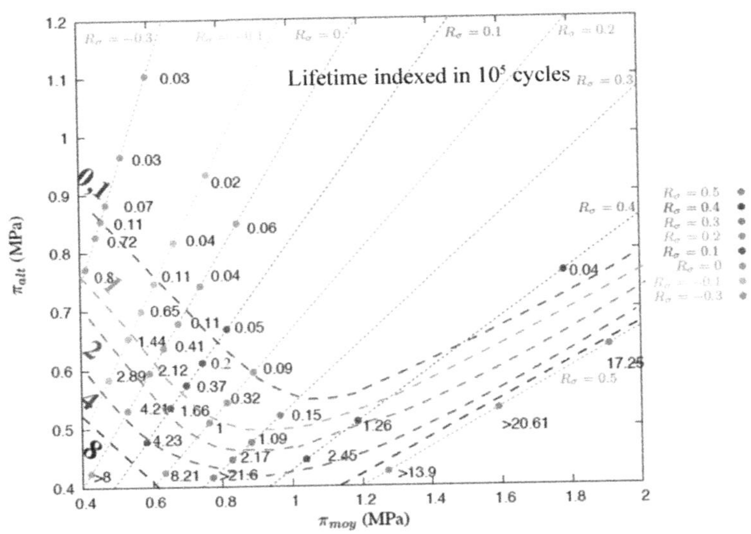

Figure 7.8. *Haigh diagram for simple tension for polychloroprene. For a color version of this figure, see www.iste.co.uk/richaud/industrialpolymers1.zip*

7.2.3.4. *Study under multiaxial loading (tension/torsion)*

Using the same specimens (Figure 7.3), a fatigue campaign under multiaxial loading was carried out. It consisted of applying a displacement in tension and torsion[1]; strain control was therefore adopted. The same loading ratio (denoted by R_ε) in tension and torsion is chosen, and three phase shifts ($\delta = 0°,\ 90°, 180°$) between these two loadings are explored. Pure tensile tests were also included to complete the campaign. Experimental conditions are summarized in Table 7.2.

δ (°)	R_ε	U_{max} (mm)	U_{min} (mm)	θ_{max} (°)	θ_{min} (°)
0°, 90°, 180°, Tension	0.2	20	4	40	8
		17	3.4	34	6.8
		15	3	30	6
	0.1	20	2	40	4
		17	1.7	34	3.4
		15	1.5	30	3
	0	20	0	40	0
		17	0	34	0
		15	0	30	0
	–0.2	20	–4	40	–8
		17	–3.4	34	–6.8
		15	–3	30	–6

Table 7.2. *Multiaxial fatigue conditions*

7.2.3.5. *Studied criteria*

Like in the uniaxial study (section 7.2.3.2), various criteria have been studied:

– the maximum on a stabilized cycle of the first main stress (π_1) locally calculated;

– the maximum on a stabilized cycle of the SED calculated both globally on the specimen and locally;

– the DED on a stabilized cycle, defined either globally or locally;

– Σ^*, the criterion based on configurational mechanics, is calculated locally;

– the locally calculated CED.

1 These tests were conducted on an electromechanical instrument ElectroForce BOSES 3300.

Some local evaluations require analytical calculations and/or modeling using a finite element model. The details of the latter can be found in Méo (2012), Poisson (2012) and Poisson et al. (2018).

From the results presented below, it is observed that the maximum principal stress (Figure 7.9) is not suitable for representing the fatigue behavior of our CR under multiaxial loading conditions, as the dispersion of results with respect to a linear trendline is too significant (a similar approach to that presented in section 7.2.3.3.1 is adopted). The partial consideration of the loading path in the calculation of the SED (Figure 7.12) leads to a reduced dispersion of the results. Finally, the criteria Σ^* (Figure 7.10), CED (Figure 7.11) and DED (Figure 7.13), which fully consider the loading path, yield much better results. Although the DED seems to provide a better description (in terms of dispersion, i.e., the value of the correlation coefficient R^2), caution is recommended to avoid establishing a hierarchy among these three criteria. On the one hand, if such a hierarchy existed, it would only be valid for the material, the specimen, and the loading conditions investigated in this study. On the other hand, for a non-dissipative material (hyperelastic, for example), the dissipation-based criterion could not be used. Finally, the criteria Σ^* and CED provide information on the direction of propagation of crack(s), which the DED, being a scalar quantity, cannot.

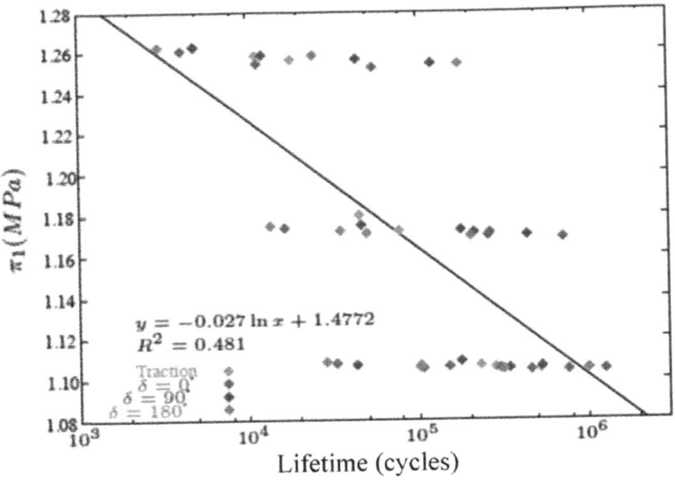

Analytical calculation

Figure 7.9. *(A) S–N curves plotted against the first principal stress. Points: experimental results; black line: linear trendline. For a color version of this figure, see www.iste.co.uk/richaud/industrialpolymers1.zip*

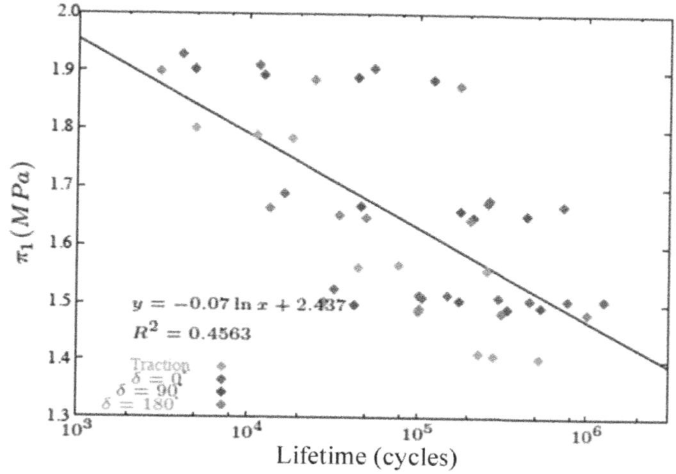

F.E. calculation

Figure 7.9. *(B) S–N curves plotted against the first principal stress. Points: experimental results; black line: linear tendency curve (continued). For a color version of this figure, see www.iste.co.uk/richaud/industrialpolymers1.zip*

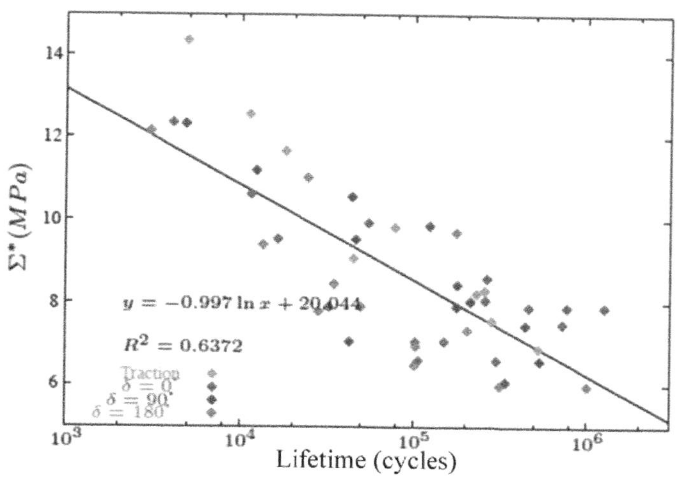

Figure 7.10. *S–N curve plotted against Σ^* (analytical calculation). Points: experimental results; black line: linear trendline. For a color version of this figure, see www.iste.co.uk/richaud/industrialpolymers1.zip*

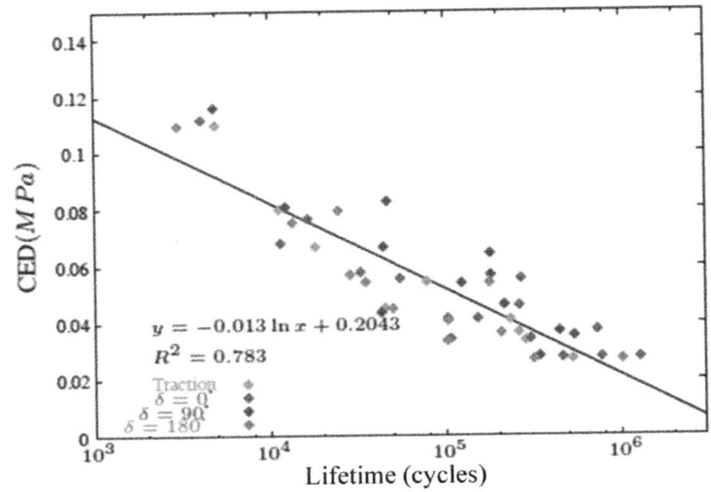

a) Analytical calculation

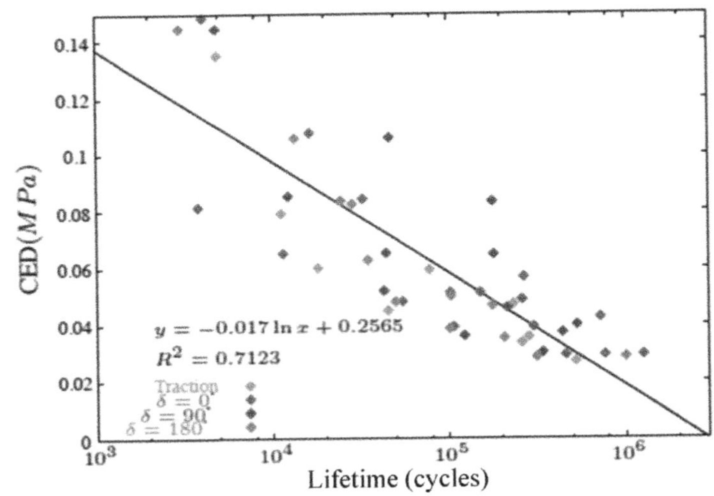

b) F.E. calculation

Figure 7.11. *S–N curve plotted against the CED. Points: experimental results; black line: linear trendline. For a color version of this figure, see www.iste.co.uk/richaud/industrialpolymers1.zip*

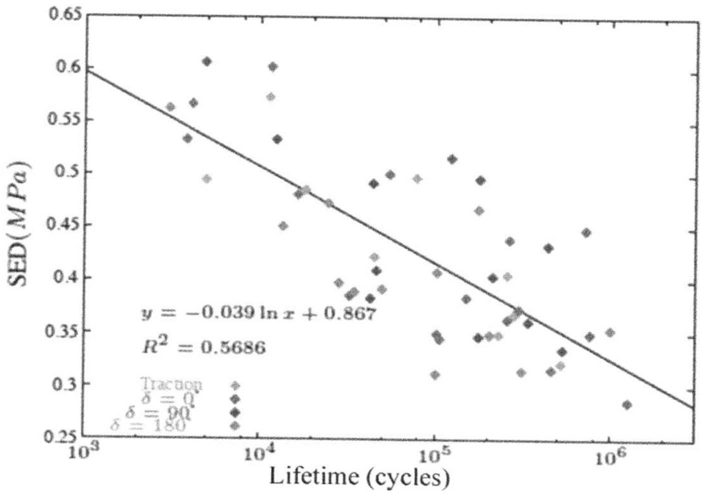

a) Analytical calculation

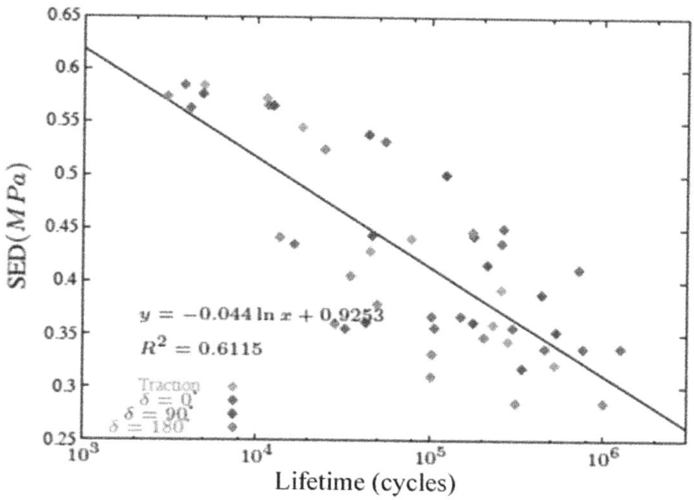

b) Analytical calculation

Figure 7.12. *(A) S–N curve plotted against SED. Points: experimental results; black line: linear trendline. For a color version of this figure, see www.iste.co.uk/richaud/industrialpolymers1.zip*

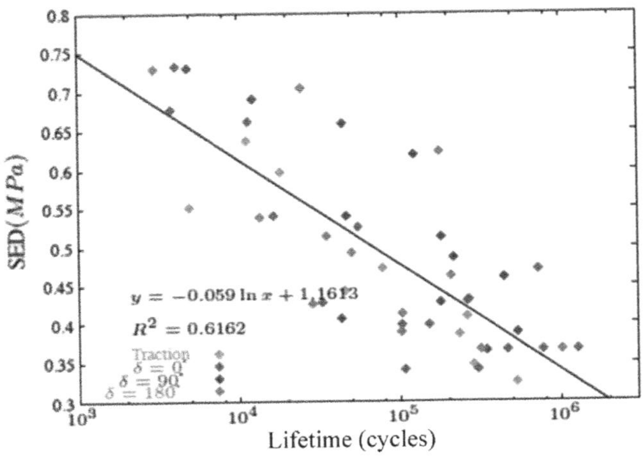

F.E. calculation

Figure 7.12. *(B) S–N curve plotted against SED. Points: experimental results; black line: linear trendline (continued). For a color version of this figure, see www.iste.co.uk/richaud/industrialpolymers1.zip*

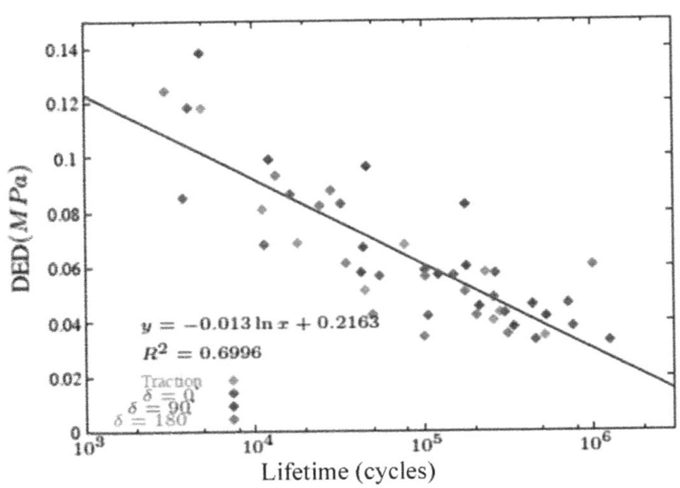

Analytical calculation

Figure 7.13. *(A) S–N curve plotted against DED. Points: experimental results; black line: linear trend line. For a color version of this figure, see www.iste.co.uk/richaud/industrialpolymers1.zip*

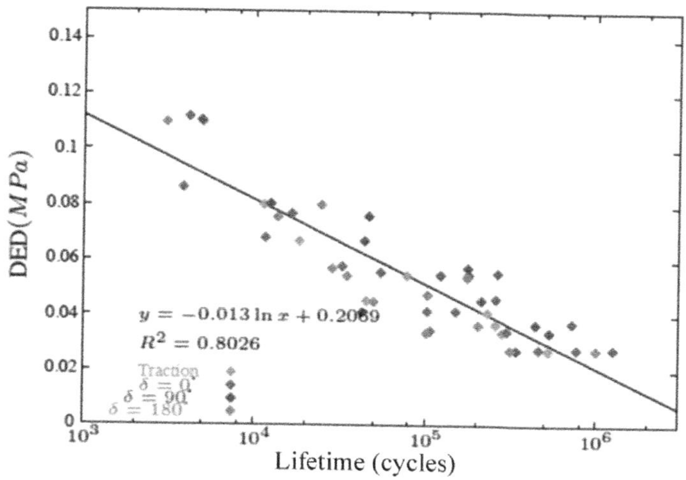

a) Analytical calculation

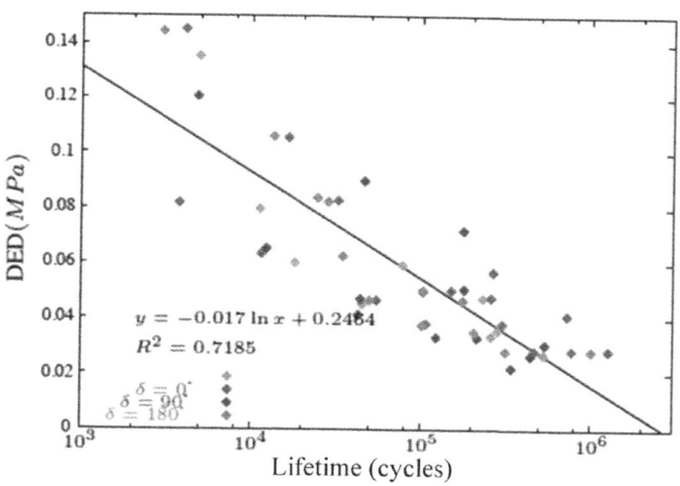

b) F.E. calculation

Figure 7.13. *(B) S–N curve plotted against DED. Points: experimental results; Black line: linear trendline (continued). For a color version of this figure, see www.iste.co.uk/richaud/industrialpolymers1.zip*

7.3. Conclusion

This chapter focuses on the characterization of elastomer fatigue behavior. Regardless of the selected approach, it is rarely specific to these materials and often relies on techniques that have already been validated on other materials (particularly metallic). However, owing to the complex nature of these materials, their environmental sensitivity, high deformability and complex thermomechanical behavior, this task remains challenging. Consequently, the authors have to adapt existing concepts or propose new, more complex ones.

Our choice was to focus on initiation-based approaches (rather than propagation or damage mechanics). As previously mentioned, the standard tools used in such approaches have been improved (e.g. configurational mechanics approaches, cracking energy density and DED), enabling an accurate representation of the fatigue response of various elastomers under complex loading. These criteria can also be readily applied to industrial components subjected to complex loading (via a finite element calculation, for example) provided that there is good knowledge of the law of behavior of the material.

7.4. References

Andriyana, A. and Verron, E. (2007). Prediction of fatigue life improvement in natural rubber using configurational stress. *International Journal of Solids and Structures*, 44(7), 2079–2092.

Arruda, E.M. and Boyce, M.C. (1993). A three-dimensional constitutive model for the large stretch behavior of rubber elastic materials. *Journal of the Mechanics and Physics of Solids*, 41(2), 389–412.

Ayoub, G., Naït-abdelaziz, M., Zaïri, F., Gloaguen, J.M. (2010). Multiaxial fatigue life prediction of rubber-like materials using the continuum damage mechanics approach. *Procedia Engineering*, 2(1), 985–993.

Bathias, C. (ed.) (2013). Fatigue of polymers and elastomers. In *Fatigue of Materials and Structures*. ISTE Ltd, London and John Wiley & Sons, New York [Online]. Available at: https://onlinelibrary.wiley.com/doi/abs/10.1002/9781118616789.ch4 [Accessed 15 July 2020].

Beatty, J.R. (1964). Fatigue of rubber. *Rubber Chemistry and Technology*, 37(5), 1341–1364.

Boukamel, A., Méo, S., Débordes, O., Jaeger, M. (2001). A thermo-viscoelastic model for elastomeric behaviour and its numerical application. *Archive of Applied Mechanics*, 71(12), 785–801.

Cadwell, S.M., Merrill, R.A., Sloman, C.M., Yost, F.L. (1940). Dynamic fatigue life of rubber. *Rubber Chemistry and Technology*, 13(2), 304–315.

Christensen, R.M. (1982). *Theory of Viscoelasticity*. Elsevier, Berlin [Online]. Available at: https://linkinghub.elsevier.com/retrieve/pii/B978012174252 2X50017 [Accessed 10 June 2020].

Chrysochoos, A. (2018). Effets de couplage et effets dissipatifs accompagnant la déformation des matériaux solides. Doctoral Thesis [Online]. Available at: https://hal.archives-ouvertes.fr/cel-02057720 [Accessed 10 July 2020].

Coleman, B.D. and Noll, W. (1974). Foundations of linear viscoelasticity. In *The Foundations of Mechanics and Thermodynamics: Selected Papers*, Noll, W. (ed.). Springer, Berlin [Online]. doi: 10.1007/978-3-642-65817-4_6.

Cruanes, C. (2015). Étude du comportement en fatigue d'un élastomère synthétique : relation entre endommagement, renforcement et auto-échauffement. PhD Thesis, Université de Tours, Tours [Online]. Available at: http://www.theses.fr/2015TOUR4012 [Accessed 14 June 2022].

Cruanes, C., Deffarges, M.P., Lacroix, F., Méo, S. (2019). Modeling of the thermo-mechanical behavior of rubbers during fatigue tests from infrared measurements. *International Journal of Fatigue* [Online]. Available at: https://hal.archives-ouvertes.fr/hal-02126778.

Delattre, A. (2014). Caractérisation et modélisation du comportement hyper-viscoelastique d'un élastomère chargé pour la simulation de pièces lamifiées élastomère-métal et étude en fatigue. PhD Thesis, Université de Tours, Tours [Online]. Available at: http://www.theses.fr/2014TOUR4039/document.

Diani, J., Fayolle, B., Gilormini, P. (2009). A review on the Mullins effect. *European Polymer Journal*, 45(3), 601–612.

Eshelby, J.D. (1951). The force on an elastic singularity. *Philosophical Transactions of the Royal Society of London. Series A, Mathematical and Physical Sciences*, 244(877), 87–112.

Eshelby, J.D. (1975). The elastic energy-momentum tensor. *Journal of Elasticity*, 5(3), 321–335.

Göktepe, S. and Miehe, C. (2005). A micro–macro approach to rubber-like materials. Part III: The micro-sphere model of anisotropic Mullins-type damage. *Journal of the Mechanics and Physics of Solids*, 53(10), 2259–2283.

Govindjee, S. and Simo, J. (1991). A micro-mechanically based continuum damage model for carbon black-filled rubbers incorporating Mullins' effect. *Journal of the Mechanics and Physics of Solids*, 39(1), 87–112.

Grandcoin Laneyrie, J. (2008). Contribution à la modélisation du comportement dissipatif des élastomères chargés : d'une modélisation micro-physiquement motivée vers la caractérisation de la fatigue. PhD Thesis, Université de la Méditerranée – Aix-Marseille II, Marseille [Online]. Available at: https://tel.archives-ouvertes.fr/tel-00326048 [Accessed 10 July 2020].

Greensmith, H.W. (1957). Rupture of rubber. IV. Tear properties of vulcanizates containing carbon black. *Rubber Chemistry and Technology*, 30(2), 584–595.

Greensmith, H.W. and Thomas, A.G. (1956). Rupture of rubber. III. Determination of tear properties. *Rubber Chemistry and Technology*, 29(2), 372–381.

Griffith, A.A. (1921). VI. The phenomena of rupture and flow in solids. *Philosophical Transactions of the Royal Society of London. Series A, Containing Papers of a Mathematical or Physical Character*, 221(582/593), 163–198.

Halphen, B. and Son Nguyen, Q. (1975). Sur les matériaux standard généralisés. *Journal de Mécanique*, 14, 39–63.

Hart-Smith, L.J. (1966). Elasticity parameters for finite deformations of rubber-like materials. *Zeitschrift für angewandte Mathematik und Physik ZAMP*, 17(5), 608–626.

Holzapfel, G.A. and Reiter, G. (1995). Fully coupled thermomechanical behaviour of viscoelastic solids treated with finite elements. *International Journal of Engineering Science*, 33(7), 1037–1058.

Holzapfel, G.A. and Simo, J.C. (1996a). A new viscoelastic constitutive model for continuous media at finite thermomechanical changes. *International Journal of Solids and Structures*, 33(20), 3019–3034.

Holzapfel, G.A. and Simo, J.C. (1996b). Entropy elasticity of isotropic rubber-like solids at finite strains. *Computer Methods in Applied Mechanics and Engineering*, 132(1), 17–44.

Hosseini, M., Méo, S., Lacroix, F., Berton, G. (2016). The effect of temperature on fatigue behavior of chloroprene rubber. In *16th International Conference on New Trends in Fatigue and Fracture*. Dubrovnik [Online]. Available at: https://hal.archives-ouvertes.fr/hal-01403202 [Accessed 15 July 2020].

Lacroix, F., Méo, S., Berton, G., Chalon, F., Tougui, A., Ranganathan, M. (2005). A local criterion for fatigue crack initiation on chloroprene rubber: Approach in dissipation. Report.

Lake, G.J. and Lindley, P.B. (1964). Cut growth and fatigue of rubbers. II. Experiments on a noncrystallizing rubber. *Journal of Applied Polymer Science*, 8(2), 707–721.

Lake, G.J. and Lindley, P.B. (1965). The mechanical fatigue limit for rubber. *Journal of Applied Polymer Science*, 9(4), 1233–1251.

Le Cam, J.B. (2005). Endommagement en fatigue des élastomères. PhD Thesis, Université de Nantes, Nantes [Online]. Available at: https://hal-univ-rennes1.archives-ouvertes.fr/tel-01448458 [Accessed 17 July 2020].

Le Cam, J.B., Huneau, B., Verron, E., Gornet, L. (2004). Mechanism of fatigue crack growth in carbon black filled natural rubber. *Macromolecules*, 37(13), 5011–5017.

Le Saux, V., Marco, Y., Calloch, S., Doudard, C., Charrier, P. (2010). Fast evaluation of the fatigue lifetime of elastomers based on a heat build-up protocol and microtomography measurements. *International Journal of Fatigue*, 32(10), 1582–1590.

Lejeunes, S., Eyheramendy, D., Boukamel, A., Delattre, A., Méo, S., Ahose, K.D. (2018). A constitutive multiphysics modeling for nearly incompressible dissipative materials: Application to thermo–chemo-mechanical aging of rubbers. *Mechanics of Time-Dependent Materials*, 22(1), 51–66.

Lemaitre, J., Chaboche, J.L., Benallal, A., Desmorat, R. (2009). *Mécanique des matériaux solides*. Dunod, Paris.

Lion, A. (1997a). A physically based method to represent the thermo-mechanical behaviour of elastomers. *Acta Mechanica*, 123(1), 1–25.

Lion, A. (1997b). On the large deformation behaviour of reinforced rubber at different temperatures. *Journal of the Mechanics and Physics of Solids*, 45(11), 1805–1834.

Lion, A. (2000). Constitutive modelling in finite thermoviscoplasticity: A physical approach based on nonlinear rheological models. *International Journal of Plasticity*, 16(5), 469–494.

Majesté, J.C. (2017). Renforcement des élastomères. *Techniques de l'Ingénieur*, AM8100 v1. doi: 10.51257/a-v1-am8100.

Marckmann, G., Verron, E., Gornet, L., Chagnon, G., Charrier, P., Fort, P. (2002). A theory of network alteration for the Mullins effect. *Journal of the Mechanics and Physics of Solids*, 50(9), 2011–2028.

Mars, W. (2001). Multiaxial fatigue of rubber. PhD Thesis, University of Toledo.

Mars, W. (2009). Computed dependence of rubber's fatigue behavior on strain crystallization. *Rubber Chemistry and Technology*, 82(1), 51–61.

Mars, W. and Fatemi, A. (2002). A literature survey on fatigue analysis approaches for rubber. *International Journal of Fatigue*, 24(9), 949–961.

Mars, W. and Fatemi, A. (2004). Factors that affect the fatigue life of rubber: A literature survey. *Rubber Chemistry and Technology*, 77(3), 391–412.

Martinez, J.M., Boukamel, A., Méo, S., Lejeunes, S. (2011). Statistical approach for a hyper-viscoplastic model for filled rubber: Experimental characterization and numerical modeling. *European Journal of Mechanics – A/Solids*, 30(6), 1028–1039.

Méo, S. (2012). Modélisation du comportement (thermo-) mécanique des élastomères et prévision de leur tenue en fatigue. PhD Thesis, Université François Rabelais, Tours [Online]. Available at: https://hal.archives-ouvertes.fr/tel-01237757.

Méo, S., Boukamel, A., Debordes, O. (2002). Analysis of a thermoviscoelastic model in large strain. *Computers & Structures*, 80(27), 2085–2098.

Miehe, C. (2000). Superimposed finite elastic–viscoelastic–plastoelastic stress response with damage in filled rubbery polymers. Experiments, modelling and algorithmic implementation. *Journal of the Mechanics and Physics of Solids*, 48(2), 323–365.

Miehe, C. and Keck, J. (2000). Superimposed finite elastic–viscoelastic–plastoelastic stress response with damage in filled rubbery polymers. Experiments, modelling and algorithmic implementation. *Journal of the Mechanics and Physics of Solids*, 48(2), 323–365.

Mooney, M. (1940). A theory of large elastic deformation. *Journal of Applied Physics*, 11(9), 582–592.

Morman, K.N. (1988). An adaptation of finite linear viscoelasticity theory for rubber-like viscoelasticity by use of a generalized strain measure. *Rheologica Acta*, 27(1), 3–14.

Mullins, L. (1969). Softening of rubber by deformation. *Rubber Chemistry and Technology*, 42(1), 339–362.

Narynbek Ulu, K., Huneau, B., Le Gac, P.Y., Verron, E. (2016). Fatigue resistance of natural rubber in seawater with comparison to air. *International Journal of Fatigue*, 88, 247–256.

Ogden, R.W. (1972). Large deformation isotropic elasticity – On the correlation of theory and experiment for incompressible rubberlike solids. *Proceedings of the Royal Society of London. A. Mathematical and Physical Sciences*, 326(1567), 565–584.

Ostoja-Kuczynski, E., Charrier, P., Verron, E., Marckmann, G., Gornet, L., Chagnon, G. (2003). Crack initiation in filled natural rubber: Experimental database and macroscopic observations. In *Third European Conference on Constitutive Models for Rubber*. CRC Press, London [Online]. Available at: https://hal.archives-ouvertes.fr/hal-01004685 [Accessed 15 July 2020].

Payne, A.R. and Whittaker, R.E. (1971). Influence of hysteresis on tensile and fatigue failure in rubbers. *Journal of Applied Polymer Science*, 15(8), 1941–1948.

Poisson, J.L. (2012). Détermination d'un critère de fatigue multiaxial appliqué à un élastomère synthétique. PhD Thesis, Université de Tours, Tours [Online]. Available at: http://www.theses.fr/2012TOUR4024/document.

Poisson, J.L., Méo, S., Lacroix, F., Berton, G., Hosséini, M., Ranganathan, N. (2018). Comparison of fatigue criteria under proportional and non-proportional multiaxial loading. *Rubber Chemistry and Technology*, 91(2), 320–338.

Reese, S. (2003). A micromechanically motivated material model for the thermo-viscoelastic material behaviour of rubber-like polymers. *International Journal of Plasticity*, 19, 909–940.

Rivlin, R.S. and Thomas, A.G. (1953). Rupture of rubber. I. Characteristic energy for tearing. *Journal of Polymer Science*, 10(3), 291–318.

Robisson, A. (2000). Comportement visco-hyperélastique endommageable d'élastomères SBR et PU : prévision de la durée de vie en fatigue. PhD Thesis, École nationale supérieure des mines, Paris.

Saintier, N. (2001). Fatigue multiaxiale dans un élastomère de type NR : mécanismes d'endommagement et critère local d'amorçage de fissure. PhD Thesis, ENMP, Paris [Online]. Available at: http://www.theses.fr/2001ENMP1089 [Accessed 10 July 2020].

Sidoroff, F. (1973). The geometrical concept of intermediate configuration and elastic finite strain. *Arch. Mech*, 25(2), 299–309.

Sidoroff, F. (1975). Variables internes en viscoélasticité, milieux avec configuration intermèdiaire. *J. Méc*, 14(4), 571–595.

Sidoroff, F. (1976). Variables internes en viscoélasticité, milieux avec plusieurs configurations intermèdiaires. *J. Méc*, 15(1), 85–118.

Tallec, P.L. and Rahler, C. (1994). Numerical models of steady rolling for nonlinear viscoelastic structures in finite deformations. *International Journal for Numerical Methods in Engineering*, 37(7), 1159–1186.

Thomas, A.G. (1955). Rupture of rubber. II. The strain concentration at an incision. *Journal of Polymer Science*, 18(88), 177–188.

Thomas, A.G. (1958). Rupture of rubber. V. Cut growth in natural rubber vulcanizates. *Journal of Polymer Science*, 31(123), 467–480.

Thomas, A.G. (1961). Rupture of rubber. VI. Further experiments of the tear criterion. *Rubber Chemistry and Technology*, 34(1), 66–75.

Verron, E. and Andriyana, A. (2008). Definition of a new predictor for multiaxial fatigue crack nucleation in rubber. *Journal of the Mechanics and Physics of Solids*, 56(2), 417–443.

Verron, E., Le Cam, J.B., Gornet, L. (2006). A multiaxial criterion for crack nucleation in rubber. *Mechanics Research Communications*, 33(4), 493–498.

Wang, B., Lu, H., Kim, G. (2002). A damage model for the fatigue life of elastomeric materials. *Mechanics of Materials*, 34, 475–483.

Wöhler, A. (1867). Wöhler's experiments on the strength of metals. *Engineering*, 4, 160–161.

Zine, A., Benseddiq, N., Abdelaziz, M.N., Hocine, N.A., Bouami, D. (2006). Prediction of rubber fatigue life under multiaxial loading. *Fatigue & Fracture of Engineering Materials & Structures*, 29(3), 267–278.

Zine, A., Benseddiq, N., Naït Abdelaziz, M. (2011). Rubber fatigue life under multiaxial loading: Numerical and experimental investigations. *International Journal of Fatigue*, 33(10), 1360–1368.

8

Influence of Polyamide 6,6 Molecular Mobility on Mechanical Behavior

Agustín RIOS DE ANDA[1,2], **Louise-Anne FILLOT**[3,4] **and Paul SOTTA**[3,5]

[1]*LPMA, Université Lyon 1, Saint-Fons, France*

[2]*ICMPE, UPEC, Thiais, France*

[3]*LPMA, CNRS, Université Lyon 1, Saint-Fons, France*

[4]*Solvay Engineering Plastics, Saint-Fons, France*

[5]*IMP, INSA, Villeurbanne, France*

8.1. Introduction

This work is part of applied research aimed at developing automobile gasoline tanks made from polyamide 6,6 (PA6,6)-based formulations.

PA6,6 is a good candidate for technical applications because of its good thermomechanical properties relative to its weight (Kohan 1995). PA6,6 is obtained by a polycondensation reaction from a stoichiometric mixture of hexamethylenediamine and adipic acid or from its nylon 6,6 salt, as shown in Figure 8.1.

PA6,6 formulations must have good barrier properties to standard gasoline and biofuels, which are gasoline containing a certain percentage of ethanol, and to the water that is present in the atmosphere, while having good mechanical properties, particularly good resistance to impacts.

Aging of Industrial Polymers 1,
coordinated by Emmanuel RICHAUD.

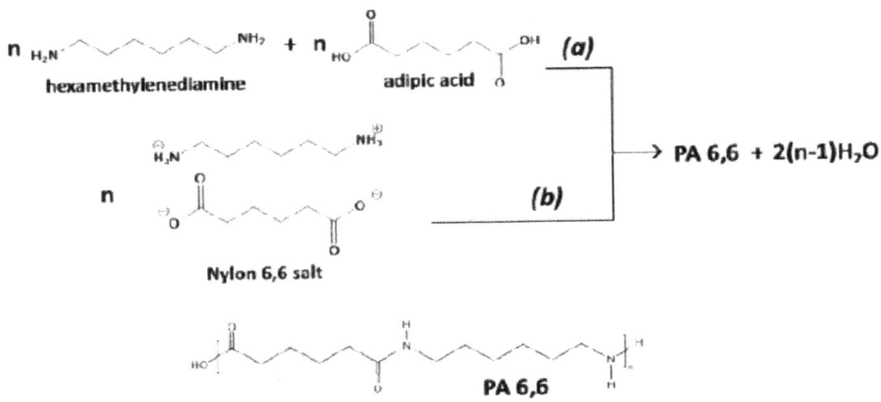

Figure 8.1. *Polycondensation reactions yielding polyamide 6,6 (a) from hexamethylenediamine and adipic acid; (b) from nylon 6,6 salt*

The scientific stakes of this work are, first, a better comprehension of the interactions between PA6,6 and a series of pure solvents of various polarities and sizes but also of ternary mixtures composed of ethanol, toluene and isooctane at various concentrations, which model the biofuels commercially available in Europe. We also want to study the effects of these solvents or mixtures of solvents on the glass transition temperature (T_g), which is a signature of the molecular mobility of the amorphous phase of this polymer. Then, the effect of additivation or addition of comonomers to pure PA6,6 on the barrier and mechanical properties, particularly the impact properties, has been studied.

It is known that polyamides have a network of H bonds formed between the amide groups of various polymer chains (Kohan 1995). This network is present in the crystalline phase, which facilitates polymer crystallization, and is also present in the amorphous phase. A schematic representation of this H-bond network is presented in Figure 8.2.

The mobility of the amorphous phase is governed by molecular movements at various time scales. In the case of PA6,6, three distinct relaxation processes have been described: the main α relaxation, which involves the movement of long chain segments (15 monomers), and two secondary β and γ relaxations that correspond to the rotation of amide groups and to vibrations of methylene chains, respectively (Owen and Bonart 1985; Brandrup and Immergut 1989; Le Huy and Rault 1994; Diaz-Calleja and Riande 2004). These transitions are schematically represented in Figure 8.3.

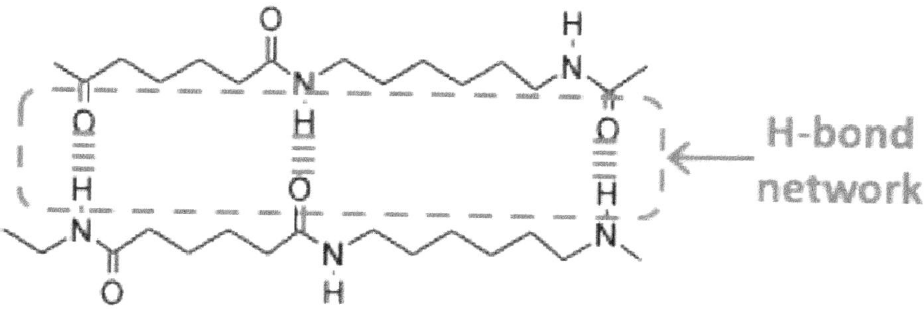

Figure 8.2. *Schematic representation of the network of hydrogen bonds between the amide groups of PA6,6. For a color version of this figure, see www.iste.co.uk/richaud/industrialpolymers1.zip*

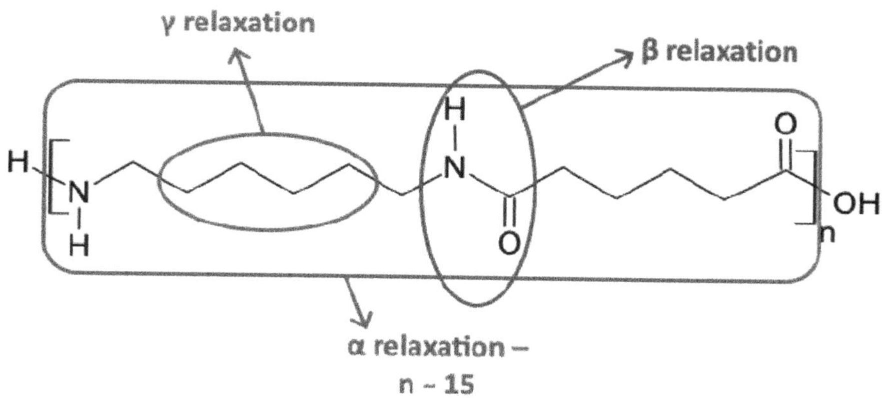

Figure 8.3. *Schematic representation of the α, β and γ relaxations of PA6,6. For a color version of this figure, see www.iste.co.uk/richaud/industrialpolymers1.zip*

Although several factors influence the molecular mobility of PA6,6, the presence of a network of interchain H bonds significantly controls or determines this mobility (Bessler and Bier 1969; Schroeder and Cooper 1976; Garcia and Starkweather 1985; Skrovanek et al. 1985a, 1985b; Hirschinger et al. 1990; Miura et al. 1990; Triggs et al. 1993).

The mechanisms of sorption (Fick et al. 1855; Starkweather 1959, 1975; Hunt 1980; Lim et al. 1999; Crank 2004; Picard et al. 2008; Sabard 2011), diffusion (Hunt 1980; Auerbach and Carnicom 1991; Mansfield et al. 1992; Yeh et al. 2000;

Alexandre et al. 2009) and permeation (Kamal and Jinnah 1984; Low et al. 2004; Alexandre et al. 2009, 2010; Eceolaza et al. 2009) of water and its plasticizing effect in PA6,6 are well known in the literature. Water is a polar molecule that is absorbed by PA6,6 because the amide groups can generate strong interactions with this solvent (Illiers 1960; Illiers and Jacobs 1960; Papir et al. 1972; Razumovskii et al. 1985; Hernández et al. 1992; Pathmanathan et al. 1992; Rotter and Ishida 1992; Le Huy and Rault 1994; Laredo and Hernández 1997; Park et al. 1997; Lim et al. 1999; Laredo et al. 2003). The literature acknowledges that water can plasticize PA6,6 because this solvent interacts with and disrupts the network of interchain H bonds. It would be interesting to study the sorption at equilibrium of a series of solvents of various sizes and polarities to better understand how the nature and/or size of a molecule conditions its interactions with PA6,6 and therefore its intake by this polymer. This work aims to study the effect that various studied solvents have on the T_g of PA6,6, as well as on secondary relaxations. Plasticization mechanisms specific to each family of (polar and apolar) solvents will be proposed.

For the sorption of solvent mixtures modeling commercial biofuels in direct connection with the industrial application of this work, which is the development of PA6,6 formulations for gasoline tanks, no information is available in the literature; therefore, we will study sorption and its effect on the molecular mobility of PA6,6 ternary mixtures that model biofuels and are composed of ethanol, toluene and isooctane at various concentrations. First, we will study the liquid/vapor equilibrium of these mixtures, which will facilitate a better understanding of the thermodynamics of these mixtures, as well as an estimation of the interactions that these solvents might have with PA6,6. The presence of these mixtures in PA6,6 may lead to a modification of these polymer properties. It is therefore important to study the sorption of these solvents and to understand the thermodynamic phenomena related to sorption. In the end, we will study the effect of these mixtures on the molecular mobility of PA6,6.

Finally, and still in relation to the context of this applied research, we explore the mechanical properties of PA6,6. The effect of the crystalline phase on the mechanical properties of polyamides has already been studied in the literature (Starkweather et al. 1956; Starkweather and Brooks 1959; Collyer 1994; Takeda et al. 1996; Bureau et al. 2002; Shan et al. 2007; Miri et al. 2009). When the crystallinity rate increases, the mechanical performances of polyamides also increase. The size of spherolites also influences these properties: when their size decreases, the Young's modulus increases. Finally, the crystalline structure also influences the mechanical properties: the more stable and denser this structure is, the better the mechanical properties.

However, PA6,6 has a relatively low crystallinity rate (40%$_{wt}$), which may suggest that the mobility of the amorphous phase has a role in the polymer properties. The effect of water on the tensile and impact properties of PA6,6 has also been studied in the literature. In this study, water will be used as a modifier of the state of molecular mobility of PA6,6. Moreover, this mobility can be modified by adding additives or comonomers to PA6,6 or by increasing the temperature of the samples. The mechanical properties that will be studied are the yield stress σ_Y, the Young's modulus E, the brittle transition temperature $T_{F/D}$ and the impact resilience. The influence of molecular mobility on these properties will be studied and discussed.

8.2. Materials and techniques

To study the sorption of a series of pure solvents and mixtures of solvents, as well as their effect on the molecular mobility of the amorphous phase and on the crystalline structure, a standard non-additivated and unloaded PA6,6 has been considered. This PA6,6, referred to as Neat PA6,6 in this work, was extruded in the form of 100-μm-thick films.

To study the effect of matrix modification on the barrier and mechanical properties of PA6,6, a series of modified materials have been designed. This series includes two additivated formulations, PA6,6 + 10%$_{wt}$ lauryl aldehide (10LA) and PA6,6 + 10%$_{wt}$ heptanaldehyde (10HA), two copolymers, PA6,6/6HIA 95/5$_{mol}$ and PA6,6/6AiSLi 95/5$_{mol}$, and a pure PA6,6 (REF). The chemical structures of the comonomers and additives are shown in Figure 8.4.

The additivated formulations 10LA and 10HA, as well as REF, were obtained by extrusion and then injected into disks and into impact and tensile bars. HIA and AISLI copolymers, after being synthesized, have also been injected into bars and extruded to obtain 300-μm-thick films.

To characterize the crystalline phase of the samples, we used DSC and X-ray diffraction. DSC was used to obtain the melting and crystallization temperatures, as well as the degrees of crystallinity, whereas X-ray diffraction was used to characterize the crystalline structures. Modulated DSC (MDSC), broadband dielectric spectroscopy (BDS) and dynamic mechanical analysis (DMA) were used to characterize the molecular mobility of the amorphous phase of the studied samples either in the dry state (MDSC, BDS and DMA) or conditioned in solvents (MDSC and BDS).

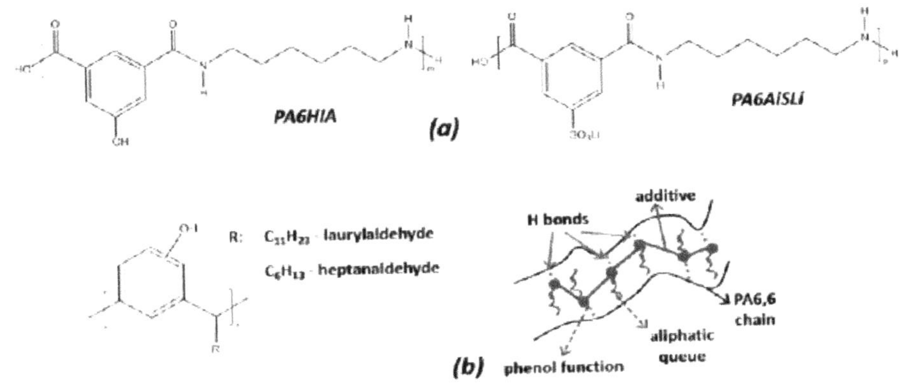

Figure 8.4. *Chemical structures (a) of PA6HIA and PA6AiSLi comonomers; (b) phenolic additives with a schematic representation of interactions with PA6,6. For a color version of this figure, see www.iste.co.uk/richaud/industrialpolymers1.zip*

Sorption by PA6,6 in a series of polar and apolar pure solvents of various sizes and polarities, as well as their effect on the molecular mobility of the polymer, have been studied. The polar solvents used are water, methanol, ethanol, ethylene glycol, 1-propanol, 2-propanol, 2-butanol and cyclopentanol. The specific solvents used are toluene, o-xylene, p-xylene, 2,3-dimethylbutane, 3-methylpentane, n-hexane and isooctane. The chemical structures of all these solvents are shown in Figure 8.5.

The sorption of these solvents was conducted at the ambient temperature (23°C), and the PA6,6 films were put in contact with each solvent and weighed regularly until sorption equilibrium was reached, which is the moment when the polymer no longer absorbs any solvent. This is schematically shown in Figure 8.6.

The sorption and the plasticizer effect on the PA6,6 of a series of ternary mixtures composed of ethanol, toluene and isooctane (referred to as EX) at various concentrations have also been studied.

First, we focused on calculating the chemical activities of these solvents in each of these mixtures. For this purpose, we used gas chromatography (GC-FID) to quantify the concentration of species in the vapor phases of these mixtures and calculated their chemical activities. This setup is shown in Figure 8.7.

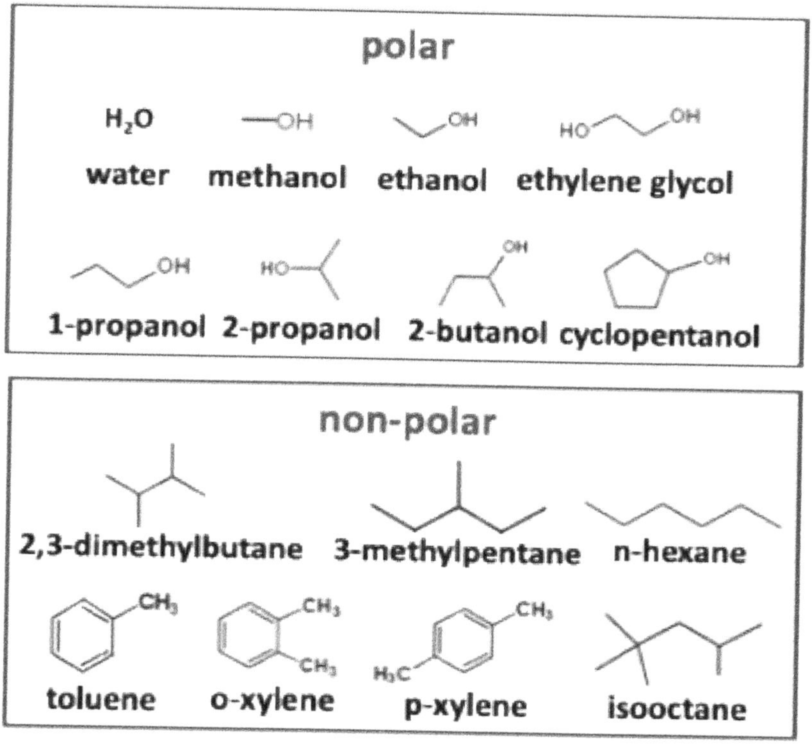

Figure 8.5. *Chemical structures of the studied solvents*

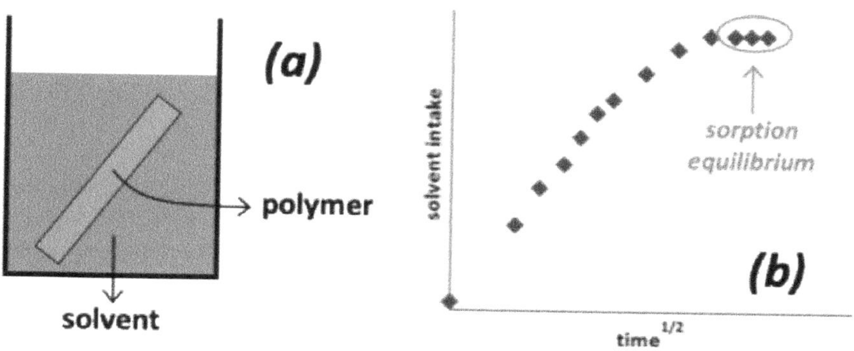

Figure 8.6. *(a) Schematic representation of a sorption test in the liquid phase; (b) sorption kinetics of a solvent as a function of the square root of time. For a color version of this figure, see www.iste.co.uk/richaud/industrialpolymers1.zip*

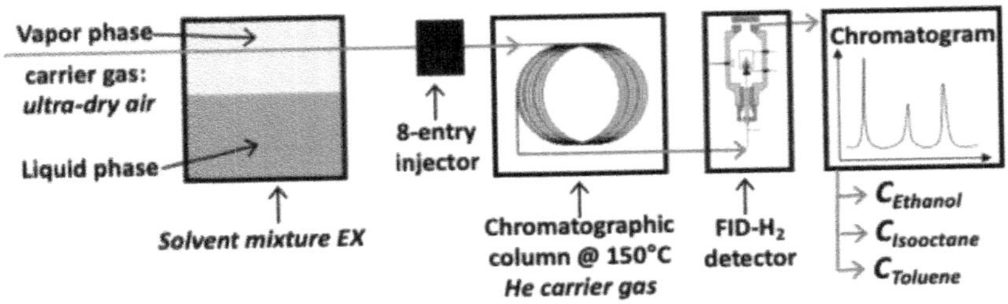

Figure 8.7. *GC-FID setup for chemical activities of species involving ternary mixtures EX. For a color version of this figure, see www.iste.co.uk/richaud/industrialpolymers1.zip*

These mixtures were then put in contact with the PA6,6 films until they reached sorption equilibrium. GC-FID was used to quantify the concentrations of ethanol, toluene and isooctane absorbed by the films by quantifying the amount of sample desorption vapors. This setup is represented in Figure 8.8(a).

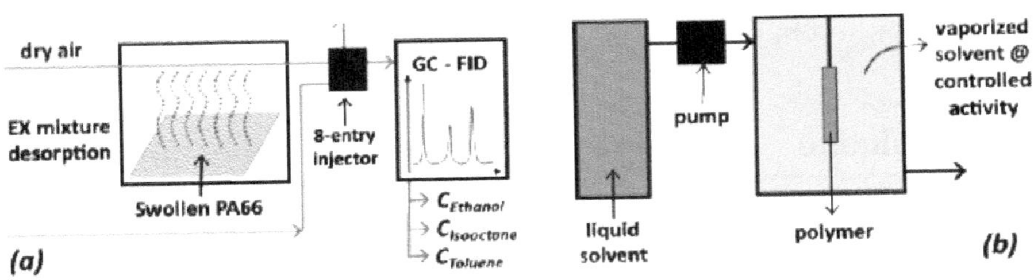

Figure 8.8. *(a) Simplified GC-FID setup for the quantification of the species absorbed by PA6,6 in contact with EX ternary mixtures; (b) schematic representation of a dynamic vapor sorption (DVS) setup. For a color version of this figure, see www.iste.co.uk/richaud/industrialpolymers1.zip*

The sorption of mixtures by PA6,6 was compared to that of pure ethanol with the same chemical activities. For this purpose, a DVS was used. This device, schematically shown in Figure 8.8(b), makes it possible to have an ethanol atmosphere with controlled activity. The films were introduced into the DVS, and their solvent intake was measured continuously until they reached sorption equilibrium.

Ethanol permeation tests were conducted. For this purpose, a sandwich setup of permeation cells was made: thinned films or plates were put in contact, on the one hand, with a tank containing this solvent and, on the other hand, with a humidity-controlled atmosphere (RH20). The tests were conducted at 40°C, and the cells were regularly weighed until a permanent permeation regime was obtained, that is, when the instantaneous solvent loss was constant over time. These tests are represented in Figure 8.9.

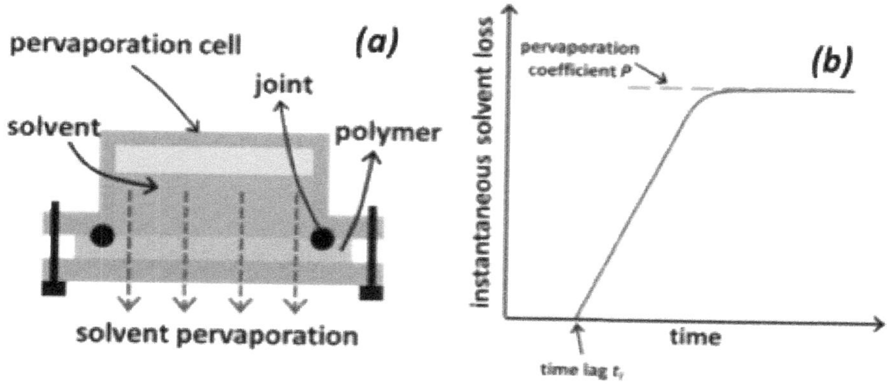

Figure 8.9. *(a) Schematic representation of a permeation test; (b) kinetics used to obtain the permeation coefficient P and the induction time t_l. For a color version of this figure, see www.iste.co.uk/richaud/industrialpolymers1.zip*

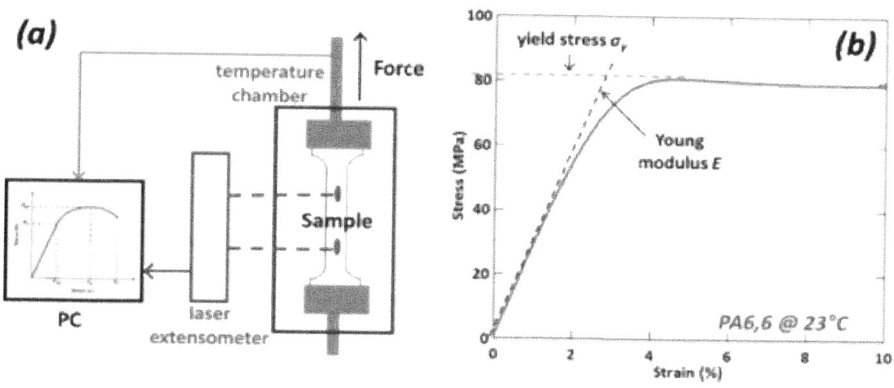

Figure 8.10. *(a) Tensile test; (b) PA6,6 stress–strain curve plotted at 23°C. For a color version of this figure, see www.iste.co.uk/richaud/industrialpolymers1.zip*

Finally, concerning the characterization of the mechanical properties, tensile tests were conducted to obtain the yield stress σ_y and the Young's modulus E. These tests, shown in Figure 8.10, were conducted for samples in the dry state at different temperatures: 23, 45, 65, 90 and 120°C. Samples conditioned at RH50 and RH100 have been studied only at 23°C.

Concerning the impact properties, the resilience of samples at various temperatures and their brittle–ductile transition temperatures $T_{F/D}$ have been obtained by instrumented Charpy impact tests, as shown in Figure 8.11. These tests are used to plot the force–time and force–displacement curves. These curves are used to calculate the resilience. Dry samples conditioned at RH50 and RH100 were characterized at temperatures ranging between –60 and 140°C.

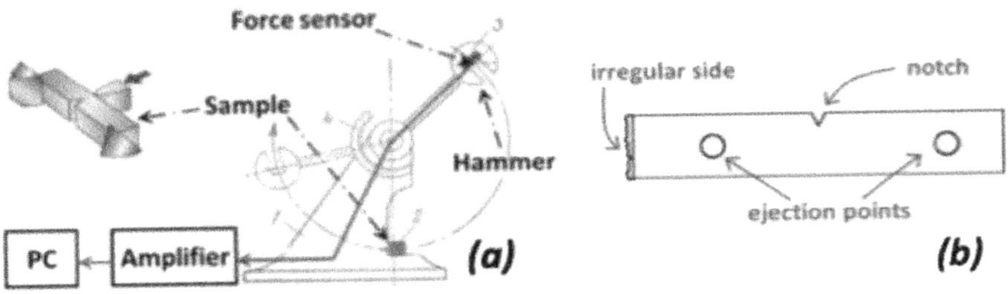

Figure 8.11. *(a) Instrumented Charpy impact test; (b) preparation of an impact specimen. For a color version of this figure, see www.iste.co.uk/richaud/industrialpolymers1.zip*

8.3. Sorption and plasticizing effect of pure polar and apolar solvents and of mixtures of solvents on pure PA6,6

The sorption of a series of pure polar and apolar solvents and their effect on the crystalline and amorphous phases of pure PA6,6 have first been studied. The solvent intake in PA6,6 for the previously mentioned series of molecules has first been considered. Notably, the intake of apolar solvents by PA6,6 does not vary as a function of size, as shown in Figure 8.12(a). In the case of polar solvents, their mass intake decreases when their size increases (Figure 8.12(a)), but when the size of polar solvents increases, the polarity of these molecules decreases. As shown in Figure 8.12(b), when the polarity of a molecule decreases, its intake in PA6,6 also decreases. This suggests that the polarity of a molecule will condition its intake in PA6,6. In the case of apolar solvents, their intake will not necessarily be conditioned by their size but rather by the weak interactions (van der Waals) of these solvents with PA6,6.

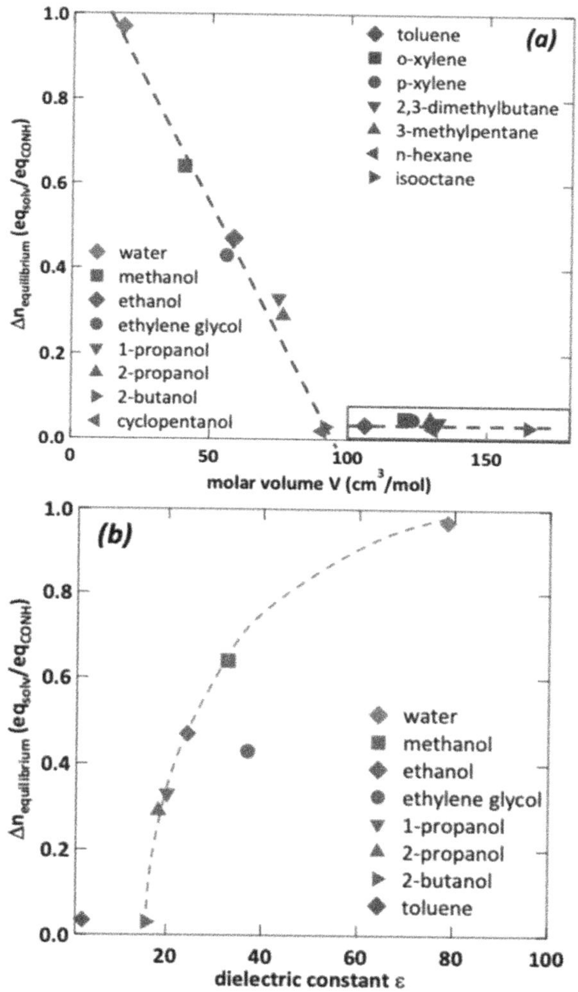

Figure 8.12. *Molar fraction at equilibrium $\Delta n_{equilibrium}$ for the studied solvents as a function of (a) their molar volume V and (b) their relative dielectric constant ε_r. For a color version of this figure, see www.iste.co.uk/richaud/industrialpolymers1.zip*

The effect of the polymer matrix on the sorption of solvents was then studied. A thermodynamic approach was proposed. The Flory–Huggins equation, which describes sorption in terms of interactions between the solvent and the polymer, was first considered. On the basis of the results yielded by this equation, ethanol should solubilize PA6,6, and the sorption of toluene should be approximately 40%$_{vol}$ for an activity of $a_{toluene} = 1$. Therefore, this equation, particularly the parameter of interaction χ, cannot describe the sorption observed for these solvents by itself (approximately 15%$_{vol}$ for ethanol and 2%$_{vol}$ for toluene).

Then, an elastic term that describes a polymer with interchain bonds that act as physical cross-linking points was added to the Flory–Huggins equation to describe the sorption of solvents in PA6,6. This elastic term takes into account the density of H bonds and a module of osmotic pressure that characterizes the stiffness of H bonds. Figure 8.13 shows that the effect of this elastic term significantly limits the solvent intake in PA6,6 compared with the standard Flory–Huggins model. As shown in the literature, other models can be used, for example, to describe water sorption in polyamide. However, these models do not consider the elasticity of the PA6,6 matrix. The effect of this elastic term on solvent intake would show that the H bonds present in the amorphous phase of the polymer act as cross-linking points and limit the solvent intake.

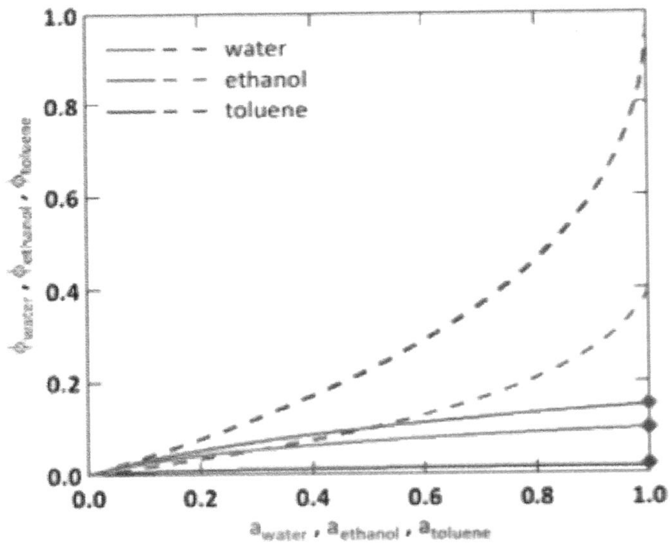

Figure 8.13. *Solvent volume fractions as a function of the activity of the solvent for ethanol, water and toluene, obtained with a "standard" Flory–Huggins model (dotted curves) and obtained with the model presented in this work (solid curves). The values measured at equilibrium are indicated by rhombuses. For a color version of this figure, see www.iste.co.uk/richaud/industrialpolymers1.zip*

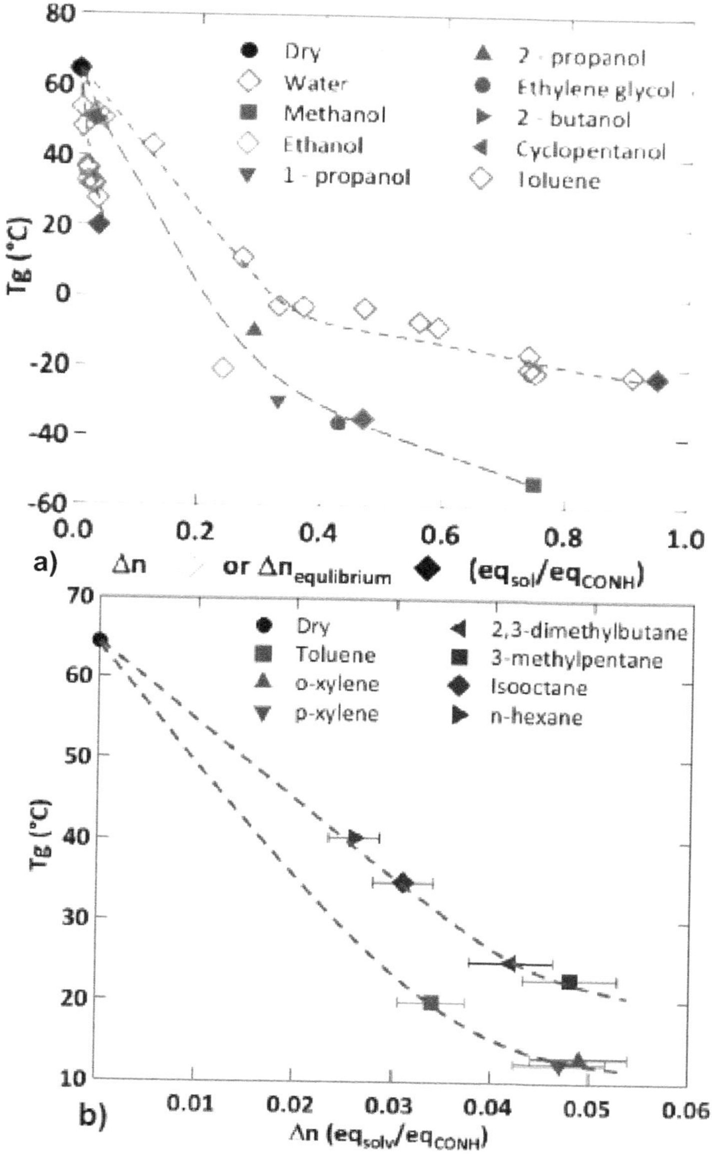

Figure 8.14. *PA6,6 T_g obtained by DSC as a function of (a) the molar fraction Δn for polar solvents and toluene (Rios de Anda et al. 2011). For a color version of this figure, see www.iste.co.uk/richaud/industrialpolymers1.zip*

Next, we focused on the effect of solvents on the molecular mobility of PA6,6, characterized by the T_g. A tendency was observed for the series of alcohols, as shown in Figure 8.14. When their PA6,6 molar intake increases, their plasticizing effect also increases. Notably, neither water nor toluene followed the same trend as alcohols did. To compare these various tendencies, the T_g values of PA6,6 for various contents of water, toluene and ethanol were measured. The following can be noted: ethanol tends to have a series of alcohols, water has a less significant plasticizing effect than these solvents for the same molar intake and toluene has the most significant plasticizing effect of all the other solvents, as shown in Figure 8.14.

Two different plasticization tendencies were also observed for the series of apolar solvents. It seems that aromatic solvents generate a more significant drop in the T_g (approximately 10°C) than aliphatic solvents do for the same molar intake. This effect will be discussed further when the hypotheses for solvent plasticization mechanisms will be described.

The decrease in the T_g caused by these solvents was compared with models of plasticization in the literature, and it was observed that none of these models properly described the plasticization tendencies of these three solvents. This is explained by the fact that these models do not consider strong interactions that may exist between a plasticizer/solvent and the polymer (Rios de Anda et al. 2011).

The effect of water, ethanol and toluene on the main α relaxation, as well as the effect of toluene on secondary β and γ relaxations, has been studied by dielectric spectroscopy. For α relaxation, the studied solvents have no effect on the width or amplitude of this relaxation. With respect to β relaxation, toluene causes a decrease in the temperature of this relaxation and a slight decrease in the activation energy associated with this molecular process. These results suggest that toluene is capable of locally modifying the movements of amide groups associated with β relaxation. Furthermore, no effect of toluene on the width or amplitude of this relaxation was observed. Finally, toluene has no notable effect on γ relaxation.

Hypotheses on the mechanisms of plasticization of water, alcohol series and apolar solvents have since been proposed. These hypotheses are schematically represented in Figure 8.15. In the case of water, this solvent can interact directly with the network of H bonds, leading to a decrease in the lifetime of these bonds, which leads to a decrease in the T_g.

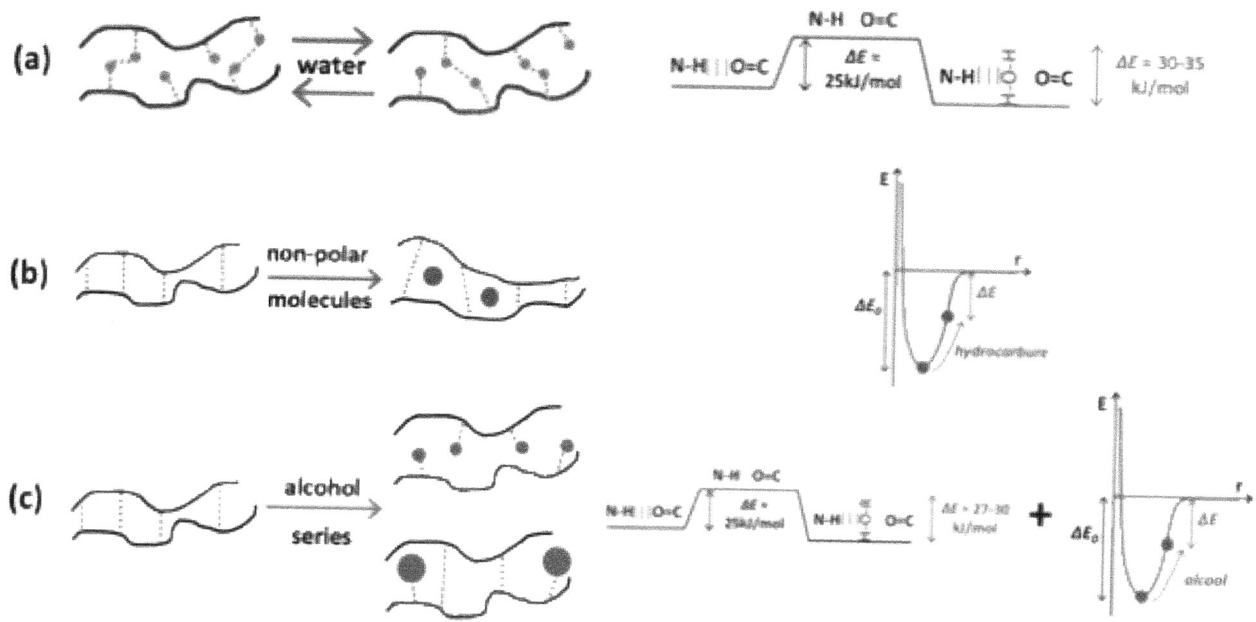

Figure 8.15. *Schematic representation of the plasticizing effects for (a) water, (b) apolar solvents and (c) a series of alcohols. For a color version of this figure, see www.iste.co.uk/richaud/industrialpolymers1.zip*

Moreover, water can form two H bonds; therefore, it may be possible that it forms bridges between PA6,6 chains. In the case of apolar molecules, these solvents cause swelling of the PA6,6 chains, which would stretch the H bonds of the system and would lead to a decrease in the lifetime of these bonds, resulting in an increase in molecular mobility. A hypothesis that would explain the various plasticization effects observed for the aromatic and aliphatic solvents would be that aromatic molecules, owing to their chemical structure and orientation in space, might lead to more significant PA6,6 swelling than aliphatic solvents. Finally, concerning the series of alcohols, these solvents would be capable of plasticizing PA6,6 by a "double effect" that would combine direct interactions with amide bonds due to their hydroxyl function and a swelling effect caused by their alkyl chain (Rios de Anda et al. 2011).

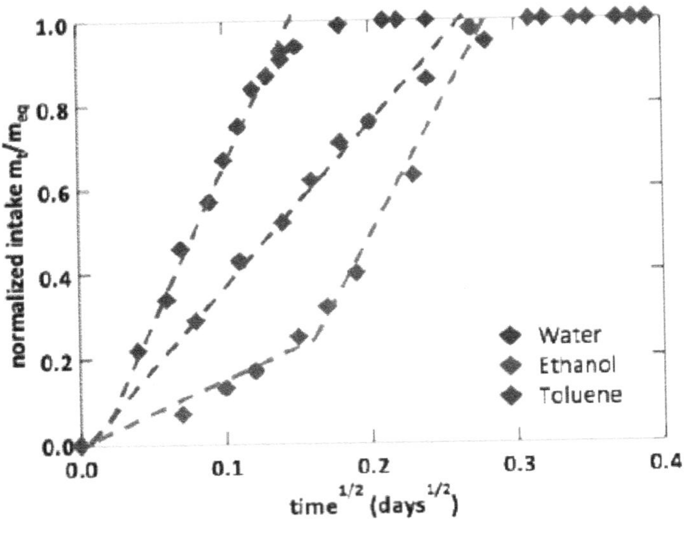

Figure 8.16. *Sorption kinetics of water, ethanol and toluene in PA6,6 at 23°C. For a color version of this figure, see www.iste.co.uk/richaud/industrialpolymers1.zip*

The sorption kinetics of water, ethanol and toluene has then been studied, as shown in Figure 8.16. Notably, toluene seems to follow Fickian kinetics a priori, and water and ethanol have a sorption regime that varies with time. Notably, water plasticizes PA6,6 very rapidly, which induces the diffusion of the solvent in a plasticized polymer according to pseudo-Fickian kinetics. Ethanol also plasticizes

PA6,6, but more slowly, which leads to the diffusion of this solvent according to an abnormal regime. Finally, toluene would not succeed in plasticizing PA6,6 during sorption, which would lead to the diffusion of this solvent according to the Fickian regime.

Then, the effect of water and ethanol on the structure of the crystalline phase of PA6,6 was studied by X-ray diffraction. As shown in Figure 8.17, after the sorption of these solvents, the crystalline structure of the polymer evolves toward the more stable structure α_I, characterized by a wider spacing of the Bragg peaks corresponding to crystallographic planes (100) and (010). Given that these solvents increase the molecular mobility of the amorphous phase, the crystalline phase may reorganize such that this phase evolves toward the most stable structure (Laurati et al. 2014). This suggests that if a solvent has an effect on the amorphous phase, then it could also have an effect on the crystalline phase.

Interestingly, the effect on the molecular mobility of a dry sample induced by an increase in temperature is similar to that generated by the presence of solvents in the PA6,6.

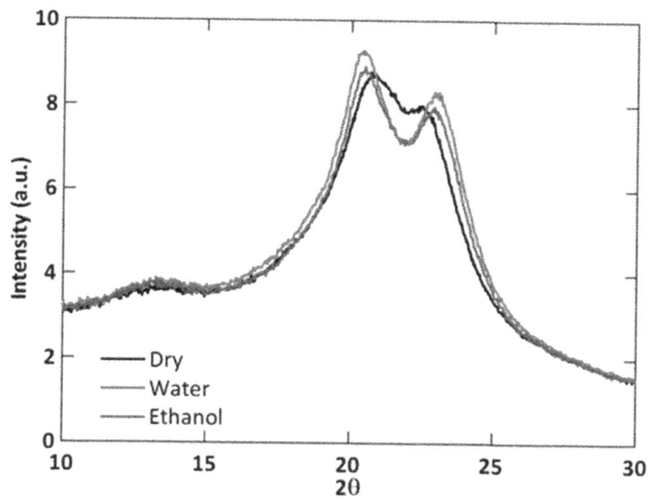

Figure 8.17. *X-ray diffraction spectra of Neat PA66 films in the dry state and after sorption of water or ethanol (Laurati et al. 2014). For a color version of this figure, see www.iste.co.uk/richaud/industrialpolymers1.zip*

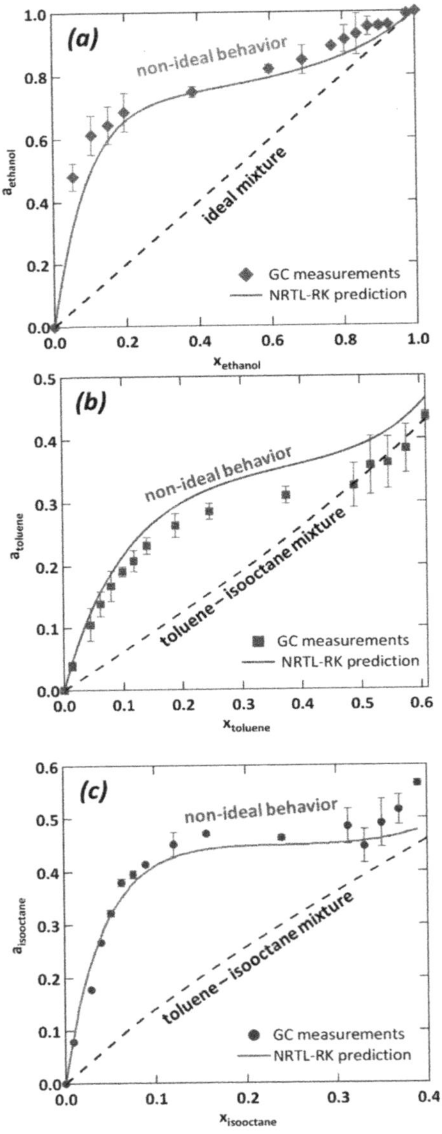

Figure 8.18. *Chemical activities a_i obtained by GC-FID for (a) ethanol, (b) toluene and (c) isooctane as a function of the molar fraction of these solvents in EX mixtures. The data obtained are compared with simulations made using the NRTL-RK model, as well as with a real toluene-isooctane mixture (dotted lines on (a) and (b)) (Hodgman 1962; Renon and Prausnitz 1968; Redlich and Kwong 1979; Rios de Anda et al. 2014)*

The next study focused on the sorption by PA6,6 of a series of ternary mixtures composed of ethanol, toluene and isooctane (EX). These mixtures are strongly non-ideal. The chemical activity of each of the species in these mixtures has therefore been determined, as it is the activity that controls the sorption of solvents. GC-FID was used to analyze the vapor phase of these mixtures and obtain the chemical activities of ethanol, toluene and isooctane. Notably, the chemical activities of these species are relatively high because of the repulsive effects between ethanol and hydrocarbons, especially for mixtures containing these low-concentration solvents, as shown in Figure 8.18 (Hodgman 1962; Renon and Prausnitz 1968; Redlich and Kwong 1979; Rios de Anda et al. 2014).

Next, we compared the ethanol intake by the PA6,6 films in contact with EX mixtures with the pure ethanol intake for the same chemical activities. Notably, the films in contact with the mixtures absorb more ethanol, as shown in Figure 8.19 (Rios de Anda et al. 2014).

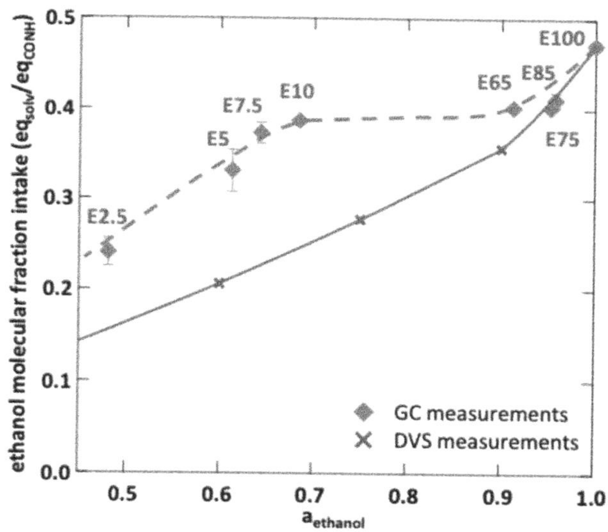

Figure 8.19. *Comparison between the ethanol molar fraction obtained by DVS at several pure ethanol activities (cross and solid curve) and the ethanol fraction for the PA66 films in contact with EX mixtures, as measured by GC-FID (points and dotted curve), with everything plotted as a function of ethanol activity (Rios de Anda et al. 2014)*

This might be explained by an effect caused by toluene, which is also present in the films, which would provoke PA6,6 swelling and would facilitate a greater absorption of ethanol by the polymer. Finally, the effect of EX mixtures on the molecular mobility of PA6,6 was studied. Notably, the decrease in T_g caused by these mixtures would be mainly due to the presence of ethanol in these samples, and even if toluene is present, this solvent does not induce an additional plasticizing effect, as shown in Figure 8.20 (Rios de Anda et al. 2014).

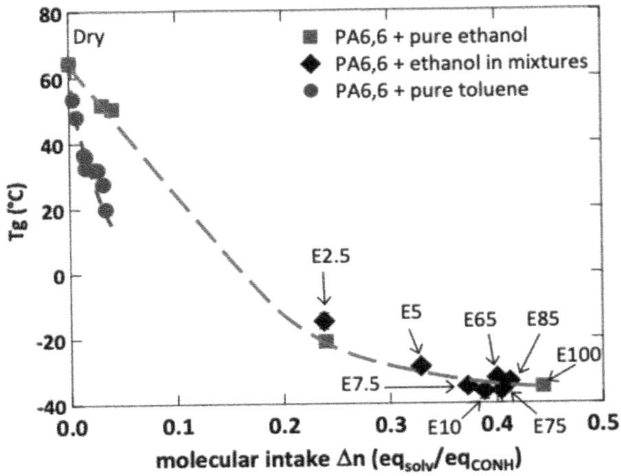

Figure 8.20. *Decrease in the T_g of PA6,6 as a function of the pure ethanol molar fraction Δn (green points) of ethanol in the EX mixtures (black dots) and in pure toluene (red dots) (Rios de Anda et al. 2014). For a color version of this figure, see www.iste.co.uk/richaud/industrialpolymers1.zip*

8.4. Properties of additivated or chemically modified PA6,6

We then investigated the effects of introducing an additive or a chemical modification of PA6,6 on the polymer structure and properties. Two additivated formulations (10HA and 10LA), two copolymers (HIA and AISLI) and a pure PA6,6 (REF) have been studied.

The effect of modifiers on the crystalline phase of PA6,6 was first studied. Notably, the 10LA and 10HA samples have a slightly higher crystallinity rate X_C, and their melting temperature T_f and crystallization temperature T_C are similar to those of the REF. The HIA and AISLI samples have crystallinity rates near the REF, but their melting and crystallization temperatures are lower, as shown in Figure 8.21. The presence of comonomers in PA6,6 may obstruct, to a certain extent, polymer crystallization.

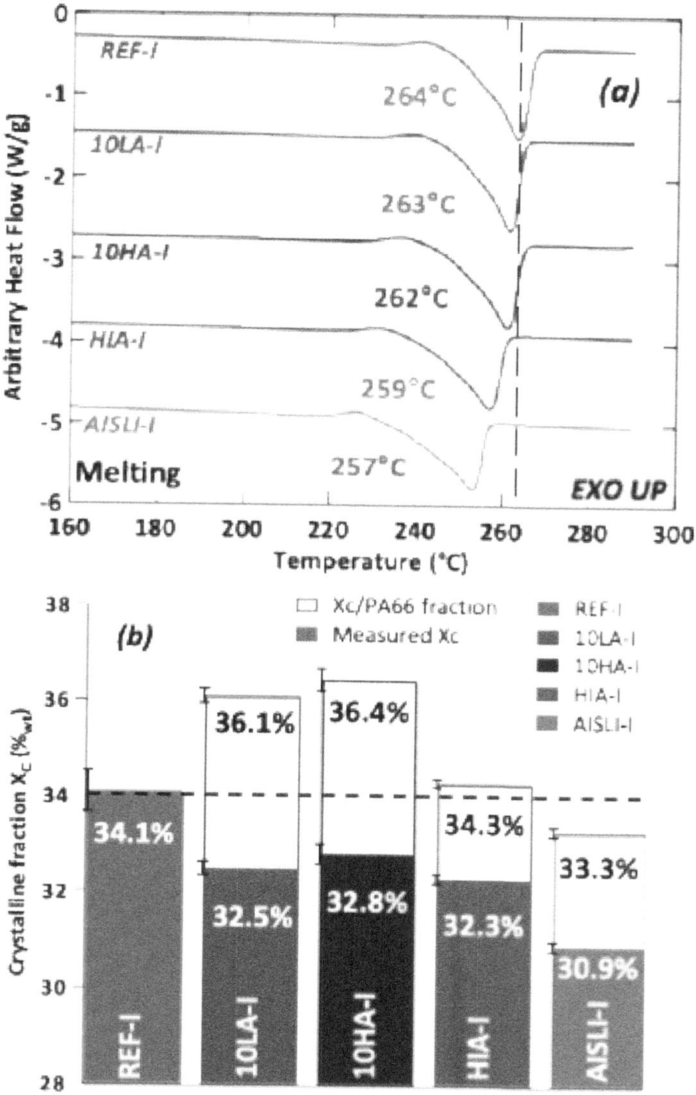

Figure 8.21. *(A) Data on the crystalline phase of the REF, 10HA, 10LA, HIA and AISLI samples. (a) DSC thermograms showing melting; (b) total crystalline fractions XC (full) and corrected XC-PA66 (empty) obtained by DSC. For a color version of this figure, see www.iste.co.uk/richaud/industrialpolymers1.zip*

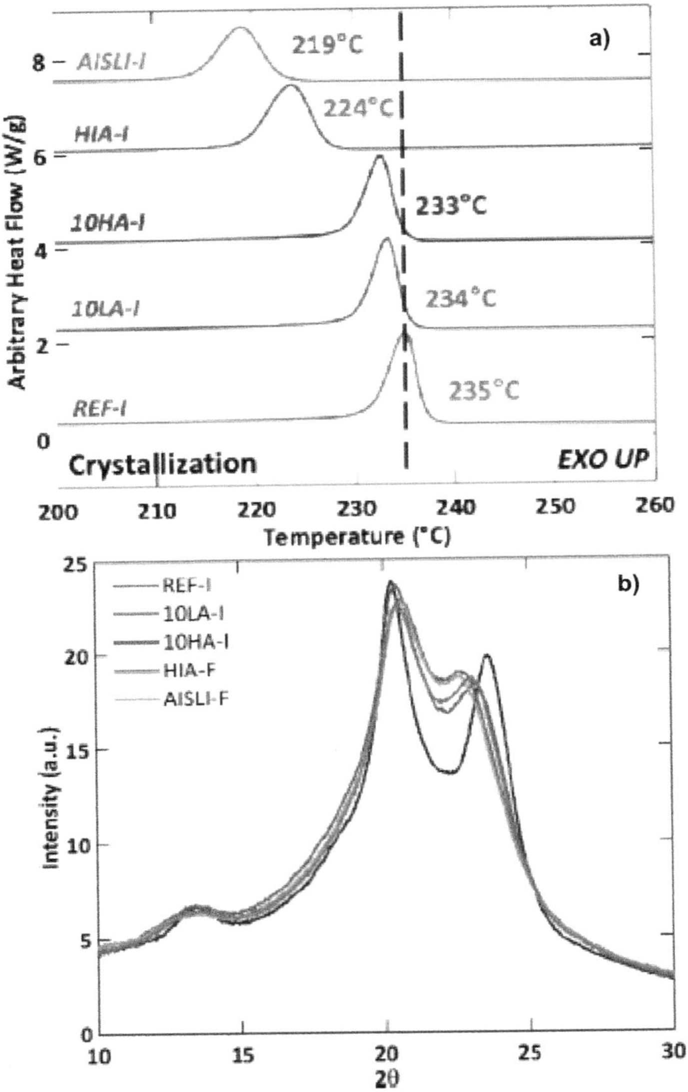

Figure 8.21. *(B) Data on the crystalline phase of the REF, 10HA, 10LA, HIA and AISLI samples. (a) DSC Thermograms showing crystallization; (b) X-ray diffraction spectra (continuation). For a color version of this figure, see www.iste.co.uk/richaud/industrialpolymers1.zip*

The effect of modifiers on the crystalline structure has also been studied by X-rays. Notably, the presence of lauryl and heptanaldehyde additives induces an evolution of the crystalline phase of PA6,6 to a more stable α_I structure. On the other hand, the presence of the 6HIA and 6AiSLi co-monomers induces a less perfect crystalline form. This effect might also be related to the process of sample implementation, as in this work we studied injected and extruded samples.

Then the molecular mobility of the amorphous phase of these samples was studied by DSC, DMA and dielectric spectroscopy. The results are shown in Figure 8.22. It seems that the heptanaldehyde additive, especially the laurylaldehyde additive, plasticizes PA6,6. For the copolymers, the HIA and AISLI samples have higher T_g values than the REF. This is due to the fact that these copolymers have aromatic groups in their chains. In the case of AISLI, lithium sulfonate functions can interact with amide groups and form stronger complexation bonds than amide–amide bonds, which would reduce the molecular mobility of chains. Moreover, the addition of 6AiSLI patterns would increase the width of the dynamic loss module E'', which might be related to a modification or to a variation of intermolecular interactions in the PA6,6 because of the presence of this comonomer (Rios de Anda et al. 2016).

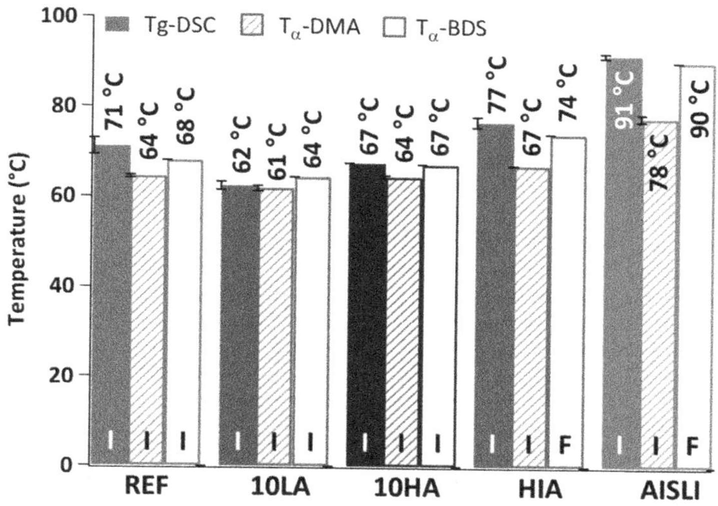

Figure 8.22. *T_g and T_α (1 Hz) obtained by DSC, DMA and BDS for the dry REF, 10HA, 10LA, HIA and AISLI formulations (Rios de Anda et al. 2016). For a color version of this figure, see www.iste.co.uk/richaud/industrialpolymers1.zip*

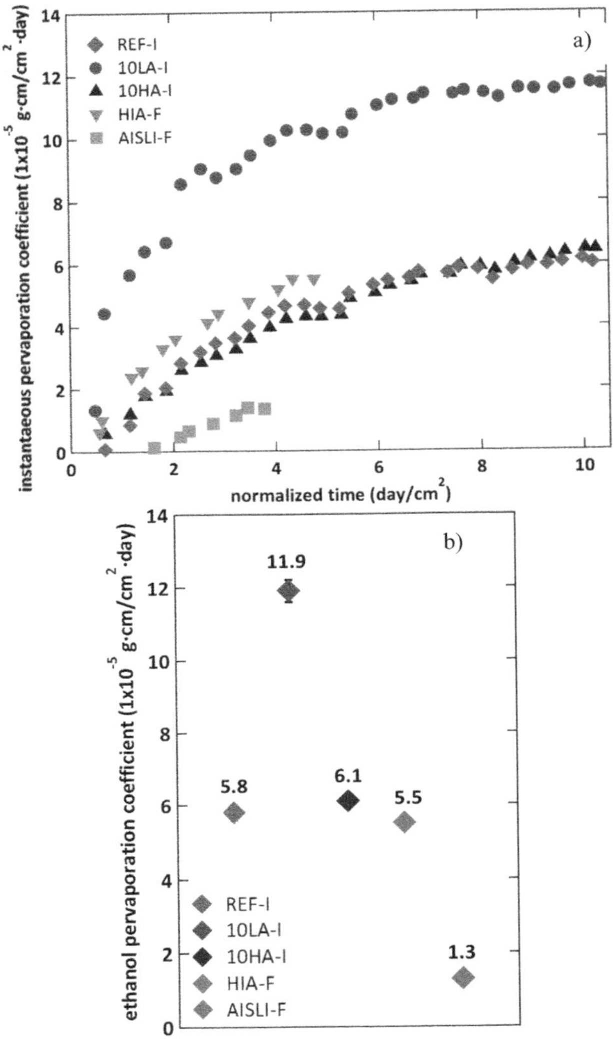

Figure 8.23. *Ethanol pervaporation coefficient of ethanol (a) instantaneously, as a function of the time normalized by the thickness of the samples; (b) in the permanent regime, obtained at 40°C for the modified formulations. For a color version of this figure, see www.iste.co.uk/richaud/industrialpolymers1.zip*

The effect of modifiers on the molecular relaxations of PA6,6 was also studied. Notably, lauryl- and heptanaldehyde additives decrease the temperature and activation energy of β relaxation, which suggests that these additives can interact with and disrupt the amide functions of PA6,6. Finally, for γ relaxation, no effect provoked by modifiers has been observed (Rios de Anda et al. 2016).

Then, the barrier properties of the modified formulations were evaluated. The sorption of water, ethanol and toluene was studied first. Notably, water is absorbed in similar quantities by all formulations, except for AISLI, which absorbs more water, probably because of the presence of lithium sulfonate functions that are highly hydrophilic. Concerning the ethanol, its intake by 10LA, HIA and AISLI is higher than that of REF. This is explained by the fact that the additives or the comonomers might promote the swelling of PA6,6, increasing the accessibility of amide groups and facilitating the further absorption of ethanol by these formulations. This hypothesis is similar to that formulated in the case of the sorption of previously described ternary mixtures. On the other hand, this same effect was not observed for 10HA, whose ethanol intake is slightly below that of REF. Moreover, no effect of modifiers on toluene intake was observed, probably because the space freed by the modifiers is not sufficiently wide to allow toluene to be further absorbed by the polyamide.

The sorption kinetics of water, ethanol and toluene were studied. Notably, water plasticizes the PA6,6 formulations very rapidly and diffuses according to an abnormal regime. In the case of ethanol, this solvent also plasticizes the studied formulations, and it diffuses similarly according to anomalous kinetics. Furthermore, the kinetics of ethanol before the plasticization of the polymer is Fickian, and its speed depends on the molecular mobility state of the formulation in a dry state: if a polymer has a low initial temperature (T_g), the diffusion of ethanol in this material would be faster than that for another polyamide whose initial T_g is higher. Finally, toluene cannot plasticize the formulations during sorption, which would lead this solvent to diffuse with a Fickian regime.

Permeability to ethanol was then assessed. The results are shown in Figure 8.23. Notably, 10LA has the highest permeability coefficient, AISLI has the lowest and the other formulations have similar coefficients. These differences could be explained by the sorption kinetics and ethanol sorption at equilibrium in these materials, since permeability depends on these two phenomena. In the case of 10LA, ethanol is absorbed to a greater extent compared to REF, and its sorption kinetics in this formulation is the fastest, which may lead to more significant permeation of this solvent. In terms of AISLI, this polymer absorbs more ethanol than REF, but its kinetics in this polymer is the slowest, which may lead to a low permeation of ethanol. 10HA and HIA, which have ethanol intakes and sorption kinetics similar to REF, also have close permeations.

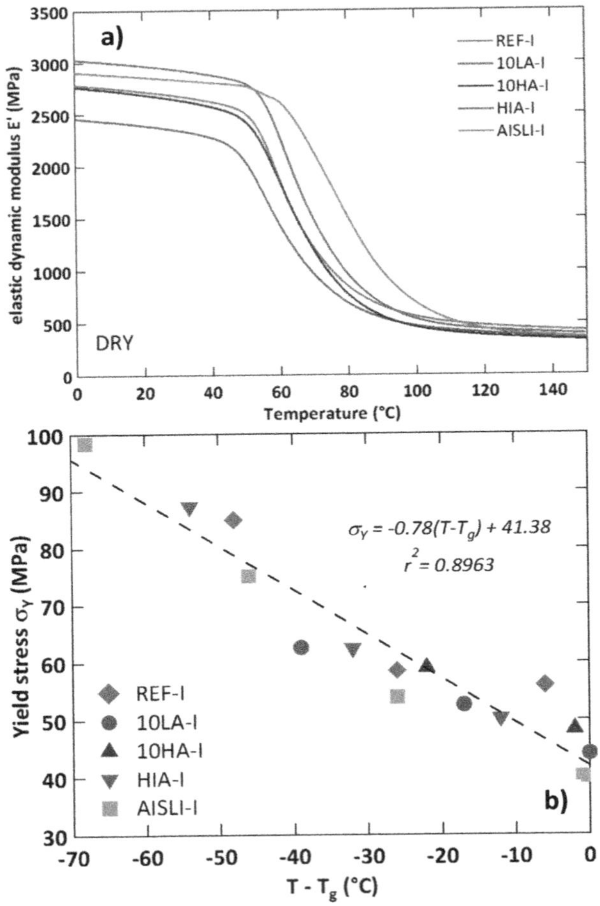

Figure 8.24. *(a) Elastic dynamic modulus E' as a function of temperature; (b) plotted yield stress σ_Y as a function of T-T_g for samples modified in the dry state (Rios de Anda et al. 2016). For a color version of this figure, see www.iste.co.uk/richaud/industrialpolymers1.zip*

The effect of solvents on the molecular mobility of these formulations was then studied. Dielectric spectroscopy revealed that water had the same plasticizing effect in all formulations. This is also observed for ethanol. With respect to toluene, this solvent has the same plasticizing effect in all formulations, except for AISLI, whose effect is twofold greater. This could be explained by the fact that toluene would be able to disrupt the amide-AISLI sulfonate interactions, which would induce an additional plasticizing effect (Rios de Anda et al. 2016).

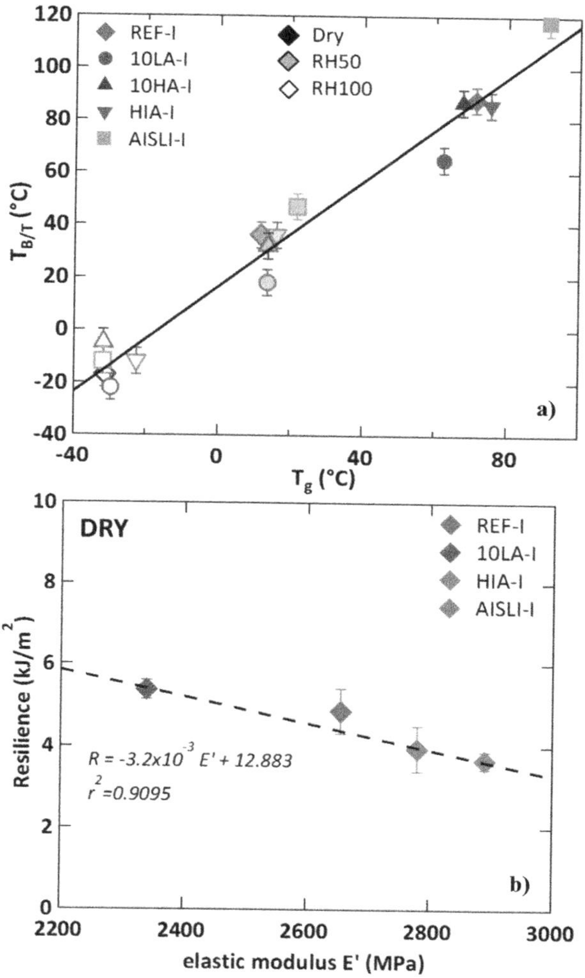

Figure 8.25. *(a) Brittle–ductile transition temperature $T_{F/D}$ plotted as a function of T_g for modified samples both in the dry state and conditioned with HR50 and HR100; (b) resilience at the brittle plateau (T-$T_{F/D}$ = –30°C) for dry samples as a function of the elastic modulus E' in the glassy state (Rios de Anda et al. 2016). For a color version of this figure, see www.iste.co.uk/richaud/industrialpolymers1.zip*

Next, we focused on the mechanical properties of these samples under tension in the dry state. In the case of the Young's modulus E and the dynamic modulus of elasticity E' of these samples, plotted as a function of temperature, Figure 8.24(a) (E') shows that the values of the modulus in the glass plateau depend on the intermolecular interactions between PA6,6 and the modifiers. 6HIA and 6AISLI rigidify PA6,6 by adding aromatic functions and strong interactions, which would increase the modulus. Phenolic additives are capable of plasticizing PA6,6 by steric effects, which would decrease the modulus of the polymer. Compared with that of T-T_g, the yield stress σ_Y of these samples yields a master curve, as shown in Figure 8.24(b). This suggests that for these samples, the yield stress σ_Y is similar for a given level of molecular mobility (Rios de Anda et al. 2016).

Then, we focused on the impact strength of these samples in the dry state and in two states of conditioning in water (HR50 and HR100) to obtain various states of molecular mobility. The brittle–ductile transition temperatures $T_{F/D}$ were obtained by instrumented Charpy impact tests. The resulting $T_{F/D}$ values were compared with the T_g values measured by DSC for these samples, as shown in Figure 8.25(a).

As shown in Figure 8.25(a), there is a global linear correlation between $T_{F/D}$ and T_g, which suggests that for a given polyamide formulation, the brittle–ductile transition temperature is mainly determined by the molecular mobility of this polymer formulation.

Notably, the values of $T_{F/D}$ for the 10LA are below those of the other formulations. This might be due to the probability that laurylaldehyde might be able to form micelles in PA6,6. These micelles would act as load concentrators, which would lead to a better distribution of the mechanical stress during impact and therefore to a better resistance to impact. Finally, the relationship between the mechanical modulus and the resilience to the brittle plateau was studied, as shown in Figure 8.25(b).

A linear dependence between the resilience and modulus can be noted: when the mechanical modulus increases, the resilience decreases (Rios de Anda et al. 2016).

8.5. Conclusion

This chapter focuses on developing a better understanding of the effect of the amorphous phase of PA6,6 on the barrier and mechanical properties. We were able to describe how the nature of a solvent conditions its sorption and its plasticizing effect on PA6,6. Notably, the nature and the interchain interactions of the polymer matrix also condition the solvent intake.

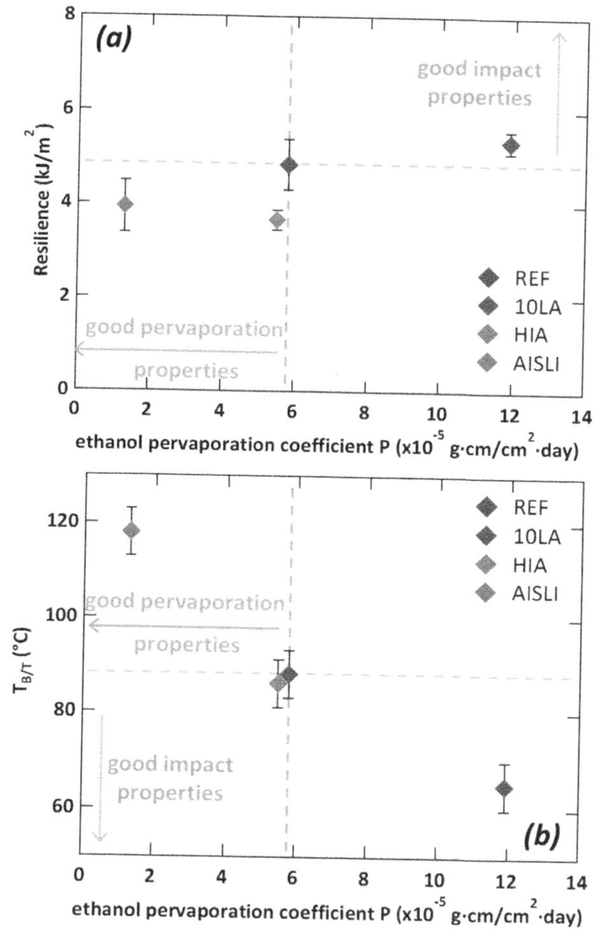

Figure 8.26. *(a) Resilience at the brittle plateau; (b) $T_{F/D}$ of dry samples as a function of the ethanol pervaporation coefficient for the studied formulations. For a color version of this figure, see www.iste.co.uk/richaud/industrialpolymers1.zip*

Moreover, a thermodynamic approach was proposed for the study and understanding of the sorption of ethanol-toluene-isooctane ternary mixtures and their plasticizing effect on PA6,6. This approach facilitates a better understanding of the thermodynamic behavior of these mixtures and the effect of this behavior on their sorption and their plasticizing effect in PA6,6. It was also shown that the mechanical properties of PA6,6 depend on or are related to the molecular mobility state of the amorphous phase of this polymer. The application part of this work, which was to obtain PA6,6 formulations for gasoline tanks with good properties of being a barrier to solvents while maintaining good resistance to impacts, was not actually complete at this stage. Notably, a formulation with good barrier properties does not have adequate impact resistance, and vice versa.

The results obtained in this work concerning the effect of solvents and modifiers on the barrier and mechanical properties of PA6,6 would enable us to better focus on a potential structural modification of this polymer to obtain a formulation featuring a good compromise between these two properties.

8.6. References

Alexandre, B., Langevin, D., Médéric, P., Aubry, T., Couderc, H., Nguyen, Q.T., Saiter, A., Marais, S. (2009). Water barrier properties of polyamide 12/montmorillonite nanocomposite membranes: Structure and volume fraction effects. *J. Memb. Sci.*, 328, 186.

Alexandre, B., Colasse, L., Langevin, D., Médéric, P., Aubry, T., Chappey, C., Marais, S. (2010). Transport mechanisms of small molecules through polyamide 12/montmorillonite nanocomposites. *J. Phys. Chem. B*, 114, 8827.

Auerbach, I. and Carnicom, M.L. (1991). Sorption of water by Nylon 66 and Kevlar 29. Equilibria and kinetics. *J. Appl. Polym. Sci.*, 42, 2417.

Bessler, E. and Bier, G. (1969). Über Wasserstoffbrücken in Polyamiden. *Makromol. Chemie*, 122(1), 30.

Brandrup, J. and Immergut, E.H. (1989). *Polymer Handbook*. John Wiley & Sons, New York.

Bureau, M.N., Denault, J., Cole, K.C., Enright, G.D. (2002). The role of crystallinity and reinforcement in the mechanical behavior of polyamide-6/clay nanocomposites. *Polym. Eng. Sci.*, 42(9), 1897.

Collyer, A.A. (1994). *Rubber Toughened Engineering Plastics*. Chapman & Hall, London.

Crank, J. (2004). *The Mathematics of Diffusion*. Oxford University Press, Oxford.

Diaz-Calleja, R. and Riande, E. (2004). Comparative study of mechanical and dielectric relaxations in polymers. *Materials Science and Engineering A*, 370, 21–33.

Eceolaza, S., Iriarte, M., Uriarte, C., Etxeberría, A. (2009). Barrier property enhancement of polyamide 6 by blending with a polyhydroxyamino-ether resin. *J. Polym. Sci. Part B: Polym. Phys.*, 47, 1625–1634.

Fick, A. (1855). Über Diffusion. *Ann. Phys, Lpz.*, 170(1), 59.

Garcia, D. and Starkweather, H.W. (1985). Hydrogen bonding in nylon 66 and model compounds. *J. Polym. Sci. Part B: Polym. Phys.*, 23, 537–555.

Hernández, R.J., Giacin, J.R., Grulke, E.A. (1992). The sorption of water vapor by an amorphous polyamide. *J. Memb. Sci.*, 65, 187–199.

Hirschinger, J., Miura, H., Gardner, K.H., English, A.D. (1990). Segmental dynamics in the crystalline phase of nylon 66: Solid-state ^{2}H NMR. Report.

Hodgman, C.D. (1962). *CRC Handbook of Chemistry and Physics*. CRC Press, Cleveland.

Hunt, D.G. (1980). Prediction of sorption and diffusion of water vapour by nylon-6,6. *Polymer*, 21, 495–501.

Illiers, K.H. (1960). Der Einfluß von Wasser auf die molekularen Beweglichkeiten von Polyamiden. *Makromol. Chemie*, 38(1), 168.

Illiers, K.H. and Jacobs, H. (1960). Der Einfluß von Quellundsmitteln auf die Wasserstoffbrücken-Bindug und das elastische Verhalten von Polyamiden und Polyurethanen. *Makromol. Chemie*, 39(1), 234.

Kamal, M.R. and Jinnah, I.A. (1984). Permeability of oxygen and water vapor through polyethylene/polyamide films. *Polym. Eng. Sci.*, 24(17), 1337.

Kohan, M.L. (1995). *Nylon Plastics Handbook*. Carl Hanser Verlag, Munich.

Laredo, E. and Hernández, M.C. (1997). Moisture effect on the low- and high-temperature dielectric relaxations in nylon-6. *J. Polym. Sci. Part B: Polym. Phys.*, 35, 2879.

Laredo, E., Grimau, M., Sánchez, F., Bello, A. (2003). Water absorption effect on the dynamic porperties of nylon-6 by dielectric spectroscopy. *Macromolecules*, 36, 9840.

Laurati, M., Arbe, A., Rios de Anda, A., Fillot, L.A., Sotta, P. (2014). Effect of polar solvents (water and ethanol) on the crystalline phase of polyamides. *Polymer*, 55(12), 2867.

Le Huy, H.M. and Rault, J. (1994). Remarks on the α and β transitions in swollen polyamides. *Polymer*, 35, 136.

Lim, L.T., Britt, I.J., Tung, M.A. (1999). Sorption and transport of water vapor in nylon 6,6 film. *J. Appl. Polym. Sci.*, 71, 197.

Low, H.Y., Liu, T.X., Loh, W.W. (2004). Moisture sorption and permeation in polyamide 6/clay nanocomposite films. *Polym. Int.*, 53, 1973.

Mansfield, P., Botwell, R., Blackband, S. (1992). Ingress of water into solid nylon 6,6. *J. Magnetic Res.*, 99, 507.

Miri, V., Persyn, O., Lefebvre, J.M., Séguela, R. (2009). Effect of water absorption on the plastic deformation behavior of nylon 6. *Eur. Polym. J.*, 45, 757.

Miura, H., Hirschinger, J., English, A.D. (1990). Segmental dynamics in the amorphous phase of nylon 66: Solid state deuterium NMR. *Macromolecules*, 23(8), 2169–2182.

Owen, A. and Bonart, R. (1985). Cooperative relaxation processes in polymers. *Polymer*, 26, 1034.

Papir, Y.S., Kapur, S., Rogers, C.E., Baer, E. (1972). Efect of orientation, anisotropy, and water on the relaxation behavior of nylon 6 from 4.2 to 300°K. *J. Polym. Sci. Part A-2*, 10, 1305.

Park, Y., Ko, J., Ahn, T.K., Choe, S. (1997). Moisture effects on the glass transition and the low temperature relaxations in semiaromatic polyamides. *J. Polym. Sci. Part B: Polym. Phys.*, 35, 807.

Pathmanathan, K., Cavaillé, J.Y., Johari, G.P. (1992). The dielectric properties of dry and water-saturated nylon-12. *J. Polym. Sci. Part B: Polym. Phys.*, 30, 341.

Picard, E., Gérard, J.F., Espuche, E. (2008). Water transport properties of polyamide 6 based nanocomposites prepared by melting blending: On the importance of the clay dispersion state on the water transport properties at high water activity. *J. Membr. Sci.*, 313, 284.

Razumovskii, L.P., Markin, V.S., Ye Zaikov, G. (1985). Sorption of water by aliphatic polyamides. Review. *Polym. Sci. U.S.S.R.*, 27(4), 751.

Redlich, O. and Kwong, J.N.S. (1979). On the thermodynamics of solutions V. An equation of state. Fugacities of gaseous solutions. *Chem. Rev.*, 44, 223.

Renon, H. and Prausnitz, J.M. (1968). Local compositions in thermodynamic excess functions for liquid mixtures. *AIChE Journal*, 14(1), 135.

Rios de Anda, A., Fillot, L.A., Rossi, S., Long, D., Sotta, P. (2011). Influence of the sorption of polar and non-polar solvents on the glass transition temperature of polyamide 6,6 amorphous phase. *Polymer Engineering and Science*, 51(11), 2129.

Rios de Anda, A., Fillot, L.A., Preda, F.M., Rossi, S., Long, D., Sotta, P. (2014). Sorption and plasticization effects of ethanol-toluene-isooctane ternary mixtures in polyamide 6,6 and induced plasticization effects. *European Polymer Journal*, 55, 199.

Rios de Anda, A., Fillot, L.A., Long, D., Sotta, P. (2016). Influence of the amorphous phase molecular mobility on impact and tensile properties of polyamide 6,6. *Journal of Applied Polymer Science*, 133(21), 43457.

Rotter, G. and Ishida, H. (1992). Dynamic mechanical analysis of the glass transition: Curve resolving applied to polymers. *Macromolecules*, 25, 2170.

Sabard, M. (2011). Effet des conditions de procédé et des nanocharges sur la morphologie et les propriétés de sorption à l'eau et à l'éthanol. PhD Thesis, Université de Lyon.

Schroeder, L.R. and Cooper, S.L. (1976). Hydrogen bonding in polyamides. *J. Appl. Phys.*, 47(10), 4310.

Shan, G., Yang, W., Yang, M., Xie, B., Feng, J., Fung, Q. (2007). Effect of temperature and strain rate on the tensile deformation of polyamide 6. *Polymer*, 48, 2958.

Skrovanek, D.J., Howe, S.E., Painter, P.C., Coleman, M.M. (1985a). On the validity of a commonly employed infrared procedure used to determine thermodynamic parameters associated with hydrogen bonding in polymers. *Macromolecules*, 18, 299.

Skrovanek, D.J., Howe, S.E., Painter, P.C., Coleman, M.M. (1985b). Hydrogen bonding on polymers: Infrared temperature studies of an amorphous polyamide. *Macromolecules*, 18, 1676.

Starkweather, H.W. (1959). The sorption of water by nylons. *J. Appl. Polym. Sci.*, 2(5), 129.

Starkweather, H.W. (1975). Some aspects of water clusters in polymers. *Macromolecules*, 8(4), 476.

Starkweather, H.W. and Brooks, R.E. (1959). Effect of spherulites on the mechanical properties of nylon 66. *J. Appl. Polym. Sci.*, 1(2), 236.

Starkweather, H.W., Moore, G.E., Hansen, J.E., Roder, T.M., Brooks, R.E. (1956). Effect of crystallinity on the properties of nylons. *J. Polym. Sci.*, 21, 189.

Takeda, N., Song, D.Y., Nakata, K. (1996). Effects of temperature and water content on impact properties of injection-molded glass nylon-6 composites. *Adv. Composite Mater.*, 5(3), 201.

Triggs, N.E., Bonn, R.T., Valentini, J.J. (1993). Do all solid amides hydrogen bond? Raman evidence to the contrary. *J. Phys. Chem.*, 97, 5535.

Yeh, J.T., Chang, S.S., Yao, H.T., Chen, K.N., Jou, W.S. (2000). The permeation resistance of polyethylene, polyethylene/polyamide, and polyethylene/modified polyamide blown tubes against unleaded gasoline. *J. Mat. Sci.*, 35, 1321.

9

Statistic Approach to Polymer Network Degradation[1]

Pierre GILORMINI
PIMM, CNRS, CNAM, ENSAM, Paris, France

9.1. Introduction

Network degradation is one of the oldest problems in polymer science. It emerged in the mid-19th century with the development of industrial applications of vulcanized rubber. The efforts to clarify the mechanisms involved had no results until the 1940s, when it was acknowledged that vulcanized rubbers are macromolecular networks, mainly thanks to Staudinger (Nobel prize, 1953) and Flory (Nobel prize, 1974), and that degradation results from free radical chain oxidation, mainly thanks to Semenov (Nobel prize, 1956).

The main characteristics of polymer network degradation by random chain scission are illustrated in Figure 9.1, where the conversion rate x is defined as the number of reactive units that have reacted, divided by the initial number of reactive units, which are assumed to be equireactive. While degradation (x increasing from 0 to 1) does not compete with cross-linking, it leads to a gradual drop in cross-linking density, with a simultaneous decrease in the modulus of elasticity G. For example, if

For a color version of all figures in this chapter, see www.iste.co.uk/richaud/industrialpolymers1.zip.

1 Writing this chapter was widely inspired by many enthusiastic discussions with Jacques Verdu, a leading expert in polymer sciences, who had a fine appreciation for jazz, literature and rugby, and who unfortunately died on November 30, 2018.

Aging of Industrial Polymers 1,
coordinated by Emmanuel RICHAUD. © ISTE Ltd 2025.

the network is initially ideal and trifunctional, with a modulus of elasticity G_0, each of the very first scission reactions suppresses three elastically active chains (EAC; while two others double their length); and as the number of reactive sites is equal to L times the number of network chains (where L is the average number of reactive sites per chain), the initial slope of the relative modulus of elasticity G/G_0 as a function of x is $-3L$. If the network was at least tetrafunctional, this slope would only be $-L$ because one scission would only suppress one single elastically active chain. This linear variation cannot persist when x increases, since an increasing part of scissions occur on more or less long and branched chains, which become pending, and can detach them from the gel, which does not modify the modulus of elasticity. The soluble phase, whose growth was initially zero, given that the first scissions could not detach parts of the gel, will then increase faster. This process accelerates to the point referred to as "degelation" reached for $x = x_d$, where the modulus of elasticity becomes zero and where the polymer becomes completely soluble. Figure 9.1 shows two experimental approaches to determine the degelation point, using either the cross-linking density, which is obtained by measuring the modulus of elasticity or by gel swelling in a solvent, or the quantity of the soluble fraction or its average weight molar mass, which is expected to reach a maximum at degelation.

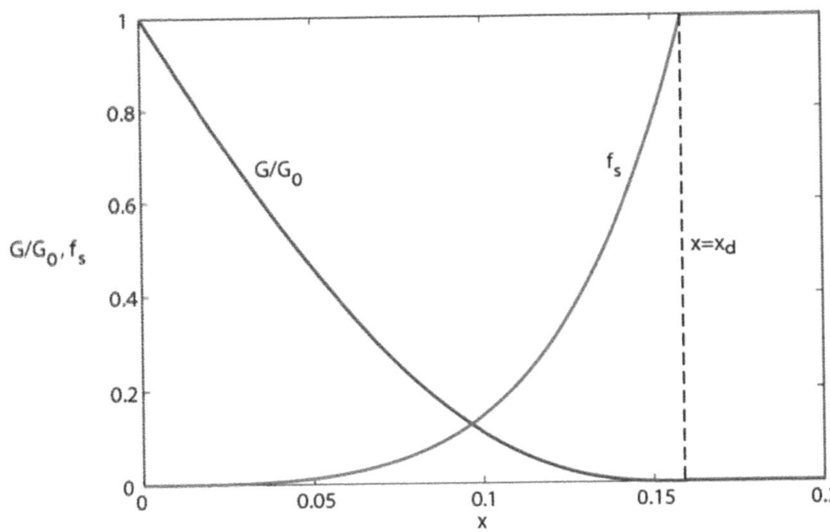

Figure 9.1. *Schematic evolutions of the relative modulus of elasticity (G/G_0, in red) and of the soluble fraction (f_s, in green) as a function of the conversion ratio x during polymer network degradation by random chain scission*

This chapter presents a theory for predicting the curves in Figure 9.1 and therefore, in particular, the degelation point. It is a probabilistic approach leading to explicit expressions and not to the differential equations to be solved. It has its place between, on the one hand, a kinetic theory that predicts the changes in the conversion rate x in time; and on the other hand, it is a theory that predicts the modulus of elasticity as a function of the cross-linking rate. This approach has already been applied to various systems, and confronted with experimental results of hydrolysis and radiolysis by Gilormini et al. (2014, 2017). These results will not be repeated here, but we would rather compare the theory to numerical simulations. This chapter covers only highly ideal systems, but hopefully the reader will draw from it the inspiration for extending the proposed approach to more complex cases.

9.2. Numerical simulation

The numerical simulation of the gradual degradation of a polymer network is conducted in three stages: a plausible network must first be built, then we should proceed to its degradation, and ultimately, the results must be analyzed to obtain relevant quantities. The network is seen here as a graph, with N vertices (the nodes of the network) connected by edges (network chains) containing, on average, L breakable bonds. Such a representation of a polymer network, reduced to its connectivity, does not locate each of its constituent entities in space and ignores potential chain entanglement; however, some of its properties can be accessed. For a perfect network of homogeneous functionality f, there is a total of $fN/2$ edges and $LfN/2$ cleavable sites.

In practice, each node is numbered from 1 to N, and in the first table, the numbers drawn at random of the two nodes at the end of each edge are defined, ensuring that the two numbers are different, that each number is only used f times and that a pair of numbers is only used once. A table containing the number of breakable bonds of each edge is then defined, considering here only uniform or Flory distributions, with, in this latter case, an intermediate draw to achieve the expected characteristics. A long table is finally defined, giving the state of all bonds, successively arranged edge by edge, in the form of Booleans initialized to the *true* state. A degraded state of the network for a conversion rate x is then reached by passing to the *false* state $xLfN/2$ randomly chosen bonds in this last table, and a chain loses its edge status as soon as one of its breakable sites reacts. The analysis of the resulting graph is at the core of the simulation. It is quite complex and probably lengthy, but it will provide various characteristics of the gel phase, including the number of EAC, as well as the composition of the soluble phase.

The number of nodes will necessarily be finite in such a simulation, and here, we will generally use 100,000 nodes after comparing results with more than one million nodes. This is why the gel phase will be defined as the largest entity connected (in number of nodes) in the system, and it is the passage by a maximum of the average weight molar mass of the soluble phase that will define the degelation in the numerical simulation.

A first analysis of the system is conducted using the SPANFO algorithm by Nijenhuis and Wilf (1978), which provides a list of all related entities that are present. This can be used to identify, on the one hand, the gel and, on the other hand, the components of the soluble phase (any entity smaller than the gel, therefore containing nodes or simple edge segments). Once the gel is identified, several operations are required to determine the number of EAC. Until the stabilization of the obtained graph, we iteratively suppress the vertices of degree 1 (from which only one edge extends and this edge is suppressed), the edge sequences connecting the vertices of degree 2 are concatenated (and the intermediate vertices are suppressed) and the loops are suppressed (the edges whose start and end are the same vertex). The number of edges of the resulting graph is by definition considered equal to the number of EAC of the gel.

9.3. Probabilistic approach to degradation

Let us consider an initially perfect network, where all nodes have the same functionality $f \geq 3$ and where the chains contain a variable number of reactive groups (or scission sites) following a distribution $s(i)$. This means that there is a fraction $s(i)$ of chains each carrying i reactive groups, with $\sum_i s(i) = 1$, where the sum applies to all values of i observed in the network. It is assumed, on the one hand, that all groups are equireactive and, on the other hand, that chain scissions do not create new groups and do not modify the reactivity of neighbor groups. The following reasoning is inspired by the probabilistic theory of networks initiated by Flory (1941) and Stockmayer (1943) and developed more recently by Macosko and Miller (1976), Dušek (1989) and Pascault et al. (2002), among many others.

Since the probability of a reactive group to not react is $1 - x$ for a conversion rate equal to x, the probability of a chain containing i reactive groups to be uninterrupted is $(1 - x)^i$. Considering the distribution of the number of reactive groups per chain, the probability that a chain does not undergo scission is as follows:

$$u(x) = \sum_i s(i)(1 - x)^i \quad [9.1]$$

The probability $g(x)$ that a chain that starts in a node is bound to the gel is as follows:

$$g(x) = u(x)\,[1 - (1 - g(x))^{f-1}] \tag{9.2}$$

Indeed, it must be intact (probability $u(x)$), and it must reach a node where at least one of the other $f - 1$ chains is bound to the gel. The probability for this latter condition to be true is equal to 1 minus the probability that none of these chains is bound to the gel, which is $(1 - g(x))^{f-1}$. The trivial solution $g(x) = 0$ of [9.2] for any value of x can be dismissed, as it would involve the absence of gel (or, it exists for $x = 0$). Therefore, after simplification by $g(x)$, it is sufficient to solve:

$$\sum_{k=1}^{f-2}(1 - g(x))^k = \frac{1}{\sum_i s(i)(1-x)^i} - 1 \tag{9.3}$$

to obtain the value of $g(x)$ in the general case. Notably, it depends on the conversion rate and network characteristics: the functionality and distribution of the number of reactive sites. It can be immediately deduced that the rate of conversion to degelation x_{d}, which corresponds to $g(x) = 0$ in equation [9.3] (as the gel has then disappeared), is given by the following solution:

$$\sum_i s(i)(1 - x_{\mathrm{d}})^i = \frac{1}{f-1} \tag{9.4}$$

Once the probability $g(x)$ is obtained from equation [9.3], the number of EAC will be deduced by first calculating the number of elastically active nodes (EAN), that is, those from which at least three chains bound to the gel emerge. As there are $\binom{f}{k}$ possibilities to define the k chains starting from a node being bound to gel and $f - k$ that are not, the probability that a node is an EAN is as follows:

$$\sum_{k=3}^{f}\binom{f}{k}\, g(x)^k\,(1 - g(x))^{f-k} \tag{9.5}$$

The number of EAC can be deduced by considering that each of them connects two EAN, counting the number of chains bound to the gel starting from an EAN. Finally, assuming that the modulus of elasticity is proportional to the number of EAC, we obtain:

$$\frac{G(x)}{G_0} = \frac{1}{f}\sum_{k=3}^{f}\binom{f}{k}\, k\, g(x)^k\,(1 - g(x))^{f-k} \tag{9.6}$$

since the initial number of EAC is $fN/2$.

The calculation of the fraction of the soluble phase involves, in principle, the molar mass of the network nodes and that of the chain segments between two breakable sites. In brief, this chapter explores only the expression of the simple case, which takes into account only the nodes of the network. Then, a node belongs to the soluble phase only if none of the f chains initially starting from it is no longer bound to the gel. Hence:

$$f_s(x) = (1 - g(x))^f \quad [9.7]$$

9.4. First example: trifunctional network with one or two reactive groups per chain

As a first very simple example, let us consider a trifunctional network with only one reactive group per chain, with $f = 3$ and $i = 1$; therefore, equation [9.4] yields $x_d = 1/2$, equation [9.3] yields $g(x) = (1 - 2x)/(1 - x)$, and equation [9.7] leads to:

$$\frac{G(x)}{G_0} = \left(\frac{1-2x}{1-x}\right)^3 \quad [9.8]$$

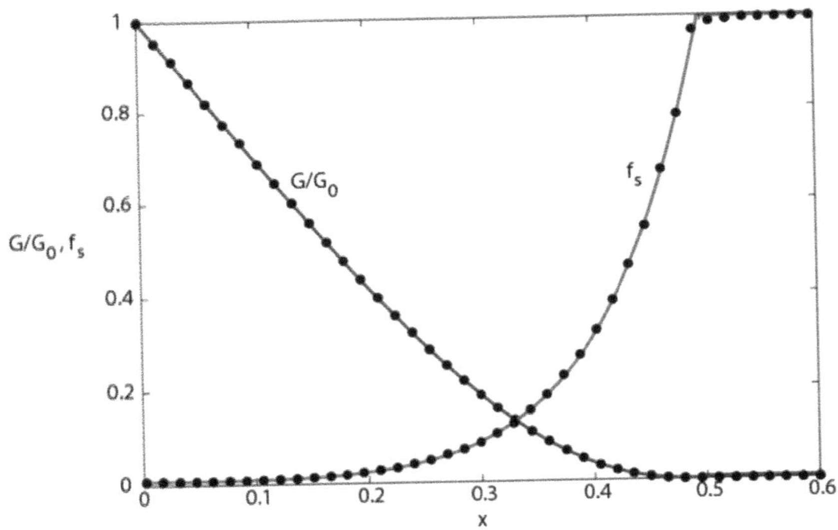

Figure 9.2. *Analytical prediction of the modulus of elasticity (red) and of the soluble fraction (green) during degradation and numerical simulation (symbols). Case of a trifunctional network with only one reactive group per chain*

These results are compared in Figure 9.2 with the results of the numerical simulation. There is total agreement. The same is true for the soluble fraction yielded by equation [9.9]:

$$f_s(x) = \left(\frac{x}{1-x}\right)^3 \qquad [9.9]$$

Galina and Lechowicz (1998) used a comparable numerical approach to study the same type of system. By conducting a single simulation on a network of 5,000 nodes, they found a degelation point at $x_d \approx 0.45$, whereas by conducting 10 simulations with 100,000 nodes, we obtained $x_d = 0.493 \pm 0.006$, which is very close to the theoretical value (1/2). This illustrates the importance of considering a sufficiently large number of nodes in numerical simulations.

If there are two reactive groups on each chain of a trifunctional network, we have $i = 2$; hence, $x_d = 1 - 1/\sqrt{2} = 0.293$, $g(x) = (1 - 4x + 2x^2)/(1 - x)^2$, and consequently:

$$\frac{G(x)}{G_0} = \left(\frac{1-4x+2x^2}{1-2x+x^2}\right)^3 \qquad [9.10]$$

that can be compared in Figure 9.3, as well as the soluble part yielded by equation [9.10]:

$$f_s(x) = \left(\frac{2x-x^2}{1-2x+x^2}\right)^3 \qquad [9.11]$$

to the results of the numerical simulation, leading once again to full agreement.

The previous results can also be obtained by reversing the direction of time in a model of polymerization or polycondensation by replacing x with $1 - x$. The first to have thought of it was probably Horikx (1956), but this approach of reverse polymerization is, however, limited to the cases considered below. It should then be noted that the rate of conversion to degelation $x_d = 1/2$, when there is only one reactive group per chain, is quite the same as that of the gel given by Flory (1941) for the polymerization of a trifunctional monomer. Equation [9.9] is strictly related to the postgel relation given by Flory (1953).

Similarly, the value of the degradation point when there are two functional groups per chain was obtained and experimentally confirmed by Argyropoulos and Bolker (1987) by changing the ratio of conversion to gel into $1 - x_d$ in the famous article by Stockmayer (1943).

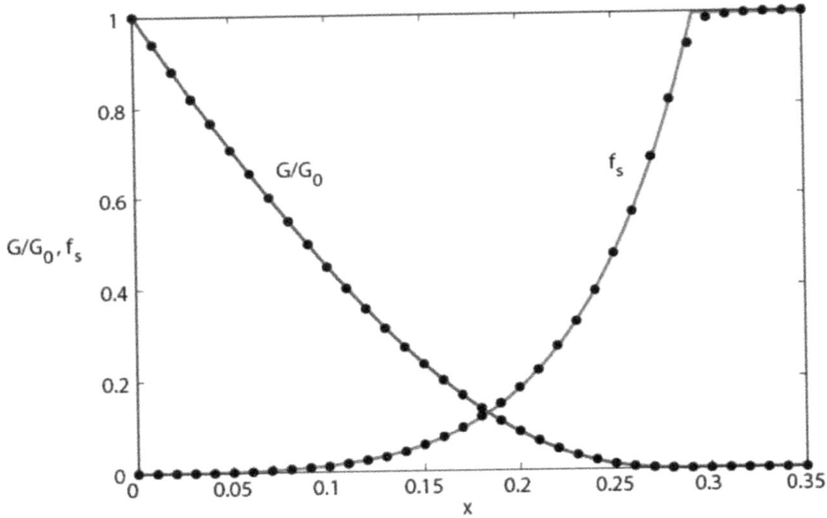

Figure 9.3. *Analytical prediction of the modulus of elasticity (red) and of the soluble fraction (green) during degradation and numerical simulation (symbols). Case of a trifunctional network with two reactive groups per chain*

9.5. Second example: trifunctional network with more than two reactive groups per chain

The proposed probabilistic theory is now applied to more complex situations. The reason for which reversing polymerization is not applicable when there are more than two reactive groups per chain is that there is competition between chain extension and bonds between chains and cross-linking nodes.

Let us first consider the case in which each chain of an initial ideal and trifunctional network has exactly L breakable sites. The only value of i for which $s(i)$ is not zero is $i = L$, and the condition of degelation of equation [9.4] is written as:

$$x_{\mathrm{d}} = 1 - \frac{1}{2^{1/L}} \qquad [9.12]$$

The change in the module before the degelation of equation [9.6] is given by:

$$\frac{G(x)}{G_0} = \left[2 - \frac{1}{(1-x)^L}\right]^3 \qquad [9.13]$$

since $g(x) = 2 - 1/(1-x)^L$, and the soluble fraction (only in terms of nodes) of equation [9.7] is given by:

$$f_s(x) = \left[\frac{1}{(1-x)^L} - 1\right]^3 \qquad [9.14]$$

These results generalize and resume the above-mentioned results for $L = 1$ or $L = 2$. For $L = 18$, for example, this leads to the curves in Figure 9.4, with degelation for $x_d = 0.038$ and in full agreement with the numerical simulation.

A comparison with Figures 9.2 and 9.3 shows the very strong effect of the number of reactive sites per chain, which significantly reduces the rate of conversion to degelation in accordance with equation [9.10].

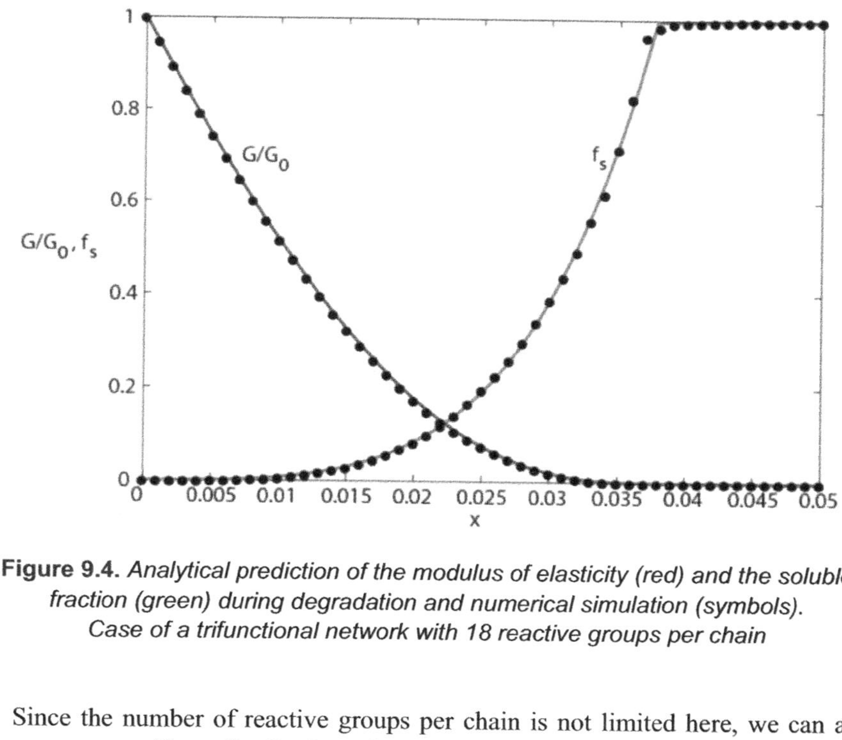

Figure 9.4. *Analytical prediction of the modulus of elasticity (red) and the soluble fraction (green) during degradation and numerical simulation (symbols). Case of a trifunctional network with 18 reactive groups per chain*

Since the number of reactive groups per chain is not limited here, we can also focus on non-uniform distributions. Let us consider, for example, a distribution of the number of reactive groups per chain following Flory's (1953) law:

$$s(i) = \frac{(L-1)^{i-1}}{L^i} \qquad [9.15]$$

with:

$$\sum_{i=1}^{\infty} s(i) = 1 \text{ and } \sum_{i=1}^{\infty} i\, s(i) = L \quad [9.16]$$

where L denotes the average number of reactive groups per chain. As equation [9.15] leads to:

$$\sum_{i=1}^{\infty} s(i)(1-x)^i = \frac{1-x}{1+(L-1)x} \quad [9.17]$$

The degelation point of equation [9.4] is given by the following remarkably simple expression:

$$x_{\mathrm{d}} = \frac{1}{L+1} \quad [9.18]$$

Since $g(x) = [1-(L+1)x]/(1-x)$, the evolution of the modulus of elasticity of equation [9.6] is as follows:

$$\frac{G(x)}{G_0} = \left[\frac{1-(L+1)x}{1-x}\right]^3 \quad [9.19]$$

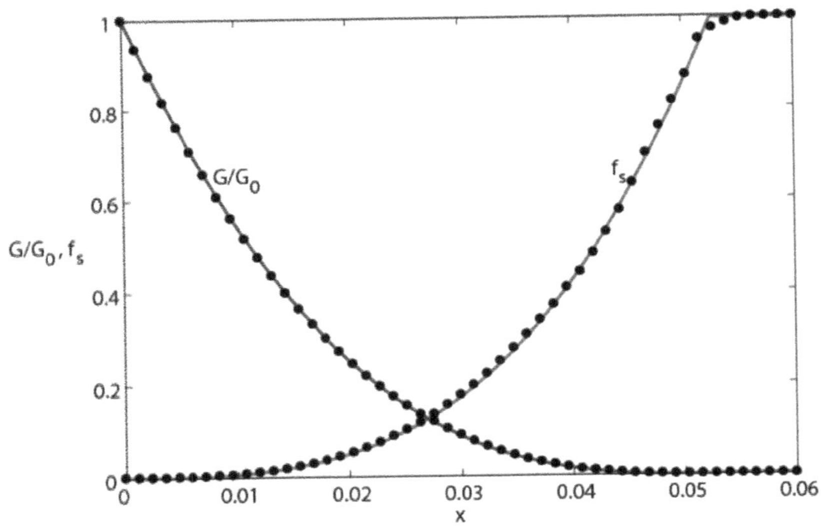

Figure 9.5. *Analytical prediction of the modulus of elasticity (red) and of the soluble fraction (green) during degradation and numerical simulation (symbols). The case of a trifunctional network with a number of reactive groups per chain following an average Flory distribution of 18*

The soluble fraction (in terms of nodes, without discounting the contribution of chain segments, to simplify) of equation [9.7] is given by:

$$f_s(x) = \left(\frac{L\,x}{1-x}\right)^3 \qquad [9.20]$$

Relations [9.19] and [9.20] are represented in Figure 9.5 for $L = 18$, as well as the result of a numerical simulation. Once more, there is full agreement between the analytical expressions that are instantaneously applied and the lengthy numerical calculations. A comparison between Figures 9.4 and 9.5 shows that degradation is slowed by dispersion, and any dispersion of the number of reactive groups per chain, irrespective of its shape, delays the degelation point (Gilormini et al. 2017). Finally, while Flory's law led us to very simple results, other types of distributions (for example, of the Poisson type) would require numerical resolutions, even to obtain the degelation point.

9.6. Third example: a tetrafunctional network

In the case of networks with tetrafunctional nodes, a chain scission transforms at low conversion two tetrafunctional nodes into trifunctional nodes; these nodes remain elastically active, that is, three EAC still start from any of them. However, using the approach proposed here, the calculations are not complex, even at high conversion, and the first step is to apply equation [9.3] with $f = 4$, which leads to a second-degree trinomial for which:

$$g(x) = \frac{3}{2} - \frac{1}{2}\sqrt{\frac{4}{u(x)} - 3} \qquad [9.21]$$

is the only root that satisfies the condition $g(x) = 1$ when $u(x) = 1$, that is, every chain belongs to gel if none undergoes scission in an ideal network. The expression for the modulus of elasticity is as follows:

$$\frac{G(x)}{G_0} = \sqrt{\frac{4}{u(x)} - 3}\left(\frac{3}{2} - \frac{1}{2}\sqrt{\frac{4}{u(x)} - 3}\right)^3 \qquad [9.22]$$

due to equation [9.6], which takes into account the presence of tetrafunctional EAN and of EAN that became trifunctional. The soluble fraction, still only in terms of network nodes, results from equation [9.7] in the following form:

$$f_s(x) = \frac{1}{16}\left(\sqrt{\frac{4}{u(x)} - 3} - 1\right)^4 \qquad [9.23]$$

Figure 9.6 illustrates these results for $L = 18$ if there are L reactive sites per chain, that is, $u(x) = (1 - x)^L$, which leads to its condition of degelation:

$$x_d = 1 - \frac{1}{3^{1/L}} \qquad [9.24]$$

Once more, the agreement with the numerical simulation is fully satisfied. When compared with Figure 9.4, increasing the network functionality delays the degelation point (which passes from $x_d = 0.038$ to $x_d = 0.059$), similarly to what a dispersion would induce.

As indicated by equation [9.5], the probability that an EAN is still tetrafunctional (obtained for $k = 4$) is equal to $g(x)^4$, whereas it is $4g(x)^3(1 - g(x))$ for it to be trifunctional ($k = 3$). As the ratio of the first probability and the second probability is zero at the degelation point, the elastically active network has therefore become trifunctional near the degelation point, which can also be noted in our numerical simulations. The evolution of the average network functionality and the different effect of scission in terms of reducing the number of EACs (section 9.1) induce a change in concavity that can be noted in the evolution of the modulus of elasticity in Figure 9.6. This was not noted in the previous figures.

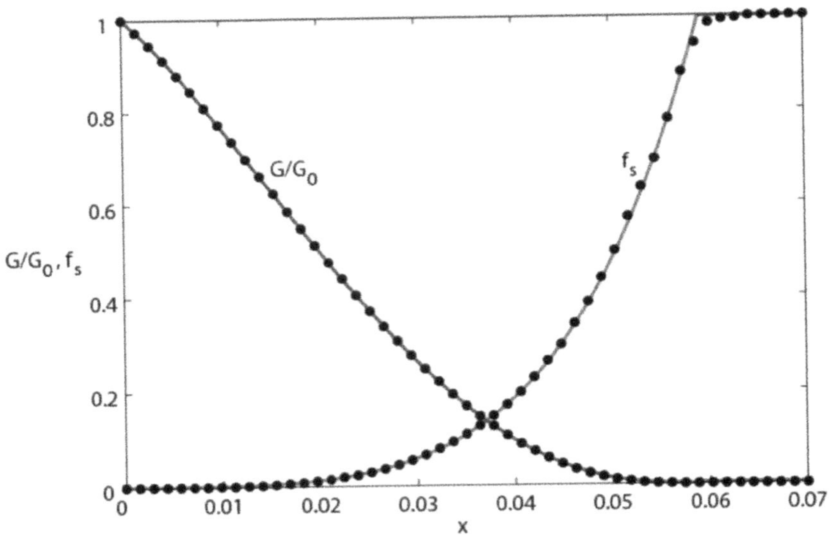

Figure 9.6. *Analytical prediction of the modulus of elasticity (red) and of the soluble fraction (green) during degradation and numerical simulation (symbols). Case of a tetrafunctional network with 18 reactive groups per chain*

The probabilistic theory proposed here highlights a very simple property of the random degradation process: for networks that are initially ideal and homofunctional, degradation always occurs when a fraction $(f-2)/(f-1)$ of the chains undergo scission, as the condition of degradation of equation [9.4] can be written as $u(x) = 1/(f-1)$ in terms of chains that are still intact, irrespective of the number and distribution of reactive groups per chain. In the trifunctional case, half of the chains should undergo scission, which requires conducting a maximum of half of the possible scissions, and even fewer, as mentioned above. This condition can already be found in Gordon's (1962) theory on cascade processes, and it was obtained in a remarkably simple manner by Flory (1941) in the context of the physics of polymers. Similarly, the degelation of an initially ideal tetrafunctional network will occur when two-thirds of the chains will undergo scission, and our numerical simulations effectively reach a value of 0.660 ± 0.005.

9.7. Conclusion

This chapter focused on a probabilistic theory of polymer networks degradation; this is illustrated in several examples and is successfully compared with numerical simulations. The latter explicitly follows the evolution of a graph that is a polymer network of sufficiently large size by means of calculations that can be very lengthy, whereas the probabilistic approach directly yields the expression of quantities of interest on the most probable state of the network for a given degree of degradation advancement. Particular attention was paid to the evolution of the modulus of elasticity of the network and to the soluble fraction gradually released during degradation. The average number of reactive groups per chain has a very strong influence on the rate of conversion to degelation. A delaying effect of the dispersion of this number was also noted, and this goes in the same direction as a higher functionality of the initial network.

The assumption made by the proposed theory is that only random chain scission is involved (no crosslinking), that all EAC contain reactive groups, that they are (and remain) equireactive and that a scission event does not create new reactive groups. Only initially homoreactive networks have been considered here, but nodes of various functionalities can also be considered. The networks were initially assumed to be ideal, but initial pending chains could also be included. For example, the number of reactive groups could be used to define an efficient ratio of conversion, as their scissions do not impact the elastically active network; alternatively, a predegraded state of the network could be considered the initial state. These development paths, and probably many others, could be explored to extend a theory whose bases are, in reality, very simple and that remains very versatile.

9.8. References

Argyropoulos, D.S. and Bolker, H.I. (1987). Gel degradation theory. 1. An experimental verification with a model trifunctional network. *Macromolecules*, 20, 2915–2922.

Dušek, K. (1989). Networks from telechelic polymers: Theory and application to polyurethanes. In *Telechelic Polymers: Synthesis and Applications*, Goethals, E. (ed.). CRC Press, Boca Raton.

Flory, P.J. (1941). Molecular size distribution in three dimensional polymers. I. Gelation. *J. Amer. Chem. Soc.*, 63, 3083–3090.

Flory, P.J. (1953). *Principles of Polymer Chemistry*. Cornell University Press, Ithaca.

Galina, H. and Lechowicz, J. (1998). An algorithm for Monte Carlo modeling of degradation of polymer networks. *Computers Chem.*, 22, 39–41.

Gilormini, P., Richaud, E., Verdu, J. (2014). A statistical theory of polymer network degradation. *Polymer*, 55, 3811–3817.

Gilormini, P., Richaud, E., Verdu, J. (2017). Radiochemical 'degelation' of polymethyl methacrylate networks. *Polymer*, 111, 130–136.

Gordon, M. (1962). Good's theory of cascade processes applied to the statistics of polymer distributions. *Proc. Roy. Soc. London A*, 268, 240–256.

Horikx, M.M. (1956). Chain scissions in a polymer network. *J. Polym. Sci.*, 19, 445–454.

Macosko, C.W. and Miller, D.R. (1976). A new derivation of average molecular weights of nonlinear polymers. *Macromolecules*, 9, 199–206.

Nijenhuis, A. and Wilf, H.S. (1978). *Combinatorial Algorithms for Computers and Calculators*. Academic Press, New York.

Pascault, J.P., Sautereau, H., Verdu, J., Williams, R.J.J. (2002). *Thermosetting Polymers*. Marcel Dekker Inc., New York.

Stockmayer, W.H. (1943). Theory of molecular size distribution and gel formation in branched-chain polymers. *Chem. Phys.*, 11, 45–55.

List of Authors

Alexis DELATTRE
Airbus
Toulouse
France

Laurent DELBREILH
GPM, CNRS
Université de Rouen Normandie
Saint-Étienne-du-Rouvray
France

Nicolas DELPOUVE
GPM, CNRS
Université de Rouen Normandie
Saint-Étienne-du-Rouvray
France

Louise-Anne FILLOT
LPMA, CNRS
Université Lyon 1
Saint-Fons
and
Solvay Engineering Plastics
Saint-Fons
France

Samuel FOREST
CNRS
Académie des sciences
Mines Paris – PSL
France

Pierre GILORMINI
PIMM, CNRS
CNAM
ENSAM
Paris
France

Florian LACROIX
LaMé
INSA
Université de Tours
France

Stéphane LEJEUNES
LMA, CNRS
AMU
Marseille
France

Stéphane MÉO
LaMé
INSA
Université de Tours
France

Yvette NGONO
CIMAP
UNICAEN
Caen
France

Jean-Louis POISSON
Baker Hugues
Houston
United States

Blandine QUÉLENNEC
GPM, CNRS
Université de Rouen Normandie
Saint-Étienne-du-Rouvray
France

Emmanuel RICHAUD
PIMM, CNRS
CNAM
ENSAM
Paris
France

Agustín RIOS DE ANDA
LPMA
Université Lyon 1
Saint-Fons
and
ICMPE
UPEC
Thiais
France

Paul SOTTA
LPMA, CNRS
Université Lyon 1
Saint-Fons
and
IMP
INSA
Villeurbanne
France

Index

G, H

I, K

M, N

O, P

Q, R

S, T

V, W

Printed and bound by CPI Group (UK) Ltd, Croydon, CR0 4YY
13/09/2026
14961071-0003